Glencoe Earth Science
Features and Benefits

Feature	Description	Pages
Dynamic Instructional Strategies	**… present clear and comprehensive coverage of middle school science.**	
	■ Each engaging chapter opener includes a *Launch Lab* and *Foldables*™.	
	■ *National Geographic Visualizing* features illustrate important concepts in middle school science.	19, 48
A Strong Reading Strand	**… encourages active reading and learning for students of all reading levels. In the *Student Edition:***	
	■ as you read gives students a preview of learning objectives and vocabulary for each section;	8, 36
	■ *Reading Checks* help students check their reading comprehension; and	23, 52
	■ *Caption questions* ask students to interpret visuals.	23, 52
	■ *Reading Essentials, An Interactive Student Textbook* is designed to help struggling readers comprehend core content. It is written at a reading level of about two to three grades below the *Student Edition.*	
Meeting the Needs of All Students	**… facilitates understanding of science concepts for students of all learning levels. In the *Teacher Wraparound Edition:***	
	■ *Differentiated Instruction* strategies help meet the needs of students with learning disabilities and physical challenges, or create opportunities to enrich and extend students' learning;	20, 48
	■ *Daily Intervention* provides intervention strategies for struggling students; and	21, 49
	■ *Identifying Misconceptions* helps uncover and address common science misconceptions.	98F
	■ The *English/Spanish Glossary*, also in the *Student Edition,* helps English-language learners comprehend science terms.	199–204
Extensive Standardized Test Practice	**… gives students the opportunity to practice for state and national exams.**	
	■ Each chapter ends with a variety of standardized test practice questions, including *Multiple Choice, Short Response/Grid In*, and *Open Ended.*	154–155
A Variety of Labs	**… gets students excited about being involved in science. The *Student Edition* provides:**	
	■ *MiniLABs,* traditional labs, and *Design Your Own, Model and Invent,* and *Use the Internet* labs; and	11, 55
	■ *Extra Try at Home Labs* provide opportunities for students to practice their science skills at home with adult supervision using materials from the kitchen, junk drawer, or backyard.	170–172
	■ *Virtual Labs* CD-ROM contains an interactive virtual lab for each chapter.	9, 55
	■ *Video Labs* (VHS) reinforce lab techniques and safety skills, offer troubleshooting tips, and give expected outcomes.	13, 35
	■ The *Science Lab Manual, Probeware Lab Manual,* and *Science Inquiry Lab Manual* provide additional opportunities to practice laboratory techniques.	
Multi-Level Review	**… presents multiple opportunities for all students to review and master content.**	
	■ Each section ends with a review that contains a *Summary* of the section's major concepts and a *Self Check* that has questions to assess student learning and practice math or science skills.	14, 43
	■ The *Study Guide* at the end of each chapter can preview, review, summarize, and visualize the chapter's main ideas.	29, 61
	■ *Study Guide* and *Reinforcement* help students grasp core content.	
Teacher Resources	**… provide innovative strategies to help new and experienced teachers.**	
	■ *Chapter Resources Fast File* ™ contains important reproducible masters.	98B
	■ Section Focus, Assessment, and Teaching transparencies accompany each chapter.	126C
	■ *Performance Assessment in the Science Classroom* has assessment guidelines, strategies, sample rubrics, and more.	
Online Resources	**… enrich the learning experience with the click of a mouse.**	
	■ For prescreened Web links, standardized test practice, self-check quizzes, chapter tests, *Vocabulary PuzzleMaker,* extra math practice, science career information, current science news, and *WebQuest* interactive projects, visit **bookh.msscience.com**.	
	■ The complete interactive *Student Edition* is available at The McGraw-Hill Learning Network Web site, **mhln.com**.	
Technology	**… provides timesaving products to help teachers creatively engage their students.**	
	■ *MindJogger Videoquizzes* (VHS & DVD) provide a game-show style interactive quiz for each chapter.	
	■ Easy to edit *Interactive Chalkboard* Microsoft® PowerPoint® presentations include step-by-step lessons, an image bank, chapter and section review questions, standardized test practice, and transparencies.	
	■ *ExamView® Pro Testmaker* CD-ROM in English or Spanish allows you to customize assessments.	
	■ *TeacherWorks* CD-ROM is your all-in-one resource center that helps you plan and organize lessons.	
	■ *StudentWorks* CD-ROM solves the heavy backpack problem.	

SAFETY SYMBOLS

SAFETY SYMBOLS	HAZARD	EXAMPLES	PRECAUTION	REMEDY
DISPOSAL	Special disposal procedures need to be followed.	certain chemicals, living organisms	Do not dispose of these materials in the sink or trash can.	Dispose of wastes as directed by your teacher.
BIOLOGICAL	Organisms or other biological materials that might be harmful to humans	bacteria, fungi, blood, unpreserved tissues, plant materials	Avoid skin contact with these materials. Wear mask or gloves.	Notify your teacher if you suspect contact with material. Wash hands thoroughly.
EXTREME TEMPERATURE	Objects that can burn skin by being too cold or too hot	boiling liquids, hot plates, dry ice, liquid nitrogen	Use proper protection when handling.	Go to your teacher for first aid.
SHARP OBJECT	Use of tools or glassware that can easily puncture or slice skin	razor blades, pins, scalpels, pointed tools, dissecting probes, broken glass	Practice common-sense behavior and follow guidelines for use of the tool.	Go to your teacher for first aid.
FUME	Possible danger to respiratory tract from fumes	ammonia, acetone, nail polish remover, heated sulfur, moth balls	Make sure there is good ventilation. Never smell fumes directly. Wear a mask.	Leave foul area and notify your teacher immediately.
ELECTRICAL	Possible danger from electrical shock or burn	improper grounding, liquid spills, short circuits, exposed wires	Double-check setup with teacher. Check condition of wires and apparatus.	Do not attempt to fix electrical problems. Notify your teacher immediately.
IRRITANT	Substances that can irritate the skin or mucous membranes of the respiratory tract	pollen, moth balls, steel wool, fiberglass, potassium permanganate	Wear dust mask and gloves. Practice extra care when handling these materials.	Go to your teacher for first aid.
CHEMICAL	Chemicals can react with and destroy tissue and other materials	bleaches such as hydrogen peroxide; acids such as sulfuric acid, hydrochloric acid; bases such as ammonia, sodium hydroxide	Wear goggles, gloves, and an apron.	Immediately flush the affected area with water and notify your teacher.
TOXIC	Substance may be poisonous if touched, inhaled, or swallowed.	mercury, many metal compounds, iodine, poinsettia plant parts	Follow your teacher's instructions.	Always wash hands thoroughly after use. Go to your teacher for first aid.
FLAMMABLE	Flammable chemicals may be ignited by open flame, spark, or exposed heat.	alcohol, kerosene, potassium permanganate	Avoid open flames and heat when using flammable chemicals.	Notify your teacher immediately. Use fire safety equipment if applicable.
OPEN FLAME	Open flame in use, may cause fire.	hair, clothing, paper, synthetic materials	Tie back hair and loose clothing. Follow teacher's instruction on lighting and extinguishing flames.	Notify your teacher immediately. Use fire safety equipment if applicable.

Eye Safety
Proper eye protection should be worn at all times by anyone performing or observing science activities.

Clothing Protection
This symbol appears when substances could stain or burn clothing.

Animal Safety
This symbol appears when safety of animals and students must be ensured.

Handwashing
After the lab, wash hands with soap and water before removing goggles.

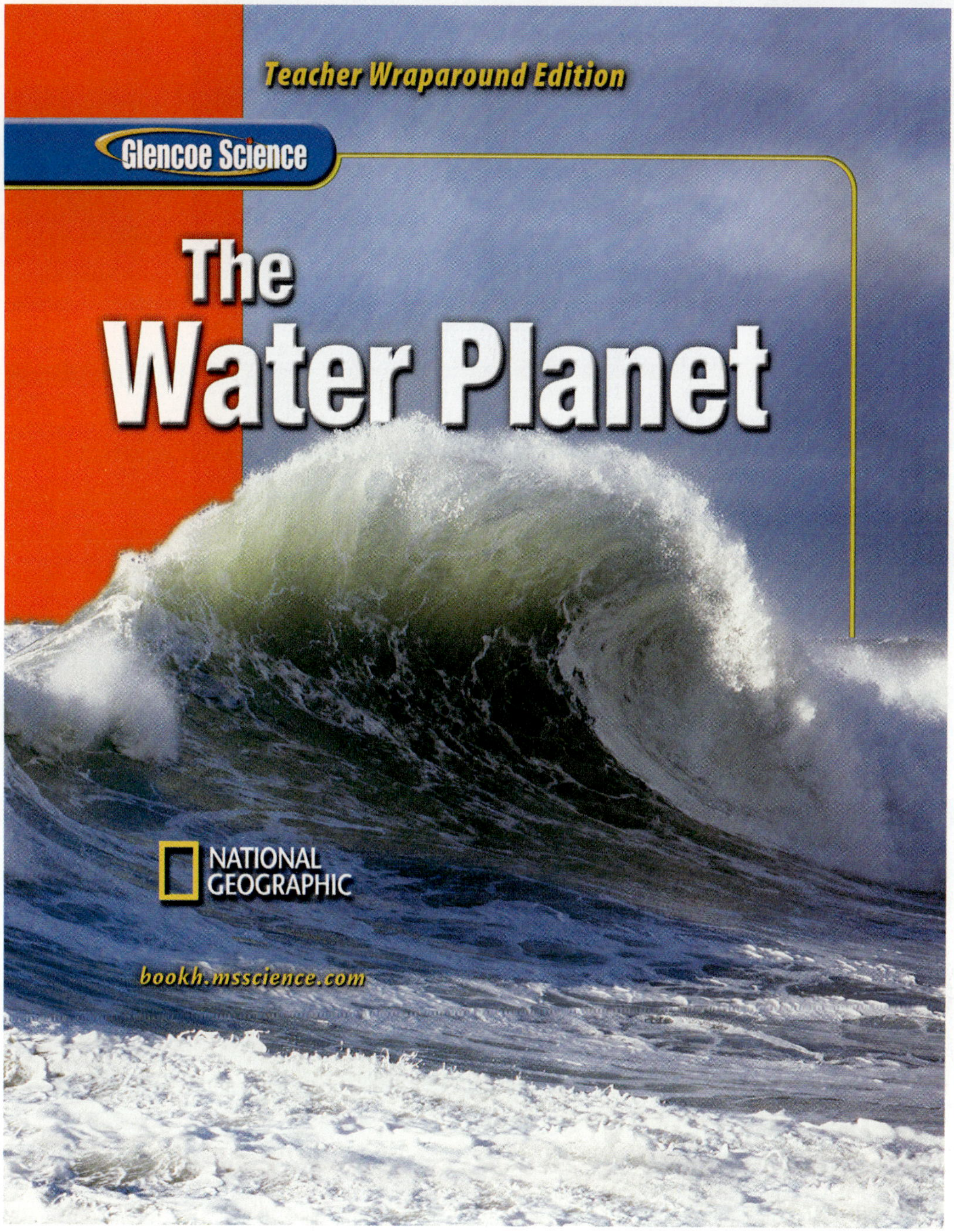

McGraw Hill Glencoe

New York, New York Columbus, Ohio Chicago, Illinois Peoria, Illinois Woodland Hills, California

Glencoe Science

The Water Planet

These crashing waves are viewed at Cape Kiwanda, on the Oregon coast. Cape Kiwanda is the smallest of three capes along the Three Capes Scenic Route (along with Cape Meares and Cape Lookout), but it's one of the best places to experience spectacular wave action. Ocean waves are a transfer of energy moving across the ocean's surface, and eventually to land.

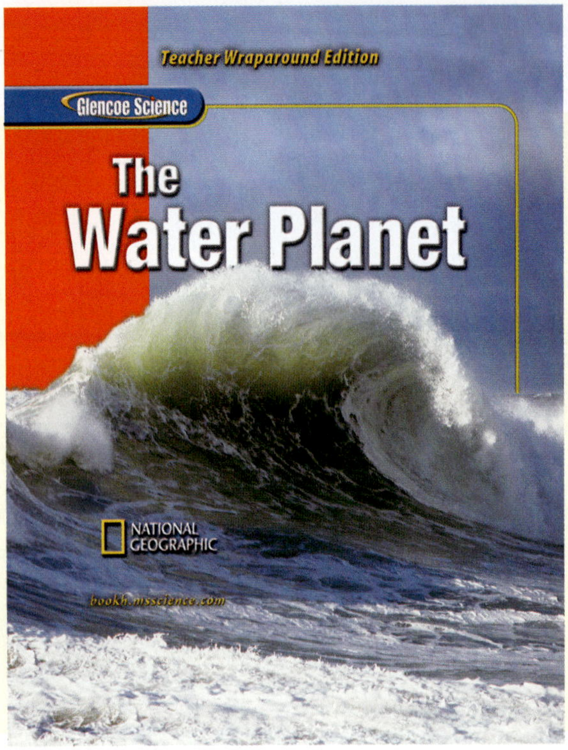

McGraw-Hill Glencoe

The McGraw-Hill Companies

Send all inquiries to:
Glencoe/McGraw-Hill
8787 Orion Place
Columbus, OH 43240-4027

ISBN 0-07-861755-3 (Student Edition)
ISBN 0-07-861756-1 (Teacher Wraparound Edition)

Printed in the United States of America.

1 2 3 4 5 6 7 8 9 10 027/043 09 08 07 06 05 04

Authors

NATIONAL GEOGRAPHIC
Education Division
Washington, D.C.

Susan Leach Snyder
Retired Earth Science Teacher
Jones Middle School
Upper Arlington, OH

Ralph M. Feather Jr., PhD
Assistant Professor
Geoscience Department
Indiana University of Pennsylvania
Indiana, PA

Dinah Zike
Educational Consultant
Dinah-Might Activities, Inc.
San Antonio, TX

Series Consultants

CONTENT

William C. Keel, PhD
Department of Physics and Astronomy
University of Alabama
Tuscaloosa, AL

MATH

Michael Hopper, DEng
Manager of Aircraft Certification
L-3 Communications
Greenville, TX

Teri Willard, EdD
Mathematics Curriculum Writer
Belgrade, MT

READING

Carol A. Senf, PhD
School of Literature, Communication, and Culture
Georgia Institute of Technology
Atlanta, GA

Rachel Swaters-Kissinger
Science Teacher
John Boise Middle School
Warsaw, MO

SAFETY

Sandra West, PhD
Department of Biology
Texas State University-San Marcos
San Marcos, TX

ACTIVITY TESTERS

Nerma Coats Henderson
Pickerington Lakeview Jr. High School
Pickerington, OH

Mary Helen Mariscal-Cholka
William D. Slider Middle School
El Paso, TX

Science Kit and Boreal Laboratories
Tonawanda, NY

Series Reviewers

Lois Burdette
Green Bank Elementary-Middle School
Green Bank, WV

Marcia Chackan
Pine Crest School
Boca Raton, FL

Mary Ferneau
Westview Middle School
Goose Creek, SC

Sharon Mitchell
William D. Slider Middle School
El Paso, TX

Joanne Stickney
Monticello Middle School
Monticello, NY

Glencoe Science 15-Book Series

Teach science your way!

With the 15 Life, Earth, and Physical Science titles in our modular series, you can select the science topics you want to cover and customize your science curriculum in any way you choose.

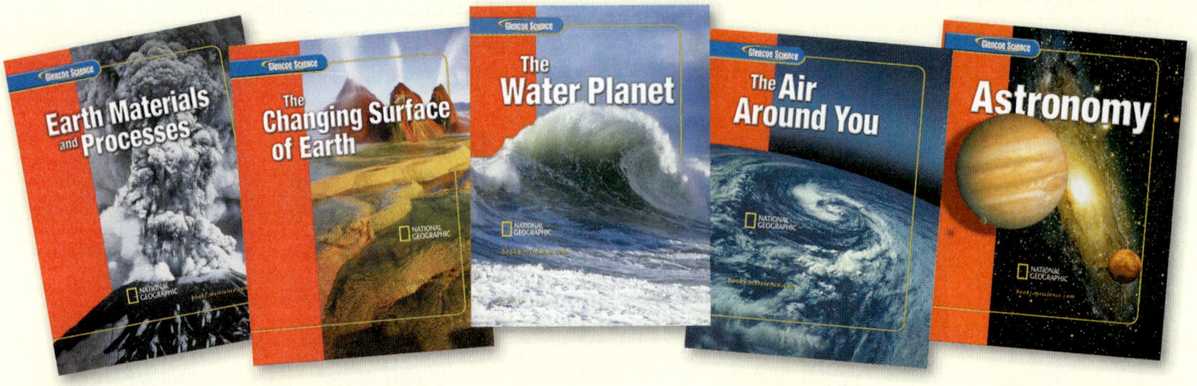

F Earth Materials and Processes
1 Minerals
2 Rocks
3 Earth's Energy and Mineral Resources
4 Plate Tectonics
5 Earthquakes
6 Volcanoes

G The Changing Surface of Earth
1 Views of Earth
2 Weathering and Soil
3 Erosional Forces
4 Water Erosion and Deposition
5 Clues to Earth's Past
6 Geologic Time

H The Water Planet
1 Water
2 Freshwater at Earth's Surface
3 Groundwater Resources
4 Ocean Motion
5 Oceanography

I The Air Around You
1 Atmosphere
2 Weather
3 Climate
4 Air Pollution

J Astronomy
1 Exploring Space
2 The Sun-Earth-Moon System
3 The Solar System
4 Stars and Galaxies

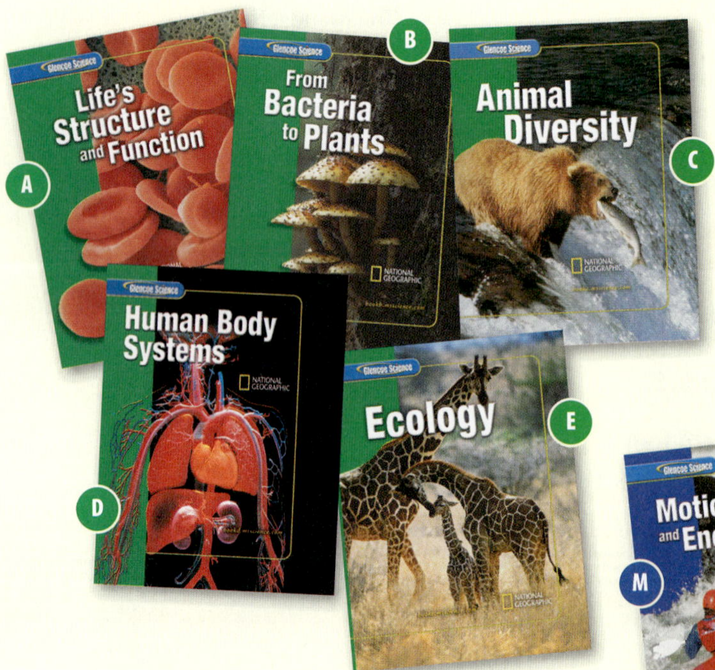

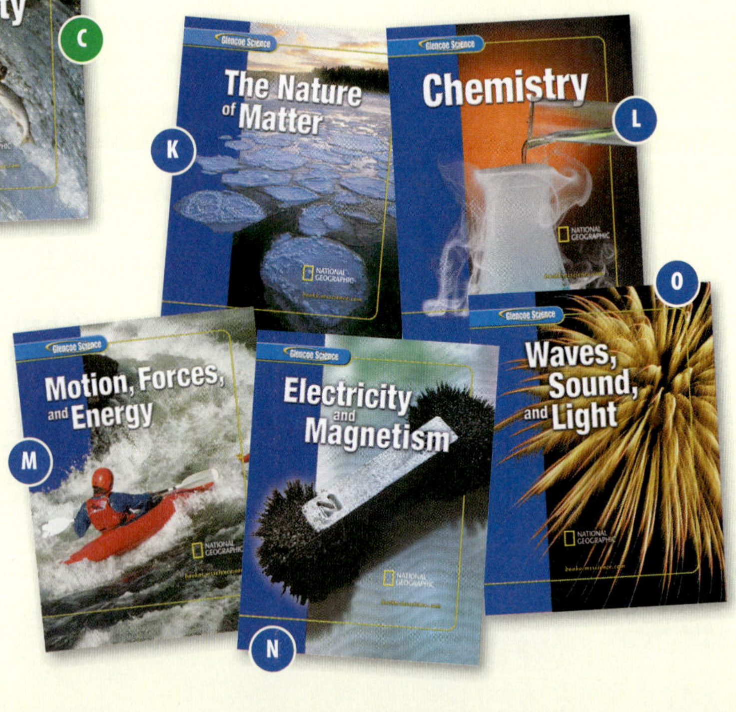

Multi-Level Review and Assessment

Each chapter provides five pages of review and testing to help you evaluate students' knowledge and ability to apply science concepts.

Section Review

- Summary pinpoints important concepts in the section.
- Skill-based questions promote critical thinking skills.

Study Guide

- Main idea summary of each section
- Concept mapping activity to help students visualize the main ideas

Chapter Review

- Using Vocabulary
- Checking Concepts
- Thinking Critically
- Performance Activities
- Applying Math

Standardized Test Practice

- Multiple Choice
- Short Response/ Grid In
- Open-Ended Questions

Dynamic Instruction

The consistent instructional strategies in each chapter strengthen students' learning—from the beginning of each chapter where students see "Chapter Preview," to the end where they have a chance to test the knowledge they have acquired and prepare for the next lesson.

Chapter Opener

- *Chapter Preview* introduces the main concepts.
- *Science Journal* promotes critical-thinking skills
- *Dinah Zike's Foldables*™ let students create interactive study guides.
- *Launch Labs* give students an opportunity to explore new ideas.

Section Opener

- *What You'll Learn* introduces main concepts.
- *Why It's Important* provides an answer to "Why do we have to learn this?"
- *Review Vocabulary* reviews a term that helps students better understand section content.
- *New Vocabulary* highlights new terms students will learn in the section.

Labs

- *Labs* allow students to design their own experiments or follow well-tested procedures, enabling them to learn and practice science processes.
- *MiniLABS* offer students quick and easy-to-do ways to clarify concepts and reinforce skills.

Assessment

Glencoe Science offers the Glencoe Assessment Advantage, a system designed to give you all the tools you need to prepare your students for success in any testing situation.

In the *Student Edition*

Section Review and **Applying Math** questions appear in every chapter.

Chapter Review questions help you evaluate students' knowledge and ability to apply science concepts.

Standardized Test Practice questions at the end of each chapter provide students with additional opportunities to practice their test-taking skills.

In the *Teacher Wraparound Edition*

Assessments located throughout the *Teacher Wraparound Edition* provide methods for assessing students' comprehension with Performance, Process, and Content exercises.

Teacher Classroom Resources

Performance Assessment in the Science Classroom

- Guidelines for assessing the performance of a task
- Reproducible activities for evaluating students
- Sample rubrics and checklists

***Fast File* Chapter Resources** provides six pages of assessement for every chapter including *Testing Concepts, Applying Concepts,* and *Writing Skills.*

Technology Support

MindJogger Videoquizzes are interactive video quizzes set in game show format. Each is designed for the full range of student learning styles.

Exam*View*® Pro Testmaker CD-ROM for Windows® and Macintosh® provides an easy way to create, edit, and customize your tests. Select your own test items by objective from two different levels of difficulty, or write and edit your own. Translate tests from English to Spanish and vice versa.

Rubrics

The following rubrics are sample scoring devices for short response and open-ended questions.

Short Response

Points	Description
2	The student demonstrates a thorough understanding of the science of the task. The response may contain minor flaws that do not detract from the demonstration of a thorough understanding.
1	The student has provided a response that is only partially correct.
0	The student has provided a completely incorrect solution or no response at all.

Open Ended

Points	Description
4	The student demonstrates a thorough understanding of the science of the task. The response may contain minor flaws that do not detract from the demonstration of a thorough understanding.
3	The student demonstrates an understanding of the science of the task. The response is essentially correct and demonstrates an essential but less than thorough understanding of the science.
2	The student demonstrates only a partial understanding of the science of the task. Although the student may have used the correct approach to a solution or may have provided a correct solution, the work lacks an essential understanding of the underlying science concepts.
1	The student demonstrates a very limited understanding of the science of the task. The response is incomplete and exhibits many flaws.
0	The student provides a completely incorrect solution or no response at all.

Time-Saving Teacher Resources

Glencoe Science provides an extensive array of support materials and resources designed to help you create and customize your science course quickly and easily.

FAST FILE Chapter Resources

For each chapter, Chapter Resources contain key reproducible masters along with additional teaching strategies, teacher support, and answer keys.

Teacher Wraparound Edition

The *Teacher Wraparound Edition* is your key to the teaching resources available. In addition to teaching strategies and suggestions, the *Teacher Wraparound Edition* provides a guide for all print and software materials available for each lesson.

Transparencies

Color Transparencies provides three types of transparencies for use while teaching each chapter. The *Section Focus Transparencies* are designed to generate interest and focus students' attention on the topic being presented in the section. The *Teaching Transparency* for each chapter addresses a major concept that will benefit from an extra visual learning aid. The *Assessment Transparency* for each chapter is set up to resemble standardized tests.

Exam*View*® Pro Testmaker CD-ROM

This CD-ROM will help you create, edit, and customize tests. In addition, it will help you create multiple versions of tests, translate tests from English to Spanish and vice versa, and build tests aligned with state standards.

Video Labs

These VHS cassettes contain step-by-step procedures for selected *Student Edition* labs. They also contain lab safety skills, teacher support, and troubleshooting advice.

TeacherWorks™

This CD-ROM is your all-in-one teacher resource center. Personalize a lesson plan, access resources from the *Teacher Wraparound Edition*, connect to the Internet, or make a to-do list. These are only a few of the many features that can assist you in the planning and organizing of your lessons.

StudentWorks™

This CD-ROM is a valuable resource for students to access content online and use online resources to continue learning chapter concepts.

Interactive Chalkboard with Image Bank

PowerPoint® Presentations

This CD-ROM brings Microsoft® PowerPoint® presentations right to your door. With the large number of graphics provided, students can use a visual approach to learning chapter content.

Virtual Labs CD-ROM Program

The Virtual Labs CD-ROM contains a collection of labs that allow students to complete labs that are too expensive, take too long to complete, or might be too dangerous in a classroom laboratory.

Science Online

This website is a portal to hundreds of prescreened Internet sites (Web links) that correlate to content in the text. Visit this link to find interactive activities that review chapter concepts and to access the *Student Edition* online.

McGraw Hill Learning Network mhln.com

mhln.com is an online teaching and learning space for teachers, students, and parents.

Differentiated Instruction

Teaching Strategies

Following each suggested assessment and activity, ability levels are supplied to accommodate all students. For a key to the Teaching Strategies designations, see the C page before each chapter.

Identifying Misconceptions

These short, diagnostic, and perscriptive lessons target common science misconceptions.

Multiple Learning Styles

Look for these italicized designations under various activities to help you target your lessons to each student's preferred learning style.

- ***Kinesthetic*** learners learn through touch, movement, and manipulating objects.
- ***Visual-Spatial*** learners think in terms of images, illustrations, and models.
- ***Interpersonal*** learners understand and work well with other people.
- ***Intrapersonal*** learners can analyze their own strengths and weaknesses and may prefer to work on their own.
- ***Linguistic*** learners write clearly and easily understand the written word.
- ***Logical-Mathematical*** learners understand numbers easily and have highly-developed reasoning skills.

Daily Interventions

Found at the end of each chapter section, this feature is designed to intercept students who are struggling and prescribe a system to help them get back on track. *Reteach* provides reinforcement of the section's concepts through visual activities.

Differentiated Instruction

These activities present various teaching strategies designed to help you meet the special needs of students with learning disabilities, physical challenges, visual impairment, and hearing impairment. *Challenge* activities provide opportunities for students who excel to engage in activities and research projects that extend the chapter's concepts. English-language learners in the classroom will also find exercises that bridge the gap between language barriers and the chapter content.

Cultural Diversity

These readings provide insights into the unique ways in which people of different ethnicities and cultural heritage have approached science. The intent of these features is to build awareness and appreciation for the global community in which we live.

Inquiry-Based Science

The call for more inquiry-based science by the *National Science Education Standards* has been met by Glencoe Science.

Glencoe Science recognizes the importance of conducting inquiry-based science activities in the classroom. The process of doing inquiry models actual science practice, encouraging problem-solving strategies and developing critical thinking skills. Inquiry gets students actively involved in the learning process by allowing them to determine materials, procedures, or the topics and questions they want to investigate.

Inquiry can range from a very structured activity for those students who need more guidance to a more open-ended approach where students lead the investigations. Glencoe Science recognizes that the inquiry activities suggested will not look the same in every classroom. We encourage teachers to modify the suggested activities in a manner that best supports your students.

Glencoe also provides teachers with *Alternative Inquiry Labs,* teaching strategies or suggestions for making existing labs more inquiry-based.

Research-Based Learning Strategies

Glencoe Science incorporates the most current and applicable educational research on science learning and follows recommendations from the American Association for the Advancement of Science and the National Science Teachers Association. The following research-based strategies can be found throughout the text.

Learning Strategies

The following research-based strategies can be found throughout the text:

- **Using Prior Knowledge** Glencoe Science encourages students to use their prior knowledge to learn information because this adds relevance to the material. Students are referred back to other parts of the text or to their own real-life experiences.
- **Practicing Important Tasks** By offering students an opportunity to practice important tasks using a variety of labs and activities in the *Student Edition, Teacher Wraparound Edition,* ancillaries and technology, Glencoe Science makes learning fun and relevant for students.
- **Using Visuals to Communicate, Organize, and Reinforce Learning** High-quality art and photos throughout the text communicate concepts more efficiently and reinforce learning, while allowing students to organize information.

White Paper

The Glencoe Science White Paper outlines the educational strategies on which this program was based. This document provides specific examples from the *Student Edition, Teacher Wraparound Edition,* ancillary program, and technology resources, highlighting extensive use of educationally sound strategies that help students learn science.

- **Motivating Students to Achieve** Active strategies and real-world experiences motivate students to achieve. Throughout Glencoe's programs, students are encouraged to apply their knowledge in ways that will motivate them to learn.
- **Developing Decoding and Reading Comprehension Strategies** Throughout the text, students are supplied with caption questions, reading checks, and other strategies to aid in comprehension.
- **Using Study Strategies** Through the use of highlighting, outlining, note-taking, summarizing, and other such strategies, students can monitor their own progress and organize information more effectively, thereby increasing their scientific literacy. These strategies are found throughout the text and ancillaries.

The use of these strategies within Glencoe Science will help teachers to achieve the goals set forth by the *National Science Education Standards.*

Field Research and Testing

Feedback from students, teachers, curriculum supervisors, department chairpersons, parents, learning specialists, and science content experts was invaluable in the development of this program. The following pre-publication and post-publication research was conducted.

Prior to Publication

- Detailed classroom teacher and curriculum supervisor **surveys** were conducted by independently contracted researchers.
- A **nationwide panel** of science teachers, curriculum supervisors, and department chairpersons provided countless hours of feedback and assistance throughout program development.
- A wide range of **educator and content reviewers** provided in-depth reviews of and suggestions for manuscripts and pre-publication versions of the program.
- **Face-to-face interviews** with science teachers provided insight into teachers' day-to-day challenges.

After Publication

- Field tests were conducted in which students and teachers used a pre-publication manuscript in the classroom.
- Follow-up interviews, observations, and surveys of Glencoe Science users provide ongoing opportunities for program development and verification of program success.

Field-Test Results

- Field-test research indicates that test scores increased among students using Glencoe Science programs.
- Nine out of ten students earned higher scores after using Glencoe programs.
- Scores improved among both male and female students.
- Scores improved among both minority and non-minority students.
- Overall, the gap between the average pre-test score and a perfect score closed by 33 percent. Stated differently, on average, **scores increased 77 percent after students used the Glencoe program.**

KS Studios

National Education Standards

Correlation of *Glencoe Science* to the National Science Education Standards.

Content Standard	Chapter and Section
(UCP) Unifying Concepts and Processes	
1. Systems, order, and organization	F1-1, F1-2, F1-3, F2-1, F2-2, F2-3, F2-4, F3-1, F3-2, F3-3, F4-1, F4-2, F4-3, F5-1, F5-2, F5-3, F6-1, F6-2, F6-3, G1-1, G1-2, G1-3, G2-1, G2-2, G2-3, G3-1, G3-2, G3-3, G4-1, G4-2, G4-3, G5-1, G5-2, G5-3, G6-1, G6-2, G6-3, H2-1, H2-2, H2-3, H2-4, H3-1, H3-2, H3-3, H4-1, H4-2, H4-3, H5-1, H5-2, H5-3, I1-1, I1-2, I1-3, I2-1, I2-2, I2-3, I3-1, I3-2, I3-3, I4-1, I4-2, I4-3, J1-1, J1-2, J1-3, J2-1, J2-2, J2-3, J3-1, J3-2, J3-3, J3-4, J4-1, J4-2, J4-3
2. Evidence, models, and explanation	F1-1, F1-2, F1-3, F2-1, F2-2, F2-3, F2-4, F3-1, F3-2, F3-3, F4-1, F4-2, F4-3, F5-1, F5-2, F5-3, F6-1, F6-2, F6-3, G1-1, G1-2, G1-3, G2-1, G2-2, G2-3, G3-1, G3-2, G3-3, G4-1, G4-2, G4-3, G5-1, G5-2, G5-3, G6-1, G6-2, G6-3, H1-1, H1-2, H1-3, H2-1, H2-2, H2-3, H2-4, H3-1, H3-2, H3-3, H4-1, H4-2, H4-3, H5-1, H5-2, H5-3, I1-1, I1-2, I1-3, I2-1, I2-2, I2-3, I3-1, I3-2, I3-3, I4-1, I4-2, I4-3, J1-1, J1-2, J1-3, J2-1, J2-2, J2-3, J3-1, J3-2, J3-3, J3-4, J4-1, J4-2, J4-3
3. Change, constancy, and measurement	F1-1, F1-2, F1-3, F2-1, F2-2, F2-3, F2-4, F3-1, F3-2, F3-3, F4-1, F4-2, F4-3, F5-1, F5-2, F5-3, F6-1, F6-2, F6-3, G1-1, G1-2, G1-3, G2-1, G2-2, G2-3, G3-1, G3-2, G3-3, G4-1, G4-2, G4-3, G5-1, G5-2, G5-3, G6-1, G6-2, G6-3, H1-1, H1-2, H1-3, H2-1, H2-2, H2-3, H2-4, H3-1, H3-2, H3-3, H4-1, H4-2, H4-3, H5-1, H5-2, H5-3, I1-1, I1-2, I1-3, I2-1, I2-2, I2-3, I3-1, I3-2, I3-3, I4-1, I4-2, I4-3, J1-1, J1-2, J1-3, J2-1, J2-2, J2-3, J3-1, J3-2, J3-3, J3-4, J4-1, J4-2, J4-3
4. Evolution and equilibrium	G6-1, G6-2, J4-3
5. Form and function	F1-1, F1-2, F1-3, F2-1, F2-2, F2-3, F2-4, F3-1, F3-2, F3-3, F4-1, F4-2, F4-3, F5-1, F5-2, F5-3, F6-1, F6-2, F6-3, G1-1, G1-2, G1-3, G2-1, G2-2, G2-3, G3-1, G3-2, G3-3, G4-1, G4-2, G4-3, G5-1, G5-2, G5-3, G6-1, G6-2, G6-3, H1-1, H1-2, H1-3, H2-1, H2-2, H2-3, H2-4, H3-1, H3-2, H3-3, H4-1, H4-2, H4-3, H5-1, H5-2, H5-3, I1-1, I1-2, I1-3, I2-1, I2-2, I2-3, I3-1, I3-2, I3-3, I4-1, I4-2, I4-3, J1-1, J1-2, J1-3, J2-1, J2-2, J2-3, J3-1, J3-2, J3-3, J3-4, J4-1, J4-2, J4-3
(A) Science as Inquiry	
1. Abilities necessary to do scientific inquiry	F1-1, F1-2, F1-3, F2-1, F2-2, F2-3, F2-4, F3-1, F3-2, F3-3, F4-1, F4-2, F4-3, F5-1, F5-2, F5-3, F6-1, F6-2, F6-3, G1-1, G1-2, G1-3, G2-1, G2-2, G2-3, G3-1, G3-2, G3-3, G4-1, G4-2, G4-3, G5-1, G5-2, G5-3, G6-1, G6-2, G6-3, H1-1, H1-2, H1-3, H2-1, H2-2, H2-3, H2-4, H3-1, H3-2, H3-3, H4-1, H4-2, H4-3, H5-1, H5-2, H5-3, I1-1, I1-2, I1-3, I2-1, I2-2, I2-3, I3-1, I3-2, I3-3, I4-1, I4-2, I4-3, J1-1, J1-2, J1-3, J2-1, J2-2, J2-3, J3-1, J3-2, J3-3, J3-4, J4-1, J4-2, J4-3
2. Understandings about scientific theory	F1-1, F1-2, F1-3, F2-1, F2-2, F2-3, F2-4, F3-1, F3-2, F3-3, F4-1, F4-2, F4-3, F5-1, F5-2, F5-3, F6-1, F6-2, F6-3, G1-1, G1-2, G1-3, G2-1, G2-2, G2-3, G3-1, G3-2, G3-3, G4-1, G4-2, G4-3, G5-1, G5-2, G5-3, G6-1, G6-2, G6-3, H1-1, H1-2, H1-3, H2-1, H2-2, H2-3, H2-4, H3-1, H3-2, H3-3, H4-1, H4-2, H4-3, H5-1, H5-2, H5-3, I1-1, I1-2, I1-3, I2-1, I2-2, I2-3, I3-1, I3-2, I3-3, I4-1, I4-2, I4-3, J1-1, J1-2, J1-3, J2-1, J2-2, J2-3, J3-1, J3-2, J3-3, J3-4, J4-1, J4-2, J4-3
(B) Physical Science	
1. Properties and changes of properties in matter	G2-1, G5-3, H1-1, H4-1, H5-2, I4-1, J4-3
2. Motions and forces	F1-2, F4-1, F4-2, F4-3, F5-1, F5-2, F6-1, F6-2, G2-2, G2-3, G3-3, I1-2, I1-3, I2-2, I3-3, J2-1, J2-2, J3-1, J4-3
3. Transfer of energy	H5-2, I1-2, I3-3, J4-3
(C) Life Science	
1. Structure and function in living systems	H2-1, H2-2, H2-3, H5-2
2. Reproduction and heredity	H5-2
3. Regulation and behavior	H5-1, H5-2
4. Populations and ecosystems	H2-1, H2-2, H2-3, H5-1, H5-2, I4-2
5. Diversity and adaptations of organisms	H2-1, H2-2, H2-3, H5-2
(D) Earth and Space Science	
1. Structure of the Earth system	F1-1, F2-1, F2-2, F2-3, F2-4, F3-1, F3-2, F3-3, F4-1, F4-2, F4-3, F5-1, F5-2, F5-3, F6-1, F6-2, G1-1, G1-2, G1-3, G2-1, G2-2, G4-1, G4-2, G4-3, G6-1, G6-2, G6-3, H2-1, H2-2, H3-1, H3-2, H3-3, H4-1, H4-2, H4-3, H5-1,H5-2, I1-1, I1-2, I1-3, I2-1, I2-2, I2-3, I3-1, I3-2, I3-3, J2-1, J2-3
2. Earth's history	F1-1, F1-2, F2-1, F2-2, F2-3, F2-4, F3-1, F4-1, F5-1, F6-1, G1-1, G1-3, G2-1, G2-3, H3-3, I2-3, I3-3
3. Earth and the solar system	F1-3, F2-1, F2-2, F2-3, F2-4, F3-1, F3-3, H4-3, J1-1, J1-2, J1-3, J2-3, J3-2, J4-2

Content Standard	Chapter and Section
(E) Science and Technology	
1. Abilities of technological design	H4-3
2. Understandings about science and technology	H4-3
(F) Science in Personal and Social Perspectives	
1. Personal health	F3-2, G2-2, I4-2
2. Populations, resources, and environments	F3-1, F3-2, F3-3, H1-3, H2-4, H3-2, H5-3, I4-1, I4-2, I4-3
3. Natural hazards	F4-1, F4-2, F4-3, F5-1, F5-2, F5-3, F6-1, F6-2, G3-3, G4-1, G4-3, I2-3, I4-1, I3-3
4. Risks and benefits	G2-1, G3-3, H3-2
5. Science and technology in society	F3-1, G5-1, G5-3, I4-3, J1-3
(G) History and Nature of Science	
1. Science as a human endeavor	F1-3
2. Nature of science	F1-2, F1-3, F4-3, F6-3, I4-3
3. History of science	F1-2, F1-3, F5-2, G5-3, H5-3, I4-3, J2-3, J3-1, J3-4, J4-3

How Glencoe Science Aligns with the National Science Education Standards

The correlations at the left and above show the close alignment between Glencoe Science and the grade-appropriate standards. Glencoe Science allows students to discover concepts within each of the content standards and gives students opportunities to make connections among the science disciplines. Hands-on activities and inquiry-based lessons reinforce the science processes emphasized in the standards.

How Glencoe Science Aligns with the NCTM Standards for Grades 6–8

Throughout Glencoe Science, each Applying Math activity provides students with the opportunity to practice and apply some of the mathematical concepts and applications described in the NCTM Standards. These activities serve to reinforce mathematical skills in real-life situations, thus preparing students to meet their needs in an ever-changing world.

Correlation of *Glencoe Science* to NCTM Standards

Math Standard	Page
1. Number and Operations	F-31, F-54, F-61, F-121, F-143, F-151, F-172, F-181, G-31, G-46, G-59, G-87, G-119, G-149, G-173, H-47, H-63, H-95, H-123, H-130, I-31, I-39, I-61, I-114, I-121, J-35, J-65, J-80, J-99, J-130
2. Algebra	F-54, F-61, F-143, F-151, F-172, F-181, G-31, G-119, G-149, H-47, H-95, H-123, H-130, I-39, I-61, I-114, I-121, J-65
3. Geometry	G-31, H-95, J-16, J-65
4. Measurement	G-31, H-95, I-91, I-121, J-35, J-65
5. Data Analysis and Probability	F-31, F-93, F-151, F-181, G-46, G-149, G-181, H-31, H-63, H-123, H-153, I-31, I-39, I-61, I-121, J-35, J-99, J-130
6. Problem Solving	F-31, G-46, G-87, G-149
7. Reasoning and Proof	F-54, F-143, F-172, G-46, G-105, G-173, H-108, H-130, I-39, J-16, J-80
8. Communication	F-117, F-147, G-83, I-87, J-31
9. Connections	F-54, F-61, F-121, F-143, F-151, F-172, F-181, G-31, G-59, G-119, H-47, H-63, H-95, H-123, H-130, H-153, I-31, I-39, I-61, I-91, I-114, I-121, J-35, J-65, J-80, J-99, J-130
10. Representation	G-31, G-149

Foldables™

Foldables™ are easy-to-make, three-dimensional, interactive graphic organizers that students create out of simple sheets of paper. These unique hands-on tools for studying and reviewing were created exclusively for Glencoe by education specialist Dinah Zike.

Research Behind Foldables™

According to research (Bransford, 1979; Corno, 1994), study strategies help students understand, organize, remember and apply new information presented in science textbooks. Some study strategies include concept mapping, highlighting, outlining, note taking, summarizing, and underlining (Peverly, Brobst, Graham & Shaw, 2003). Glencoe Science offers Dinah Zike's Foldables™ Study Organizers as an organizational tool and study guide for students.

Foldables™

- Build prereading skills
- Encourage active reading and writing
- Summarize content for review

FOLDABLES™ Study Organizer

Primary and Secondary Succession Make the following Foldable to help you illustrate the main ideas about succession.

STEP 1 **Fold** a vertical sheet of paper in half from top to bottom.

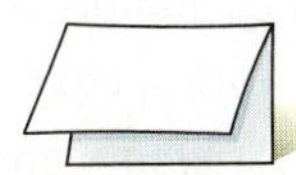

For more ideas on how to incorporate Foldables™ into your lessons consult **Dinah Zike's *Teaching Science with Foldables™***

Educational Partnerships

NATIONAL GEOGRAPHIC

Some topics in the chapter either require or benefit from a larger, more detailed visual explanation. The National Geographic Society has created *Visualizing* features that call out an important concept from the chapter and illustrate it in a way that will inform, excite, and motivate your students.

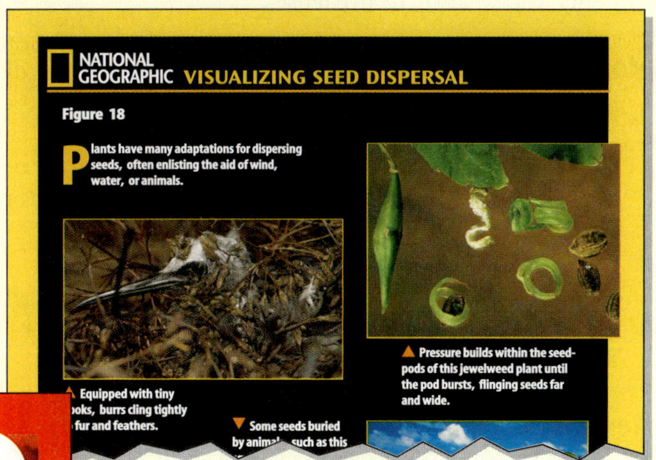

TIME

TIME magazine brings science topics and history together to further explain the chapter's main ideas and show how science relates to real life.

Safety in the Laboratory

All activities are designed to minimize dangers in the laboratory. Careful laboratory planning and management by both the instructor and the student are essential to a safe laboratory experience. **Local, state, and federal laboratory safety laws and regulations must be strictly followed.** The information provided here is one of the many resources to which you can refer for information about laboratory safety.

Classroom and Laboratory Preparation

1. Store equipment properly and securely and other thing.
   a. Clean and dry all equipment before storing.
   b. Protect electronic equipment and microscopes from dust, humidity, and extreme temperatures.
   c. Number, catalog, and organize equipment.
2. Ensure adequate work space for each student.
3. Ensure adequate classroom and storeroom ventilation.
4. Explain and post safety and evacuation guidelines along with expectations of conduct.
5. Ensure that all safety equipment is functioning properly and is clearly visible.
6. Provide hot plates as a heat source whenever possible. If gas burners are used, know where the central gas supply shutoff valve is located.
7. Ensure that each workstation has a GFCI-protected electrical source.
8. Provide safety goggles consistent with ANSI Standard Z87.1 for each student, including students who wear corrective lenses.

Before Each MiniLAB or Lab

1. Arrange the lab in such a way that equipment and supplies are clearly labeled and easily accessible.
2. Have available only equipment and supplies needed to complete the assigned investigation.
3. Review the procedure with students, emphasizing any caution statements or safety symbols that appear.
4. Be sure all students know the proper procedures to follow if an accident should occur.

After the MiniLAB or Lab

1. Be certain that students have returned all equipment and disposed of broken glassware and chemicals properly.
2. Be sure that all hot plates and electrical connections are off.

Storage of Chemicals

Be sure to store all chemicals properly. The following are guidelines commonly used. Your school, city, county, or state may have additional requirements for handling chemicals. It is the responsibility of each teacher to become informed of the rules or guidelines in effect in his or her area.

1. Separate chemicals by reaction type. Strong acids should be stored together. Likewise, strong bases should be stored together and should be separated from acids. Oxidants should be stored away from easily oxidized materials, and so on.
2. Be sure all chemicals are stored in labeled containers indicating contents, concentration, source, date purchased (or prepared), any precautions for handling and storage, and expiration date.
3. Hazardous chemicals require special storage containers and conditions. Be sure to know which chemicals those are and the accepted practices for your area. Some substances must be stored outside the building.

Disposal of Chemicals

Local, state, and federal laws regulate the proper disposal of chemicals. These laws should be consulted before chemical disposal is attempted. Although many substances encountered in the science classroom can be flushed down the drain with plenty of water, it is not safe to assume that this is always true.

DISCLAIMER

Glencoe Publishing Company makes no claims to the completeness of this discussion of laboratory safety and chemical storage. The material presented is not all-inclusive, nor does it address all of the hazards associated with handling, storing, and disposing of chemicals, or with laboratory management.

Preparation of Solutions

It is important to use safe laboratory techniques when handling all chemicals. Always check the MSDS (Material Safety Data Sheet) for each chemical before using it in the classroom. Many substances might appear harmless, but might be toxic, corrosive, or very reactive. Chemicals should never be ingested. Use proper techniques to smell any chemical, wear safety goggles and an apron in the laboratory, and observe the following precautions.

1. **Dilution of Acids and Bases** When diluting acids with water, always add the acids to the water. Never add water to acids. When sulfuric acid and sodium hydroxide are added to water, a large amount of thermal energy is released. Use extra care when handling these substances.

2. **Poisonous and Corrosive Liquids or Vapors** Use a fume hood if possible. Examples include hydrochloric acid, acetic acid, nitric acid, and ammonium hydroxide.

3. **Poisonous and Corrosive to Eyes, Lungs, and Skin** Examples include acids, bases, silver nitrate, iodine, and potassium permanganate.

Bromthymol blue: Add 0.5 g bromthymol blue powder to 500 mL distilled water to make a BTB stock solution. Dilute 40 mL BTB stock solution to 2 L with distilled water. Solution should be bright blue. If not, add one drop of NaOH at a time, swirling to mix. Check color.

Hydrochloric acid (HCL) solution: To make a 5% solution, add 13.6 mL concentrated HCl to 73 mL water while stirring. To make a 0.1*M* solution, add 1 mL concentrated hydrochloric acid to 100 mL water while stirring.

Iodine solution/Iodine stain: Dilute 1 part Lugol's solution with 15 parts water.

Lugol's solution: Dissolve 10 g potassium iodide in 100 mL distilled water. Then add and dissolve 5 g iodine. Store in dark bottle. Keeps indefinitely.

Phenolphthalein indicator: From a drug store, buy a package of any laxative that contains phenolphthalein. To make 1% solution, mash 4 tablets and pour the powder into 10 mL of rubbing alcohol. Let mixture soak for 15 minutes. Pour liquid into and store in a dropper bottle.

Potassium permanganate: For a 0.01*M* solution of potassium permanganate, dissolve 0.15 g $KMnO_4$ in 100 mL water.

Red cabbage concentrate: Put 5 leaves of red cabbage in a pot. Add 1 L of water, bring to a boil, and simmer until water turns a deep purple. Pour liquid through a strainer or piece of cheesecloth into a storage bottle. Keep refrigerated.

Salt solution: For a 3.5% salt (NaCl) solution that simulates the concentration of ocean water, dissolve 35 g of salt (NaCl) in 965 mL of water. For a 1% solution (weak), dissolve 1 g of salt (NaCl) in 99 mL of water. For a 6% solution, dissolve 6 g of salt (NaCl) in 94 mL of water.

Silver nitrate solution: To make a 10% solution, put 5 g of silver nitrate in 50 mL of distilled water.

Sugar solution: Add 1 tablespoon of sugar to 1 cup of warm water in a deep jar or flask. Stir to dissolve.

Sodium hydroxide (dilute): To make a 1% solution, dissolve 1 g NaOH in 99 mL of water.

Equipment and Materials List

Refer to the Chapter Organizer in front of each chapter for a list of equipment and materials used for each laboratory activity in the chapter.

Consumables			
Material	**Launch Lab (Chapter)**	**MiniLAB (Chapter-Section)**	**Lab (Chapter-Section)**
algae cultures			2-2
aluminum foil			4-3
chalk			4-3
cloth			3-3
coffee filter		3-2	
coffee, hot chocolate, or powdered drink		3-2	
food coloring	4	1-1, 2-4, 4-2	3-3
gravel		3-1	
gravel, aquarium			3-3
ice	4		1-1
nitrate solution			2-2
nitrate-phosphate solution			2-2
pH paper or pH meter			2-4
phosphate solution			2-2
potato, small, uncooked			4-3
salt		4-2	4-3
sand		3-1	
straw			3-1, 4-3
string			2-4
tubing, flexible		2-1	
water, pond, lake, or ocean		5-2	
water, salt			3-3
Nonconsumables			
alkalinity kit			2-4
baby food jars, small			2-2
bottle with cap, tall, soft-plastic			3-1
bottles, sample, plastic			2-4
bowl, large glass			4-3
box, clear-plastic			3-3
bucket		2-1	2-4
clamp for thermometer			1-1
cork		4-3	
dissolved oxygen kit			2-4
drinking glass	1		
glass		2-4, 3-2, 4-2	
hammer			3-1
hot plate			1-1
nail or awl			3-1
nitrate test kit			2-4
pan, large, flat		2-1	
pennies	1		
protractor			3-1
pump			3-3
ring stand			1-1
rope			2-4
spoon		4-3	
spring	5		
storage box, clear plastic		4-2, 4-3	4-3
stream table		2-1	
teaspoon			4-3
tray tables		5-1	

McGraw Hill Glencoe

New York, New York   Columbus, Ohio   Chicago, Illinois   Peoria, Illinois   Woodland Hills, California

HOW TO...

Use Your Science Book

Why do I need my science book?

Have you ever been in class and not understood all of what was presented? Or, you understood everything in class, but at home, got stuck on how to answer a question? Maybe you just wondered when you were ever going to use this stuff?

These next few pages are designed to help you understand everything your science book can be used for . . . besides a paperweight!

Before You Read

- **Chapter Opener** Science is occurring all around you, and the opening photo of each chapter will preview the science you will be learning about. The **Chapter Preview** will give you an idea of what you will be learning about, and you can try the **Launch Lab** to help get your brain headed in the right direction. The **Foldables** exercise is a fun way to keep you organized.
- **Section Opener** Chapters are divided into two to four sections. The **As You Read** in the margin of the first page of each section will let you know what is most important in the section. It is divided into four parts. **What You'll Learn** will tell you the major topics you will be covering. **Why It's Important** will remind you why you are studying this in the first place! The **Review Vocabulary** word is a word you already know, either from your science studies or your prior knowledge. The **New Vocabulary** words are words that you need to learn to understand this section. These words will be in **boldfaced** print and highlighted in the section. Make a note to yourself to recognize these words as you are reading the section.

As You Read

- **Headings** Each section has a title in large red letters, and is further divided into blue titles and small red titles at the beginnings of some paragraphs. To help you study, make an outline of the headings and subheadings.
- **Margins** In the margins of your text, you will find many helpful resources. The **Science Online** exercises and **Integrate** activities help you explore the topics you are studying. **MiniLabs** reinforce the science concepts you have learned.
- **Building Skills** You also will find an **Applying Math** or **Applying Science** activity in each chapter. This gives you extra practice using your new knowledge, and helps prepare you for standardized tests.
- **Student Resources** At the end of the book you will find **Student Resources** to help you throughout your studies. These include **Science, Technology,** and **Math Skill Handbooks,** an **English/Spanish Glossary,** and an **Index.** Also, use your **Foldables** as a resource. It will help you organize information, and review before a test.
- **In Class** Remember, you can always ask your teacher to explain anything you don't understand.

FOLDABLES™ Study Organizer

Science Vocabulary Make the following Foldable to help you understand the vocabulary terms in this chapter.

STEP 1 **Fold** a vertical sheet of notebook paper from side to side.

STEP 2 **Cut** along every third line of only the top layer to form tabs.

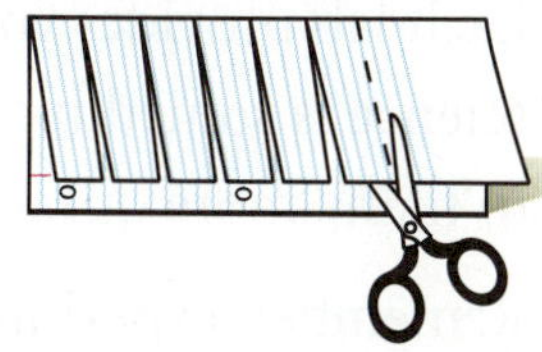

STEP 3 **Label** each tab with a vocabulary word from the chapter.

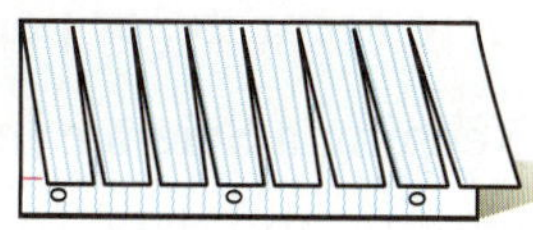

Build Vocabulary As you read the chapter, list the vocabulary words on the tabs. As you learn the definitions, write them under the tab for each vocabulary word.

In Lab

Working in the laboratory is one of the best ways to understand the concepts you are studying. Your book will be your guide through your laboratory experiences, and help you begin to think like a scientist. In it, you not only will find the steps necessary to follow the investigations, but you also will find helpful tips to make the most of your time.

- Each lab provides you with a **Real-World Question** to remind you that science is something you use every day, not just in class. This may lead to many more questions about how things happen in your world.
- Remember, experiments do not always produce the result you expect. Scientists have made many discoveries based on investigations with unexpected results. You can try the experiment again to make sure your results were accurate, or perhaps form a new hypothesis to test.
- Keeping a **Science Journal** is how scientists keep accurate records of observations and data. In your journal, you also can write any questions that may arise during your investigation. This is a great method of reminding yourself to find the answers later.

Before a Test

Admit it! You don't like to take tests! However, there *are* ways to review that make them less painful. Your book will help you be more successful taking tests if you use the resources provided to you.

- Review all of the **New Vocabulary** words and be sure you understand their definitions.
- Review the notes you've taken on your **Foldables,** in class, and in lab. Write down any question that you still need answered.
- Review the **Summaries** and **Self Check questions** at the end of each section.
- Study the concepts presented in the chapter by reading the **Study Guide** and answering the questions in the **Chapter Review.**

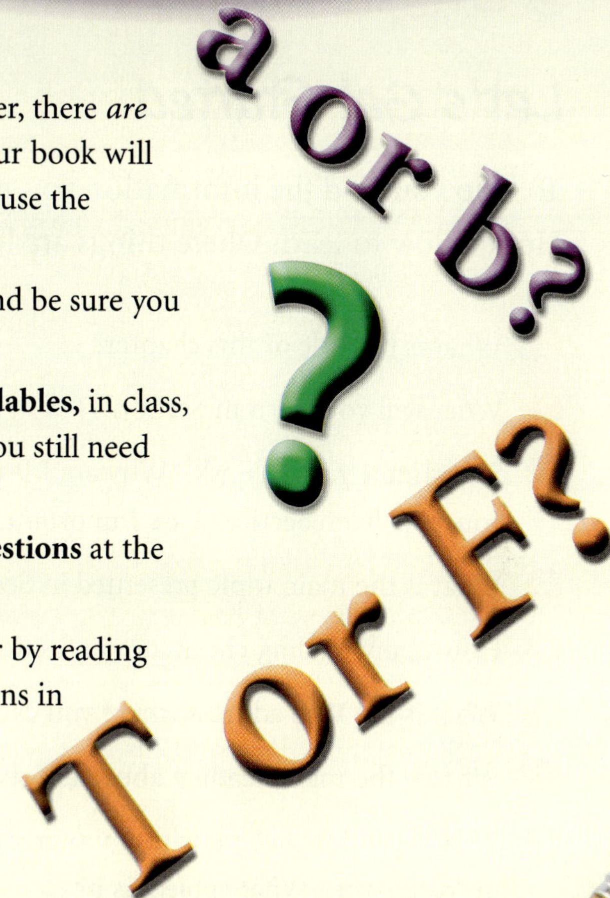

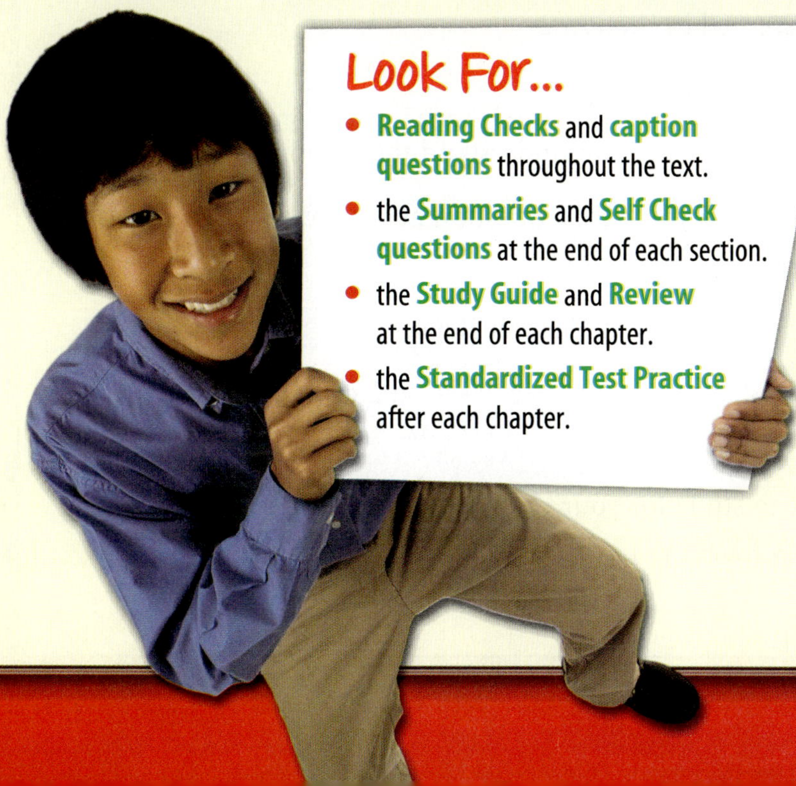

Look For...

- **Reading Checks** and **caption questions** throughout the text.
- the **Summaries** and **Self Check questions** at the end of each section.
- the **Study Guide** and **Review** at the end of each chapter.
- the **Standardized Test Practice** after each chapter.

Scavenger HUNT

Let's Get Started

To help you find the information you need quickly, use the Scavenger Hunt below to learn where things are located in Chapter 1.

1. What is the title of this chapter?
2. What will you learn in Section 1?
3. Sometimes you may ask, "Why am I learning this?" State a reason why the concepts from Section 2 are important.
4. What is the main topic presented in Section 2?
5. How many reading checks are in Section 1?
6. What is the Web address where you can find extra information?
7. What is the main heading above the sixth paragraph in Section 2?
8. There is an integration with another subject mentioned in one of the margins of the chapter. What subject is it?
9. List the new vocabulary words presented in Section 2.
10. List the safety symbols presented in the first Lab.
11. Where would you find a Self Check to be sure you understand the section?
12. Suppose you're doing the Self Check and you have a question about concept mapping. Where could you find help?
13. On what pages are the Chapter Study Guide and Chapter Review?
14. Look in the Table of Contents to find out on which page Section 2 of the chapter begins.
15. You complete the Chapter Review to study for your chapter test. Where could you find another quiz for more practice?

Teacher Advisory Board

The Teacher Advisory Board gave the authors, editorial staff, and design team feedback on the content and design of the Student Edition. They provided valuable input in the development of the 2005 edition of ***Glencoe Science.***

John Gonzales
Challenger Middle School
Tucson, AZ

Rachel Shively
Aptakisic Jr. High School
Buffalo Grove, IL

Roger Pratt
Manistique High School
Manistique, MI

Kirtina Hile
Northmor Jr. High/High School
Galion, OH

Marie Renner
Diley Middle School
Pickerington, OH

Nelson Farrier
Hamlin Middle School
Springfield, OR

Jeff Remington
Palmyra Middle School
Palmyra, PA

Erin Peters
Williamsburg Middle School
Arlington, VA

Rubidel Peoples
Meacham Middle School
Fort Worth, TX

Kristi Ramsey
Navasota Jr. High School
Navasota, TX

Student Advisory Board

The Student Advisory Board gave the authors, editorial staff, and design team feedback on the design of the Student Edition. We thank these students for their hard work and creative suggestions in making the 2005 edition of ***Glencoe Science*** student friendly.

Jack Andrews
Reynoldsburg Jr. High School
Reynoldsburg, OH

Peter Arnold
Hastings Middle School
Upper Arlington, OH

Emily Barbe
Perry Middle School
Worthington, OH

Kirsty Bateman
Hilliard Heritage Middle School
Hilliard, OH

Andre Brown
Spanish Emersion Academy
Columbus, OH

Chris Dundon
Heritage Middle School
Westerville, OH

Ryan Manafee
Monroe Middle School
Columbus, OH

Addison Owen
Davis Middle School
Dublin, OH

Teriana Patrick
Eastmoor Middle School
Columbus, OH

Ashley Ruz
Karrar Middle School
Dublin, OH

The Glencoe middle school science Student Advisory Board taking a timeout at COSI, a science museum in Columbus, Ohio.

Contents

Contents

Nature of Science: Exploring the Depths of the Water Planet—2

chapter 1

Water—6

In each chapter, look for these opportunities for review and assessment:

- Reading Checks
- Caption Questions
- Section Review
- Chapter Study Guide
- Chapter Review
- Standardized Test Practice
- Online practice at bookh.msscience.com

chapter 2

Freshwater at Earth's Surface—34

chapter 3

Groundwater Resources—66

Student Resources

Cross-Curricular Readings/Labs

available as a video lab

NATIONAL GEOGRAPHIC VISUALIZING

TIME SCIENCE AND Society

TIME SCIENCE AND HISTORY

Oops! Accidents in SCIENCE

Science and Language Arts

SCIENCE Stats

Launch LAB

Mini LAB

Mini LAB Try at Home

One-Page Labs

Two-Page Labs

Labs/Activities

Design Your Own Labs

Use the Internet Labs

Applying Math

Applying Science

INTEGRATE

Science Online

Standardized Test Practice

Content Details

The NATURE OF SCIENCE

Exploring the Depths of the Water Planet

Introduction

This feature introduces students to the history of oceanography by capturing their interest and then guiding them through some significant discoveries about the characteristics of oceans and the organisms that live in them. This feature also introduces the concept that ideas in science change as new discoveries are made. In some cases, with the collection of new data and scientific research, old ideas can be overturned. In other instances, observations made and data collected support previously stated theories.

1 Motivate

Have students make a list of everything they know about oceans. Students could make columns with headings such as Life in the Ocean, Physical Characteristics of Ocean Water (temperature, waves, tides, currents, depth), and The Seafloor. A heading of Miscellaneous could be used to list any other facts they know about oceans. Encourage students to make as many subheadings as they can, such as information about the biology of the organisms they list. Have students review their lists, and point out that the information they have listed was not always known. Encourage students to give examples of the type of equipment used to gather information about oceans.

The NATURE OF SCIENCE

History of Science

Exploring the Depths of the Water Planet

Figure 1 This diver is taking samples to learn more about sediments on the ocean floor.

Figure 2 Some of the major events in ocean exploration and discovery are shown in the time line.

Humans have been exploring and collecting data from Earth's oceans for thousands of years. Despite all that people have learned about oceans, almost 90 percent of Earth's vast oceans have yet to be visited by humans.

Many early civilizations regularly navigated long distances on the ocean relying on knowledge of the stars and ocean currents. By 800 B.C., the Phoenicians and Greeks had learned to navigate the Mediterranean Sea. The Vikings, Arabs, and Chinese also were sailors. When the compass, a Chinese invention, became standard sailing equipment in the thirteenth century, sailors were able to navigate more precisely.

Ferdinand Magellan's voyage around the world (1519–1522) proved a theory that had been stated by ancient Greeks 2,000 years earlier—Earth is round. This accurate idea of Earth's shape and new instruments allowing vessels to calculate latitude and longitude allowed explorers to begin mapping the shapes of Earth's oceans and landmasses. By the late 1700s, Antarctica was the only area on Earth that remained uncharted by Europeans.

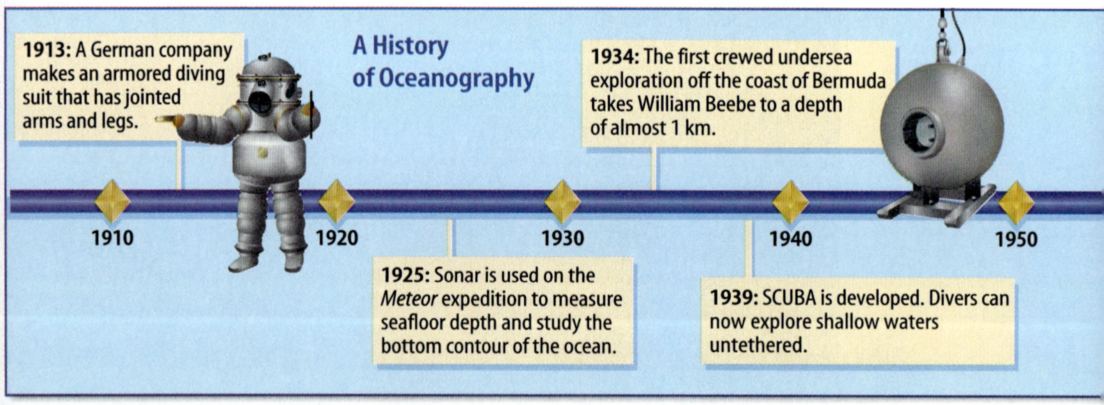

Visual Learning

Figure 2 The timeline in Figure 2 shows some of the events in oceanography during the Twentieth century. Have students study the timeline and explain how the events relate to the different science disciplines of biology, geology, and physical science. For example, undersea exploration by Beebe is related to biology because he observed organisms living at depth. The discovery of hydrothemal vents is related to biology, geology, and physical science.

Ocean Discoveries

As people continued to explore the oceans, many discoveries were made about what was in the oceans and how currents functioned. The discovery of a stalked sea lily—an organism that had previously only been seen in fossils that were millions of years old—supported Darwin's earlier theory that the ocean could give clues about Earth's evolutionary past. The discovery of this organism, recovered from a depth of more than 3 km, further fueled the growing demand for a large-scale expedition to study the deep sea.

Broad-based undersea exploration began with the voyage of the *Challenger* from 1872 to 1876. This British vessel circled Earth conducting soundings (measurements of depth) with weighted wire, gathering samples from the deep ocean, and recording information on salinity, temperature, and water density. Although data from the voyage provided a wealth of new knowledge, the voyage is best known for the revelation that the ocean floor is mountainous and for the discovery of about 4,700 species of marine life. Since many of the organisms identified on the voyage were collected from a depth of 1 km or greater, the *Challenger* research also helped to disprove an earlier theory that life only can exist as deep as 0.5 km below the ocean's surface.

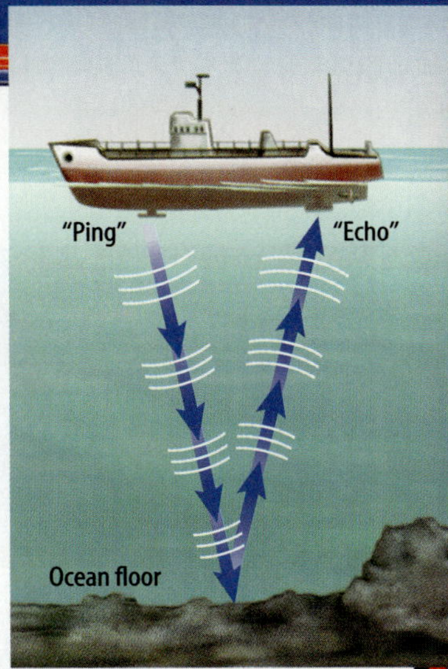

Figure 3 Scientists use sonar to gain a clearer picture of the depth and the contours of the ocean floor.

Sonar

From 1925 to 1927, the German vessel *Meteor* crisscrossed the Atlantic Ocean with an electronic echo sounder, or sonar—Sound Navigation Ranging. Analysis of these and later sonar data helped to give a clearer picture of the undersea world. The data were used to develop accurate maps of the rough ocean floor and to identify three-dimensional objects.

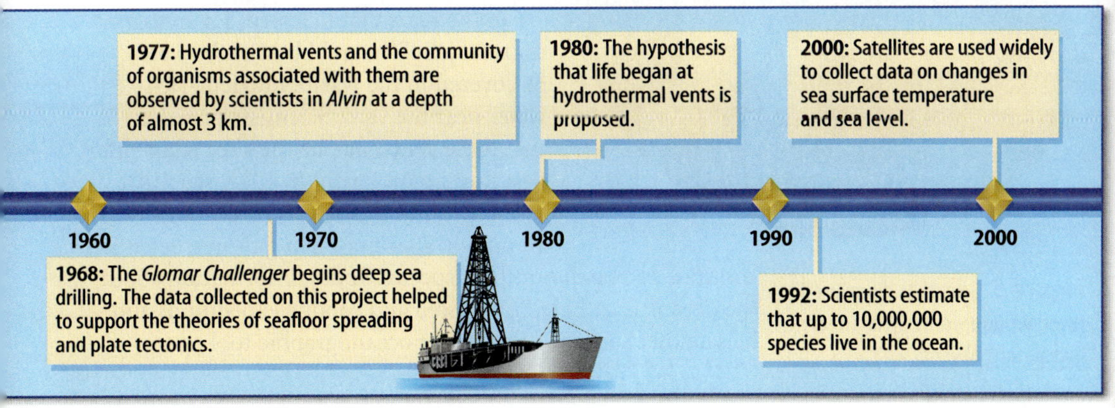

2 Teach

Content Background

Although the number of organisms that live in the waters of and on the floor of the deep sea are scarce compared to the number of organisms living in other areas of the ocean, the quest to gather information about these elusive organisms continues. Organisms that live in the abyss exist under the relatively constant environmental conditions of high pressure, cold temperatures, and complete darkness. Many of these organisms have small eyes or none at all. Decomposers such as bacteria are found in deep-sea sediments. Different types of worms, sea stars, and sea cucumbers consume sediment, removing organic material as their source of food. Other organisms such as glass sponges and crinoids live as filter feeders attached to the bottom. Fish, such as the tripod fish, can also be found living in the abyss.

Discussion

What type of information can satellites provide scientists about oceans? Possible answers: provide information on surface temperature, which can be used to study surface currents such as the Gulf Stream, and also can provide information about an El Nino event; collect data about primary production and phytoplankton blooms; record changes in ice coverage; measure wave height and changes in sea level

Curriculum Connection

Geography Have students use a world map to locate which ocean corresponds with the following latitude and longitude coordinates:

40N, 50W Atlantic Ocean
70S, 20E Antarctic Ocean
10S, 100E Indian Ocean
20N, 160W Pacific Ocean

THE NATURE OF SCIENCE H ◆ 3

Use Science Words

Word Origin Have students use a dictionary to find the origin of the word *ocean*. Possible answer: The word *ocean* is derived from the Latin *oceanus,* which comes from the Greek word *okeanos* meaning "a river that encircles Earth." When the Greeks first ventured outside the Mediterranean Sea almost 3,000 years ago, they presumed that currents were only found in rivers. When they discovered a southward current flowing outside of the Mediterranean, they thought that the water beyond them, which was really the Atlantic Ocean, was just a large river that they could not see across. *Okeanos* was the name the Greeks gave this so-called river. LS **Linguistic**

Activity

Have students read an account of a scientist's experiences aboard a deep-sea submersible. After completing some background reading, have small groups of students work together to write and perform a skit about a journey to the depths of the ocean. Students can focus on the mechanics of operating a submersible as well as what may be seen (including organisms, hydrothermal vents, mid-ocean ridges, etc.) while on the journey. LS **Interpersonal**

Figure 4 *Alvin* has allowed scientists to see some of the deepest and darkest parts of the ocean.

Exploration Technology

Seven years later, William Beebe and Otis Barton became the first humans to observe ocean life from a depth of almost 1 km. They watched from inside their diving vessel called a bathysphere as it was lowered from a ship on a cable seven eighths of an inch thick. The 1939 development of SCUBA—Self-Contained Underwater Breathing Apparatus—allowed divers to explore shallow waters. In 1960, the quest for depth was complete when Jacques Piccard and Don Walsh reached the Challenger Deep—the deepest point in the ocean at about 11 km below the surface—in an improved diving vessel, the bathyscaphe *Trieste.*

Scientific curiosity didn't stop at the bottom of the ocean. In 1968, a team of scientists aboard the *Glomar Challenger,* a ship designed to drill and collect samples from the deep ocean floor, began to drill through the oceanic crust. The goal was to obtain samples from the boundary between the crust and mantle of Earth. The drilling, sonar data, and sediment samples provided support for the then-controversial theory of seafloor spreading which suggested that new seafloor is formed along an underwater system of ridges.

Figure 5 Scientists discovered these unique worms living near hydrothermal vents.

Discovering New Species

Meanwhile, the U.S. Navy began developing crewed and robotic crafts to explore the depths of the ocean. In 1977, the U.S. Navy's *Alvin*—a deep-sea craft—was used to view a hydrothermal vent.

In the years that followed, 1.5-m worms, unusual jellyfish, and blind crabs were discovered in the warm, nutrient-rich water near the vent. Scientists proposed that these vents may be the sites where life first formed on Earth. In 1992, scientists estimated that the oceans hold about 10 million species of living things. Scientists continue to search out these species. Today, information about the oceans is gathered by crewed and robotic expeditions and by satellite. Sonar is still a major oceanographic tool.

Science Journal

Giant Squid Have students research information about the elusive giant squid. Encourage them to record in their science journals any information on recent discoveries, as well as what is known about the biology and life cycle of giant squids. LS **Linguistic**

The History of Oceanography

Knowledge of the oceans has grown in leaps and bounds, yet much is still to be discovered. Understanding the oceans is key to understanding Earth's ecosystem and its history.

Changing Courses

Science is a long process of understanding nature better. Today, scientists know that the oceans are changing shape and are full of life. Explorations of the depths have given surprising results. Each of these surprises leads to further research and exploration.

It might be hard to imagine that people once thought Earth was flat and that the ocean floor was a flat, underwater wasteland. It's important to remember that as science makes discoveries, old ideas can be overturned. These overturned ideas, however, are the beginnings of scientific research. In the process of disproving them, scientists learn more about Earth. Scientists work in the hope that later scientists will be able to use their observations to make further advances.

Building on the Past

Although some scientific ideas may change, the foundations of today's science consist of information that is now considered ancient. Without the compass and the invention of a clock in 1736 that could keep accurate time at sea, precise navigation using latitude and longitude would not have been possible. Scientists don't just disprove what others have done, they also build on their successes. So, over time, the understanding of the world increases.

Figure 6 The chart of the Gulf Stream that Franklin published in 1777 is remarkably accurate when compared to today's satellite images.

Franklin's chart was developed with the help of his cousin, Timothy Folger, and other sea captains who sailed in the North Atlantic.

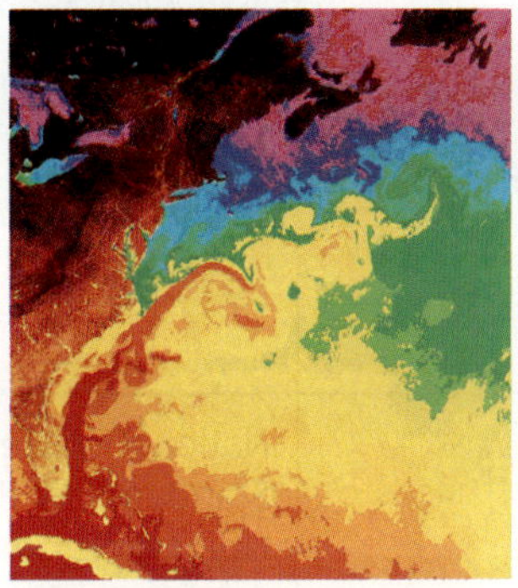

The warm water carried by the Gulf Stream shown here in red and orange, stands out against the surrounding colder water in this satellite image.

You Do It

In the 1920s, sonar helped scientists to picture the ocean floor before they could see it firsthand. Today, more sophisticated forms of sonar are used by the military, by researchers, and even by doctors. Research the history of sonar. How does it work? How has it been refined since its invention? How is it used today?

Fun Fact

Organisms have been found living even in the deepest area of the ocean. Sea anemones, sea cucumbers, worms, and at least one type of crustacean are known to live at a depth of 11,000 m in the Mariana Trench.

Make a Model

Have students use clay to make models of the seafloor. Students can shape the clay to show the contours of an ocean basin including the continental shelf and slope, abyssal plains, mid-ocean ridges, seamounts, volcanoes, and trenches. Students can reference chapters found in the text for more information on the features of an ocean basin. Remind students that the contours of ocean basins are known as a result of analyzing sonar data. LS **Kinesthetic**

3 Assess

Use the questions in You Do It to assess understanding. Answers should relate to the use of sonar and should be as specific as possible.

You Do It

Explain to students the basic concept of how sonar works. A sound wave travels through water until it reaches the ocean floor, where it bounces back to the surface. The speed of the wave is known and the travel time is recorded. With these variables known, the depth of the seafloor can be calculated using the following formula: Depth = (speed × time)/2.

chapter 1 Organizer

<table>
<tr><th rowspan="2">Section/Objectives</th><th colspan="2">Standards</th><th rowspan="2">Labs/Features</th></tr>
<tr><th>National</th><th>State/Local</th></tr>
<tr><td>Chapter Opener</td><td>See pp. 9T–10T for a Key to Standards.</td><td></td><td>Launch Lab: What is cohesion?, p. 7
Foldables, p. 7</td></tr>
<tr><td>Section 1 The Nature of Water
3 sessions
1.5 blocks
1. Explain that water exists on Earth in three states.
2. Describe several unique properties of water.
3. Explain that water is a polar molecule.</td><td>National Content Standards: UCP.2, UCP.3, UCP.5, A.1, A.2, B.1</td><td></td><td>Science Online, p. 9
Integrate Life Science, p. 10
MiniLAB: Examining Density Differences, p. 11
Applying Science: How does water behave in space?, p. 13
Integrate Life Science, p. 14
Lab: Discovering Latent Heat, p. 15</td></tr>
<tr><td>Section 2 Why is water necessary?
2 sessions
1 block
4. Explain that water is essential for life on Earth.
5. Describe ways that society uses water.
6. Describe methods for conserving freshwater.</td><td>National Content Standards: UCP.2, UCP.3, UCP.5, A.1, A.2</td><td></td><td>MiniLAB: Predicting Water Use, p. 18
Visualizing Shipping, p. 19</td></tr>
<tr><td>Section 3 Recycling Water
3 sessions
1.5 blocks
7. Identify Earth's water reservoirs.
8. Describe sources of freshwater on Earth.
9. Explain how water is recycled.</td><td>National Content Standards: UCP.2, UCP.3, UCP.5, A.1, A.2, F.2</td><td></td><td>Integrate History, p. 23
Lab: Conserving Water, p. 26
Science and Society: Not a Drop to Drink, p. 28</td></tr>
</table>

Glencoe Exclusive! TeacherWorks™ All-In-One Planner and Resource Center

Lab Materials	Reproducible Resources	Section Assessment	Technology
Launch Lab: water, drinking glass, several pennies	**Chapter *FAST FILE* Resources** Foldables Worksheet, p. 17 Directed Reading Overview, p. 19 Note-taking Worksheets, pp. 33–35	GLENCOE'S ASSESSMENT ADVANTAGE	TeacherWorks includes: • Interactive Teacher Edition • Lesson Planner with calendar • Access to all program blacklines • Correlations to standards • Web links
MiniLAB: water, 2 beakers, food coloring, pipette, refrigerator **Lab:** hot plate, ring stand, 400-mL beaker, crucible tongs, foam cup, clamp for thermometer, Celsius thermometer, 250-mL graduated cylinder, crushed ice	**Chapter *FAST FILE* Resources** Transparency Activity, p. 44 MiniLAB, p. 3 Lab Activity, pp. 9–12 Enrichment, p. 30 Reinforcement, p. 27 Directed Reading, p. 20 Lab Worksheet, pp. 5–6 **Mathematics Skill Activities,** p. 9	**Portfolio** Differentiated Instruction, p. 12 **Performance** MiniLAB, p. 11 Applying Science, p. 13 **Content** Section Review, p. 14	Section Focus Transparency Virtual Labs CD-ROM Guided Reading Audio Program Interactive Chalkboard CD-ROM
MiniLAB: bathtub/shower combination, water, meterstick	**Chapter *FAST FILE* Resources** Transparency Activity, p. 45 MiniLAB, p. 4 Lab Activity, pp. 13–16 Enrichment, p. 31 Reinforcement, p. 28 Directed Reading, p. 21 **Cultural Diversity,** p. 33	**Portfolio** Curriculum Connection, p. 18 **Performance** MiniLAB, p. 18 Applying Science, p. 21 **Content** Section Review, p. 21	Section Focus Transparency Guided Reading Audio Program Interactive Chalkboard CD-ROM
Lab: calculator *Need materials?* Contact Science Kit at 1-800-828-7777 or www.sciencekit.com on the Internet.	**Chapter *FAST FILE* Resources** Transparency Activity, p. 46 Enrichment, p. 32 Reinforcement, p. 29 Directed Reading, pp. 21, 22 Transparency Activity, pp. 47–48 Lab Worksheet, pp. 7–8 **Reading and Writing Skill Activities,** p. 19	**Portfolio** Reteach, p. 25 **Performance** Applying Math, p. 25 **Content** Section Review, p. 25	Section Focus Transparency Teaching Transparency Guided Reading Audio Program Interactive Chalkboard CD-ROM Video Lab

End of Chapter Assessment

GLENCOE'S ASSESSMENT ADVANTAGE

Blackline Masters	Technology	Professional Series
Chapter *FAST FILE* Resources Chapter Review, pp. 37–38 Chapter Tests, pp. 39–42 **Standardized Test Practice,** pp. 7–10	MindJogger Videoquiz Virtual Labs CD-ROM Exam*View*® Pro Testmaker TeacherWorks CD-ROM Interactive Chalkboard CD-ROM	**Performance Assessment in the Science Classroom (PASC)**

Transparencies

Section Focus

Section 1 Section Focus Transparency — Denting the Water

If you look closely at this water strider's feet, you can see that it is standing on top of the water. Water striders can stand on the surface of water without breaking through.

1. How can you tell this water strider is on top of the water, rather than floating in it?
2. What are the three different forms that water can take? Give some examples of each one in nature.
3. Does ice sink or float? What does this indicate about ice?

L2

Section 2 Section Focus Transparency — Artificial Oasis

This community is located in a desert. The plants and animals native to the desert can tolerate the very dry conditions. Do you think the grass you see in the photo is native to the desert?

1. What does the word *resource* mean?
2. How does this community maintain green grass in such an arid environment?
3. How do you use water during a typical day?

L2

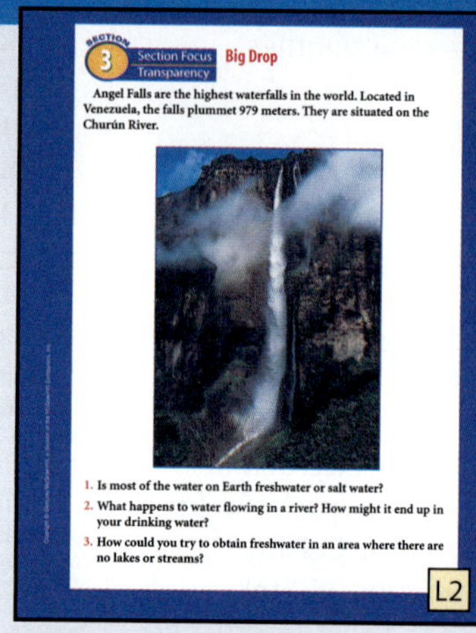
Section 3 Section Focus Transparency — Big Drop

Angel Falls are the highest waterfalls in the world. Located in Venezuela, the falls plummet 979 meters. They are situated on the Churún River.

1. Is most of the water on Earth freshwater or salt water?
2. What happens to water flowing in a river? How might it end up in your drinking water?
3. How could you try to obtain freshwater in an area where there are no lakes or streams?

L2

This is a representation of key blackline masters available in the Teacher Classroom Resources. See Resource Manager boxes within the chapter for additional information.

Assessment

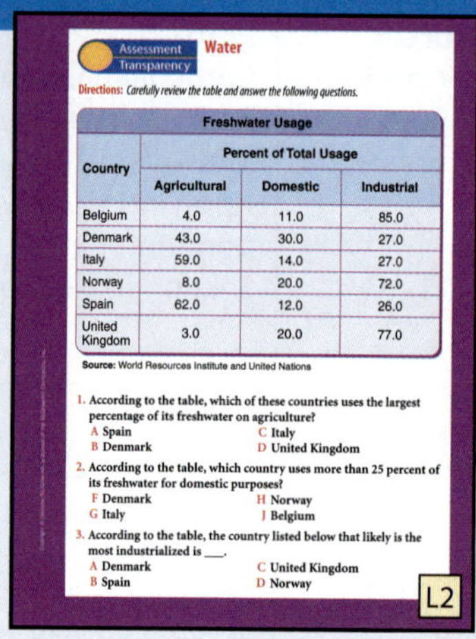
Assessment Transparency — Water

Directions: Carefully review the table and answer the following questions.

Freshwater Usage

Country	Percent of Total Usage		
	Agricultural	Domestic	Industrial
Belgium	4.0	11.0	85.0
Denmark	43.0	30.0	27.0
Italy	59.0	14.0	27.0
Norway	8.0	20.0	72.0
Spain	62.0	12.0	26.0
United Kingdom	3.0	20.0	77.0

Source: World Resources Institute and United Nations

1. According to the table, which of these countries uses the largest percentage of its freshwater on agriculture?
A Spain C Italy
B Denmark D United Kingdom
2. According to the table, which country uses more than 25 percent of its freshwater for domestic purposes?
F Denmark H Norway
G Italy J Belgium
3. According to the table, the country listed below that likely is the most industrialized is ___.
A Denmark C United Kingdom
B Spain D Norway

L2

Teaching

Section 3 Teaching Transparency — The Water Cycle

L2

Key to Teaching Strategies

The following designations will help you decide which activities are appropriate for your students.

- L1 Level 1 activities should be appropriate for students with learning difficulties.
- L2 Level 2 activities should be within the ability range of all students.
- L3 Level 3 activities are designed for above-average students.
- ELL ELL activities should be within the ability range of English Language Learners.
- COOP LEARN Cooperative Learning activities are designed for small group work.
- LS Multiple Learning Styles logos, as described on page 6T, are used throughout to indicate strategies that address different learning styles.
- P These strategies represent student products that can be placed into a best-work portfolio.
- PBL Problem-Based Learning activities apply real-world situations to learning.

Hands-on Activities

Student Text Lab Worksheet

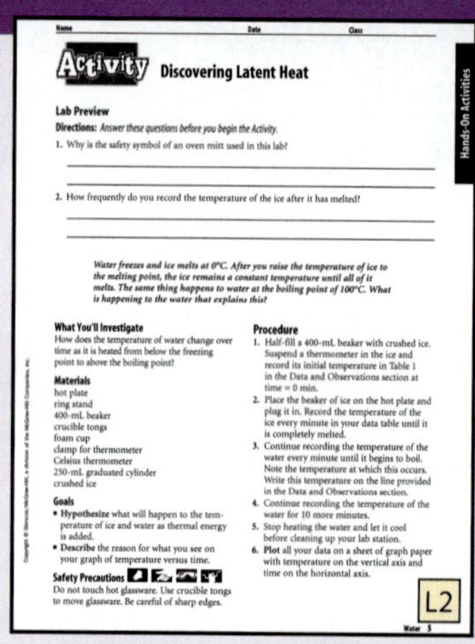
Activity — Discovering Latent Heat

Lab Preview

Directions: Answer these questions before you begin the Activity.

1. Why is the safety symbol of an oven mitt used in this lab?
2. How frequently do you record the temperature of the ice after it has melted?

Water freezes and ice melts at 0°C. After you raise the temperature of ice to the melting point, the ice remains a constant temperature until all of it melts. The same thing happens to water at the boiling point of 100°C. What is happening to the water that explains this?

L2

Laboratory Activities

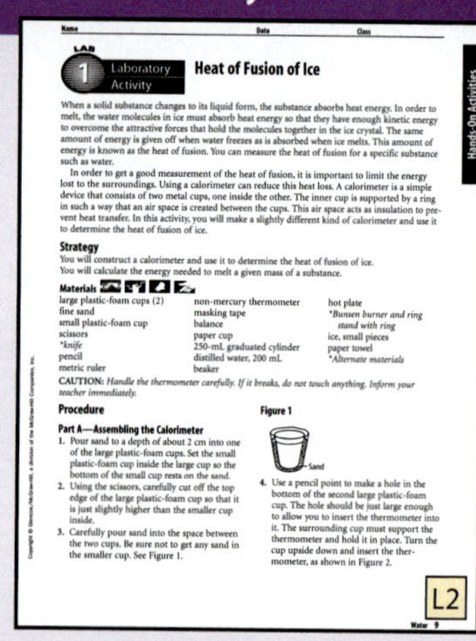
Lab 1 Laboratory Activity — Heat of Fusion of Ice

L2

Meeting Different Ability Levels

Content Outline

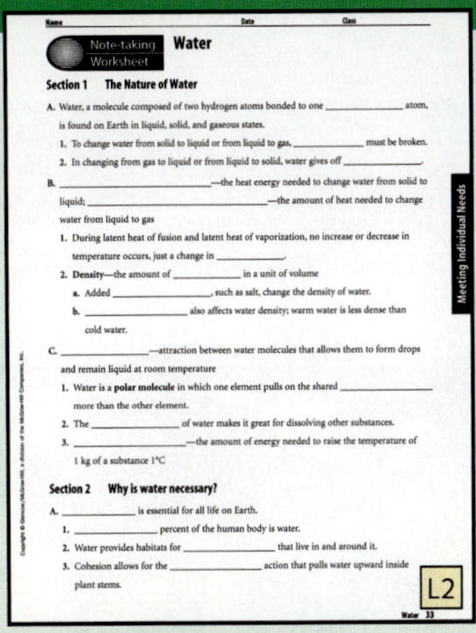

Reinforcement

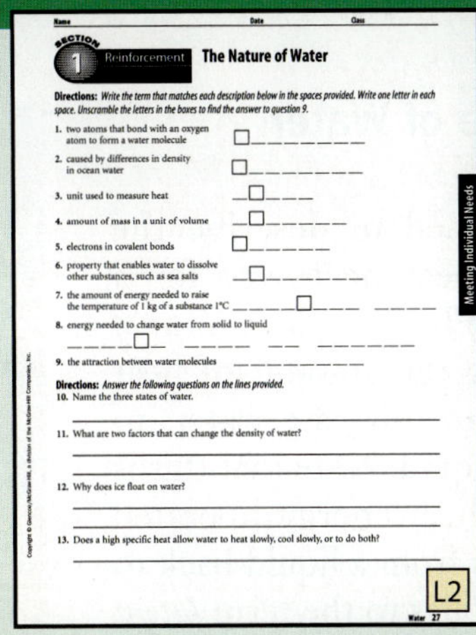

Enrichment

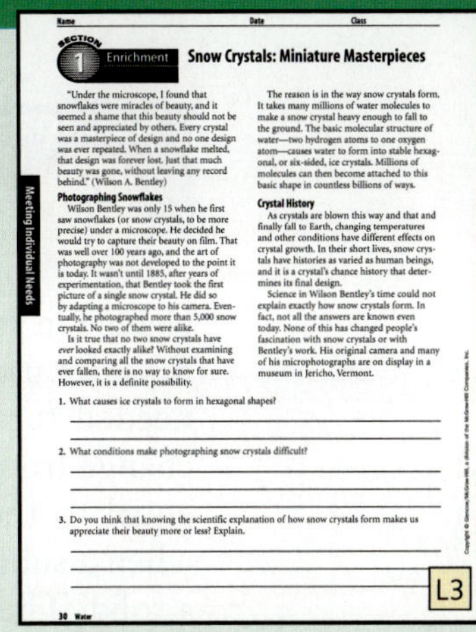

Directed Reading (English/Spanish)

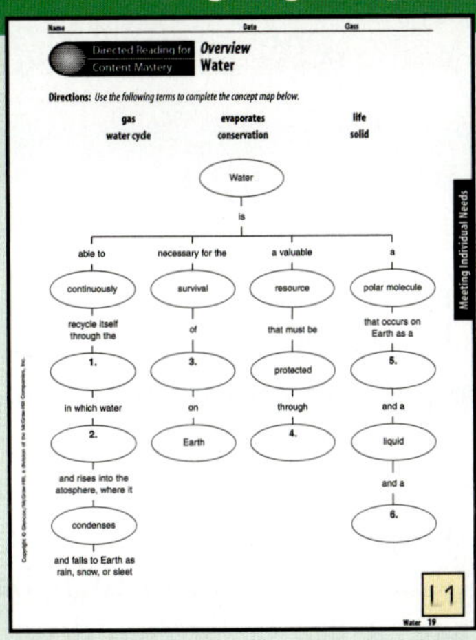

Study Guide

Reading Essentials

Assessment

Test Practice Workbook

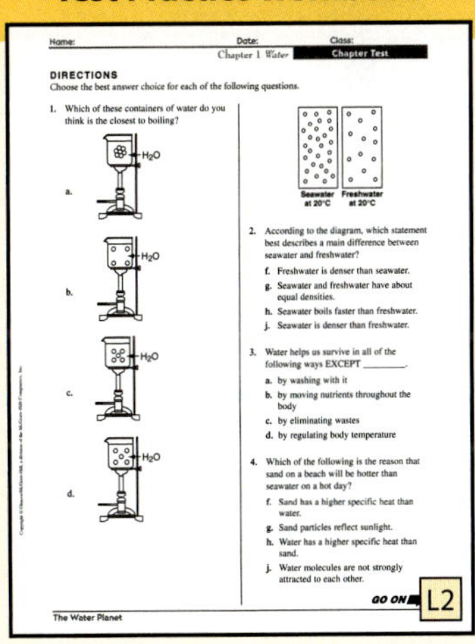

Chapter Review

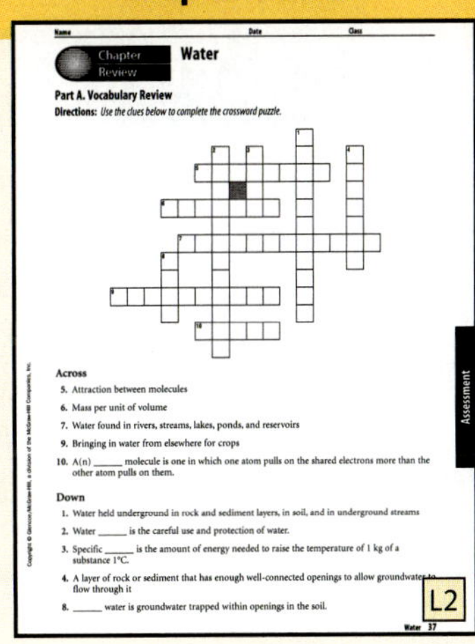

Chapter Tests

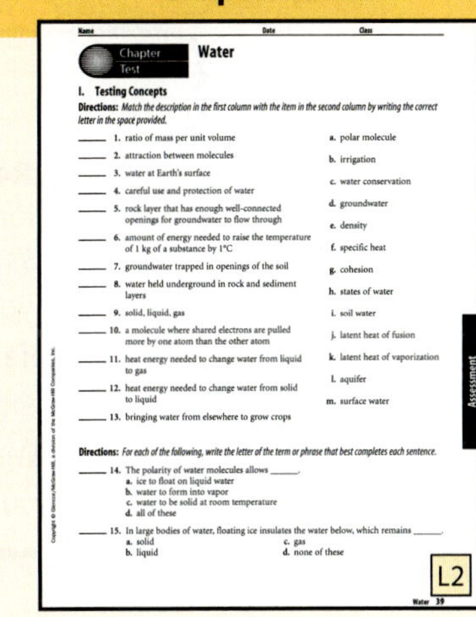

Science Content Background

section 1 The Nature of Water

Latent Heat

Terms are used to describe the amount of heat necessary for a change in state to occur. The term *latent heat of fusion* is used to describe the amount of heat needed for a substance, in this case water, to change from a solid to a liquid. Note that this is equal to the amount of heat energy liberated when a substance changes from a liquid back to a solid. The same rule applies to the term *latent heat of vaporization*, which is used to describe the amount of heat needed for a change in state from a liquid to a gas. This is equal to the amount of heat liberated when the state changes from a gas back into a liquid. Water and other substances also can undergo *sublimation*, the term used to describe a phase change from a solid directly to a gas. Water vapor is released directly from snow or ice during this process.

section 2 Why is water necessary?

Water and Life

Because water is essential to life, many plants and animals that live in desert areas, where there is little available water, have adaptations that allow them to reduce water loss or store water. Camels are among several desert animals that can reduce

Jose Fuste Raga/CORBIS

the amount of moisture lost through respiration. When exhaled air that has been warmed within the body is passed over the cooler mucus membranes that line the nasal passages, water condenses. This water is subsequently reabsorbed by the body. Another adaptation to desert life is that camels store all of their body fat in humps. That fat can be broken down to release water as needed. This fat-water reserve can allow camels to survive for up to 10 days without drinking.

chapter content resources

Internet Resources

For additional content background, visit bookh.msscience.com to:

- access your book online
- find references to related articles in popular science magazines
- access Web links with related content background
- access current events with science journal topics

Print Resources

Glossary of Hydrology edited by W. E. Wilson and J. E. Moore, American Geological Institute

Water and the Environment by Vandas, S. J., and others, American Geological Institute

Principles of Hydrology by Ward, R. and Robinson, M., McGraw-Hill Higher Education, 1999

> **Teacher to Teacher**
> Marie Renner
> Diley Middle School
> Pickerington, Ohio
>
> "I use a problem-based learning activity to help students internalize the topic of water conservation. I create a newspaper article that alerts townspeople (students) to a water shortage. 'Residents' are invited to a town meeting where the students role-play various townspeople. During the meeting, students brainstorm and discuss what they already know about the problem, what they need to know to propose a solution, and possible solutions that consider economics and practicality."
>
> *Marie Renner*

The kangaroo rat, also found in desert habitats, survives exclusively on the water that is contained in the seeds that it consumes. This small mammal can go a lifetime without actually drinking any water. A type of beetle living in the Namib desert in western Africa climbs to the tops of sand dunes every morning in order to collect the drops of condensation that form on the outside of its body as day breaks.

Many desert plants have adaptations that allow them to store large amounts of water through long periods of time. Some succulent plants, such as cacti, can take in up to 700 L of water during a rare rainstorm. Some plants produce a toxin that is added to the water in order to discourage animals from using them as a water source.

section 3 Recycling Water

Oceans

Removing salt from ocean water is done through a process called desalination. This can be done by several different methods. Salt can be removed from water through distillation, which involves boiling the salt water, collecting and cooling the steam, and filtering the resulting freshwater condensation. Although distillation is the most commonly used method of desalination worldwide, one disadvantage is that the high heat of vaporization of water forces the input of high amounts of heat energy to continue the boiling process.

Using solar power to heat ocean water and remove the salt is another alternative. Solar humidification is similar to distillation in that the condensation of evaporated water is collected. In reverse osmosis water is forced through a semi-permeable membrane under high pressure. As the water passes through the membrane, the salt is left behind. All three of these methods, as well as others, are being used in different areas of the world, including the Middle East, West Africa, and in the United States in Texas, Florida, and California to provide necessary freshwater. Saudi Arabia has the world's largest desalination plant, which delivers approximately 100 million liters of freshwater a day to surrounding areas.

Water in the Atmosphere

Absolute humidity is a measure of the amount of water vapor the air contains. *Relative humidity* is the term used to describe the ratio of the amount of moisture the air actually contains compared to the amount it can hold at a given temperature. For example, if the relative humidity is 30%, then the air is holding only about one third of the total moisture it can possibly hold. The amount of water vapor the air can hold is temperature dependent. As the temperature cools, the amount of water vapor the air can hold decreases. If the temperature of the air decreases to the point at which the air becomes saturated with water vapor, water begins to condense. If this cooling occurs near Earth's surface, the outcome is fog or dew. If the cooling occurs as an air mass rises into the atmosphere, clouds will form.

chapter 1

Chapter Vocabulary

density, p. 11
cohesion, p. 12
polar molecule, p. 12
specific heat, p. 14
irrigation, p. 17
water conservation, p. 20
groundwater, p. 23
soil water, p. 23
aquifer, p. 23
surface water, p. 24

Science Journal Paragraphs should address both residential and production uses.

INTERACTIVE CHALKBOARD with Image Bank

PowerPoint® Presentations

This CD-ROM is an editable Microsoft® PowerPoint® presentation that includes:

- a pre-made presentation for every chapter
- interactive graphics
- animations
- audio clips
- image bank
- all new section and chapter questions
- Standardized Test Practice
- transparencies
- pre-lab questions for all labs
- Foldables directions
- links to bookh.msscience.com

chapter 1

Water

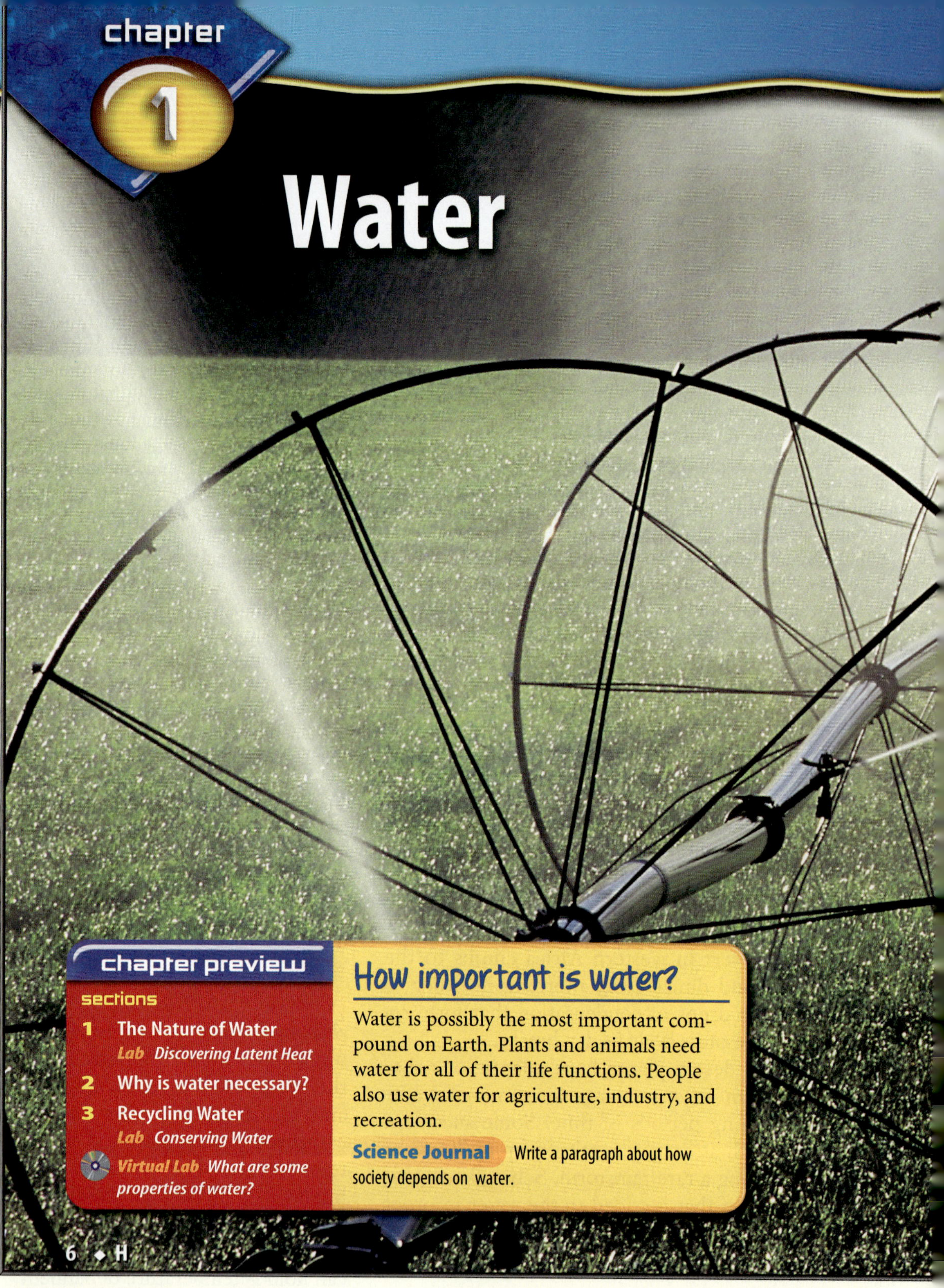

chapter preview

sections

How important is water?

Water is possibly the most important compound on Earth. Plants and animals need water for all of their life functions. People also use water for agriculture, industry, and recreation.

Science Journal Write a paragraph about how society depends on water.

6 ◆ H

Theme Connection

Systems and Interactions The special properties of water are discussed, including that it exists in all three states on Earth. The polarity of the water molecule is responsible for some of its unique properties, such as cohesion. Students will also learn that water, which is necessary for life on Earth, is constantly recycled through the water cycle.

About the Photo

Irrigation More water is used for agriculture than for any other purpose. The photo shows an aerial irrigation system. Such systems commonly are used to irrigate grain fields in the Great Plains region of the United States.

Start-Up Activities

Launch LAB

What is cohesion?

Water is an amazing substance. It has many unusual properties. One of these properties is cohesion, which is the ability of water molecules to attract each other. Observe cohesion as you do this lab.

1. Fill a drinking glass to the rim with water.
2. Slowly and carefully slip pennies into the glass one at a time.
3. After adding several pennies, observe the surface of the water at eye level.
4. **Think Critically** In your Science Journal, draw a picture of the water in the glass after the pennies were added. How might cohesion affect the water's surface?

FOLDABLES Study Organizer

Water Make the following Foldable to help you understand the vocabulary terms in this chapter.

STEP 1 **Fold** a vertical sheet of notebook paper from side to side.

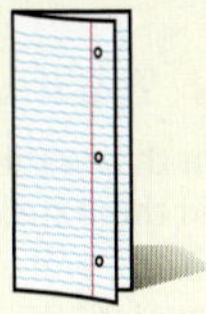

STEP 2 **Cut** along every third line of only the top layer to form tabs.

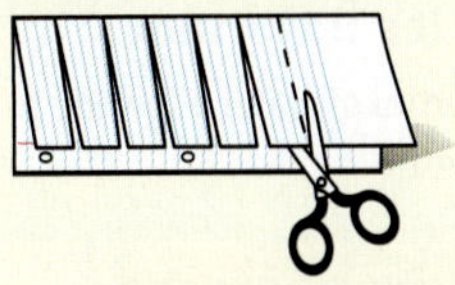

Build Vocabulary As you read the chapter, list the vocabulary words on the tabs. As you learn the definitions, write them under the tab for each vocabulary word.

Science Online | Preview this chapter's content and activities at bookh.msscience.com

H ◆ 7

Launch LAB

Purpose Use the Launch Lab to introduce students to one property of water—cohesion. L2 LS **Visual-Spatial**

Preparation Explain that a slight negative charge on the oxygen side of water molecules and a slight positive charge on the hydrogen side of water molecules cause the cohesion between molecules.

Materials clear drinking glass, water, pennies

Teaching Strategy A narrow glass will work best for this activity.

Think Critically

Pictures should show the water with a convex surface above the glass. Paragraphs should explain how the water molecules attract one another enough to overcome the tendency to spill over the rim of the glass.

Assessment

Content Have students think of other ways to show the cohesiveness of water molecules and perform a demonstration for the class. For example, they might balance a needle on the surface of water or show how water beads up on a surface. Use **Performance Assessment in the Science Classroom,** p. 95.

FOLDABLES Study Organizer **Dinah Zike Study Fold**

Student preparation materials for this Foldable are available in the Chapter *FAST FILE* Resources.

section 1

The Nature of Water

as you read

***What* You'll Learn**

- **Explain** that water exists on Earth in three states.
- **Describe** several unique properties of water.
- **Explain** that water is a polar molecule.

***Why* It's Important**

The properties of water allow life to exist on Earth.

Review Vocabulary

states of matter: the physical forms in which all matter naturally exists on Earth, most commonly as a solid, a liquid, or a gas

New Vocabulary

- density
- cohesion
- polar molecule
- specific heat

Forms of Water

Think about the water you use every day. When you drink a glass of water, you use it in liquid form. When you put ice in a drink, you're using it in solid form. Even when you breathe, you are using water. Along with the oxygen, nitrogen, and other gases from Earth's atmosphere, you inhale gaseous water with every breath you take.

The fact that water exists on Earth as a liquid, a gas, and a solid is one of its unique properties. Water is a simple molecule composed of two hydrogen atoms bonded to one oxygen atom. Yet water has several unusual properties that make it important here on Earth. Indeed, without water, this planet would be a different place from what is seen in **Figure 1.**

Reading Check *Which two elements occur in molecules of water?*

Figure 1 Water is abundant on Earth's surface.
Identify *the places where you see water in this photograph.*

1 Motivate

INTERACTIVE CHALKBOARD PowerPoint® Presentations

Bellringer

Section Focus Transparencies also are available on the Interactive Chalkboard CD-ROM. L2 ELL

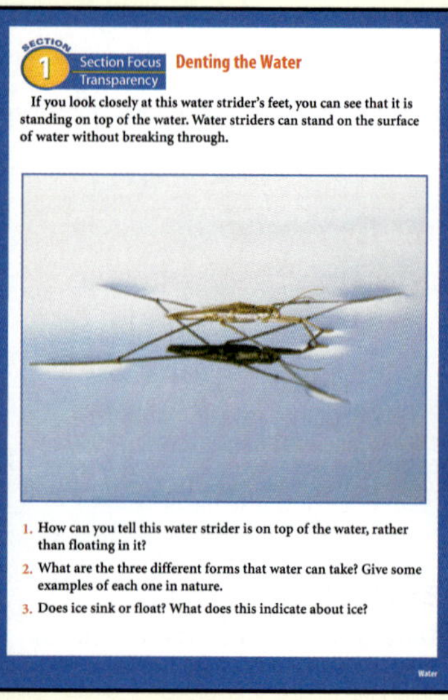

SECTION 1 Section Focus Transparency **Denting the Water**

If you look closely at this water strider's feet, you can see that it is standing on top of the water. Water striders can stand on the surface of water without breaking through.

1. How can you tell this water strider is on top of the water, rather than floating in it?
2. What are the three different forms that water can take? Give some examples of each one in nature.
3. Does ice sink or float? What does this indicate about ice?

Tie to Prior Knowledge

Water's States Ask students to recall the different forms in which they have seen water. Students should mention liquid water and solid ice; gaseous water vapor is invisible. Tell students that water is unique because it can exist naturally on Earth in all three states: solid, liquid, and gas. Then tell them they will discover in this section that water has unique properties that help make life on Earth possible.

Reading Check

Answer hydrogen and oxygen

Section 1 Resource Manager

Chapter *FAST FILE* Resources

Transparency Activity, p. 44
Directed Reading for Content Mastery, pp. 19, 20
Note-taking Worksheets, pp. 33–35
MiniLAB, p. 3
Lab Activity, pp. 9–12
Enrichment, p. 30
Reinforcement, p. 27
Lab Worksheet, pp. 5–6

Reading and Writing Skill Activities, p. 3

Figure 2 Water molecules in the liquid state are held close together by weak bonds. When water changes to steam, the molecules move farther apart. It takes energy to break weak bonds and separate the molecules.

Changing Forms of Water You have seen water change states. Ice melts on the sidewalk, and the puddle evaporates. Ice generally melts at 0°C. However, liquid water can become gas at many temperatures. Water evaporates from oceans, lakes, and rivers to enter Earth's atmosphere. Water also changes to gas when it boils at 100°C.

Water molecules are connected by weak bonds. In order to change from solid to liquid or to change from liquid to gas, bonds must be broken. Breaking bonds requires energy, as shown in **Figure 2.**

When the state changes go the other direction, water gives off energy. The same amount of heat needed to change liquid to gas, for example, is given off when the gas changes back to liquid.

Latent Heat

Would you try to boil a pan of water over a candle? Of course not. Clearly a candle does not give off enough heat. Each molecule of water is attracted weakly to other water molecules. All those attractions mean that you need a high amount of heat to boil the water in your pan—definitely more than a candle's worth. Changing the state of water, either from liquid to gas or from solid to liquid, takes more energy than you might think.

INTEGRATE Chemistry The heat energy needed to change water from solid to liquid is called the latent heat of fusion. Heat can be measured using a unit called the joule. It takes about 335 joules to melt a single gram of ice at 0°C. On the other hand, 335 joules of heat will escape when a single gram of water freezes into ice at 0°C. It might surprise you to know that the temperature does not change while the freezing or melting is going on. During freezing and melting, energy is changing the state of the water but not the temperature.

Science Online

Topic: Latent Heat

Visit bookh.msscience.com for Web links to information about how water changes from one state to another.

Activity Make a table showing which state changes require heat and which give off heat.

2 Teach

Caption Answer

Figure 1 oceans, glaciers, and clouds

Visual Learning

Figure 2 Ask students to look at the illustrations of molecules in liquid and gaseous water. Have them contrast the way the molecules are spaced. The molecules in the liquid state are closer together than the molecules in the gaseous state.

Activity

Evaporation For a quick example of the way water changes form, have students pour water over one of their hands and then wave that hand in the air. Ask them to describe what they feel and explain why it occurs. As students wave their hands in the air, their hands feel cooler because the water is evaporating and the heat energy needed for this to occur is being withdrawn from their hands.

Use Science Words

Word Meaning Have students use a dictionary to learn the meaning of the word *latent.* Ask them to relate this meaning to the words use in *latent heat.* The word *latent* means present but not obvious or visible. Latent heat is stored and not noticed unless a change of state occurs.

Virtual Labs

Water *What are some properties of water?*

Differentiated Instruction

Physically Challenged To enable these students to participate in the evaporation activity described above, spray water on one of their hands, and then direct the breeze from a small handheld fan toward it. Ask students to describe what they feel and to explain why this is occurring. Their hands feel cooler because evaporation removes heat.

SECTION 1 The Nature of Water H ◆ 9

INTEGRATE Life Science

Protecting Crops As the water freezes, the heat given off by the water is absorbed by the citrus. This helps keep the citrus from being damaged by the cold temperatures.

Reading Check

Answer The latent heat of vaporization goes into changing the water's form.

Quick Demo

Oil and Water

Materials large, glass jar with lid; water; vegetable oil

Estimated Time five minutes

Procedure Fill a glass jar halfway with water. Pour a layer of vegetable oil on top of the water. Explain to students that the vegetable oil floats on the water because it is less dense. Screw a lid on the jar and shake. Show the students that the immiscible vegetable oil will rise to again form a layer on top of the water.

Fun Fact

The bonds that form between hydrogen atoms and oxygen atoms to form water are covalent bonds. Strong covalent bonds form when atoms share electrons.

Text Question Answer

A desert home could use a "swamp" cooler. Warm air passes over water soaked pads. Evaporation of the water cools the air, which is then blown into the house.

Figure 3 Temperature does not change when water boils. The heat is used to change the water's form.

Time Requirements These processes take time. Water won't freeze the instant it goes into the freezer, and ice won't melt immediately when you place an ice cube on the counter. Ice is a stable form of water. A large amount of heat loss must occur to make ice in the first place. After water is frozen, it takes far more energy to melt the ice than it does to heat the resulting liquid water to almost boiling.

INTEGRATE Life Science

Protecting Crops Citrus farmers often protect their crop on cold nights by spraying water on the orange or grapefruit trees. If the temperature drops below freezing, the water freezes. Explain in your Science Journal how freezing water could keep citrus trees warmer.

Heat of Vaporization It takes even more heat energy to change liquid water to gas, or water vapor. The amount of heat needed to change water from liquid to gas is called the latent heat of vaporization. Each gram of liquid water needs 2,260 joules of heat to change to water vapor at 100°C. Likewise, each gram of water vapor that changes back to liquid at 100°C releases 2,260 joules of heat. During both of these processes, no increase or decrease in temperature occurs, just a change in form, as shown in **Figure 3.**

Reading Check *Why doesn't the temperature of water change when it boils?*

You might have experienced latent heat of vaporization if you've ever felt a chill after getting out of a swimming pool. When you first emerged from the pool, your skin was covered with some water. As the water evaporated into the air and became water vapor, it took heat from your body and made you feel cold. Can you think of a way that evaporating water could be used to cool other things, such as a desert home?

Science Journal

Freezing Problems Find out if students have ever placed a glass bottle containing water or a soft drink in the freezer to cool quickly, only to forget about it and find later that the bottle had broken or the top was forced off by the expansion of the ice as it froze. Have students write a brief report in their Science Journals about areas in and around their homes that would suffer similar consequences if the temperature dropped below freezing. Have them also suggest ways to avoid the problem. Possible problem areas include outside faucets, cooling systems in cars, and heating system pipes. Shut off water flow to protect outside faucets. Use antifreeze in cars. Maintain heat in heating systems or around pipes to prevent freezing. L2 ELL LS **Linguistic**

Density Which has more mass—a kilogram of plastic foam or a kilogram of lead? They're the same, of course, but you will need a much bigger container to hold the plastic foam. The volume of the lead will be smaller because lead has more mass for its size than plastic foam. In other words, the lead has greater density. **Density** is the amount of mass in a unit of volume. The density of pure water is 1.0 g/cm^3 at 4°C. Adding another substance to the water, such as salt, changes the density. Freshwater will float on top of denser salt water, just as olive oil floats atop the denser vinegar in salad dressing. This situation is found in nature where rivers flow into the ocean. The freshwater stays on top until waves and currents mix it with the seawater.

Reading Check *What is density?*

Temperature also affects the density of water. As freshwater heats up above 4°C, the water molecules gain energy and move apart. In the same volume of water, warm water has fewer molecules than cold water does. Therefore, warm water has lower density than cold water and will float on top of it, as shown in **Figure 4.** You might have experienced this while swimming in a pond or lake during the summer. The water on top is fairly warm. However, if you dive down, you suddenly feel the colder, denser water below. The difference in density between warm and cold water has an important effect in the ocean. These differences cause currents in the water.

Figure 4 When warm water is introduced into cold water, it floats on top of the cold water.
Explain *why the warm water floats on the cold water.*

Mini LAB

Examining Density Differences

Procedure

1. Place 200 mL of **water** in a **beaker** and chill in a **refrigerator.** Remove the beaker of water and allow it to sit undisturbed for at least one minute.
2. Place about 30 mL of **hot tap water** in a **small beaker.**
3. Place 5 drops of **food coloring** into the hot water to make it easy to see.
4. Use a **pipette** to slowly place several milliliters of the hot water into the bottom of the beaker of cold water. Be careful not to stir the cold water.

Analysis

1. How does the hot water behave when added to the cold water?
2. A few minutes after the hot water has been added to the cold water, observe the container. Explain the cause of what you observe.

Mini LAB

Purpose Students will model density currents in water. L2
LS **Visual-Spatial**

Materials beakers (2), water, food coloring, refrigerator, pipette

Teaching Strategies

- The temperature difference between the hot and cold water should be as great as possible.
- Instruct students to observe the beaker containing hot water from the side when adding the cold water.

Safety Precautions Caution students to use care with the hot water.

Analysis

1. The hot water rises to the top of the cold water.
2. Gradually, the colored water mixes with the cold water.

Assessment

Performance Provide students with two liquids unknown to them. Ask them to explain how they would determine which of the liquids is denser. Possible answer: Both liquids could be placed in one container. The denser liquid would form a layer on the bottom of the container. Use **Performance Assessment in the Science Classroom,** p. 89.

Reading Check

Answer the amount of mass in a unit of volume

Caption Answer

Figure 4 The warm water is less dense.

Active Reading

Cause and Effect Chart This strategy is used to focus on cause-effect relationships. In the center, students write the topic they are trying to understand. To the left, they write the apparent cause of the topic. On the right, they write the apparent effects. Have students design a Cause and Effect Chart about how density differences affect water movement.

Teacher FYI

Density of Freshwater The maximum density of freshwater occurs at 4°C. Above this temperature, increasing molecular movement causes the water molecules to be farther apart. Hence, density decreases. Below 4°C, hydrogen bonding increases the distance between water molecules until ice is formed.

SECTION 1 The Nature of Water H ◆ 11

Reading Check

Answer cohesion

Caption Answer

Figure 5 impedes evaporation at water surface; creates surface tension; promotes droplet formation of water suspended in air; facilitates capillary action

Discussion

Water Cohesion How would the landscape of Earth be different if water molecules did not have cohesion and vaporized at room temperature? Students should infer that the landscape would be drier and would not contain bodies of surface water, such as lakes and rivers.

Fun Fact

The water molecule is very common in all living things. For example, people are about 65% water.

IDENTIFYING Misconceptions

Freezing of Liquid Water Many students and some adults think that pure water always will freeze at exactly 0°C. It is common for water to be cooled below its freezing point before turning to ice. For example, liquid cloud droplets can have temperatures as low as −39°C. These droplets resist freezing because of the surface tension associated with the droplets. The droplets are said to be supercooled. Above −40°C, freezing nuclei with a structure similar to that of ice are needed for these droplets to freeze in a cloud.

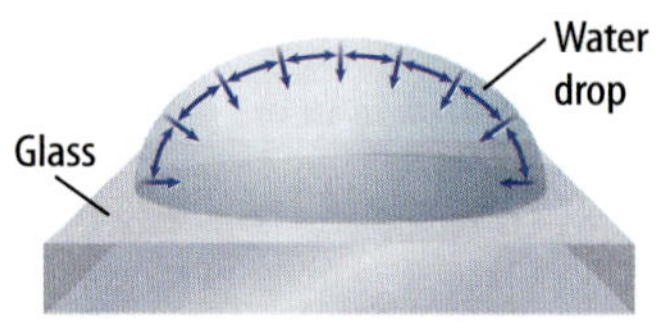

Figure 5 Cohesion causes water to form drops on a window.
List *some other effects of cohesion.*

A Polar Molecule

You might have noticed how water can bulge from the end of a graduated cylinder for a moment before it finally comes pouring out. That's due to the special property of water called cohesion. **Cohesion** is the attraction between water molecules. It's what allows water to form into drops, as shown in **Figure 5.** Cohesion also helps keep water liquid at room temperature. If not for cohesion, water molecules would quickly evaporate into the air. Molecules like carbon dioxide and nitrogen, which are close in mass to water molecules, completely vaporize at room temperature. If water molecules behaved this way, Earth would be a different, far drier place.

Reading Check *What property of water allows it to form into drops?*

Here's how cohesion works. As you know, a water molecule is made of two hydrogen atoms and one oxygen atom. These atoms share their electrons in covalent bonds. But the oxygen atom pulls more powerfully on the negatively charged electrons than the hydrogen atoms do. This gives the oxygen end of the molecule a partial negative charge and the hydrogen side of the molecule a partial positive charge. The molecule then acts like a tiny magnet, attracting other water molecules into weak bonds. This is shown in **Figure 6.** Because of this behavior, the water molecule is considered a polar molecule. A **polar molecule** has a slightly positive end and a slightly negative end because electrons are shared unequally. As you soon will see, this polarity of the water molecule explains several of water's unique properties.

Figure 6 Water molecules have weak charges at each end. These weak charges attract opposites and cause the molecules to bond.

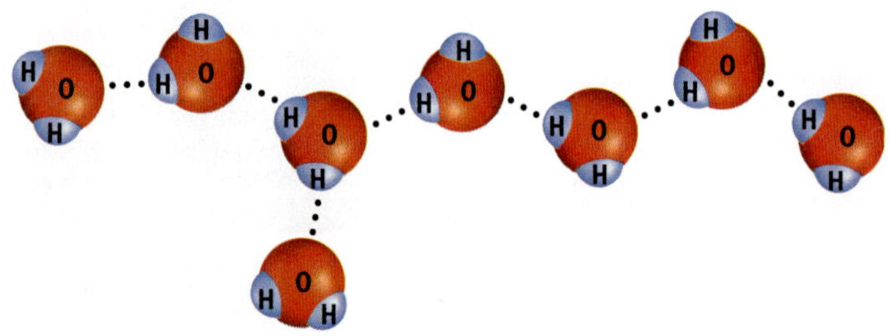

Differentiated Instruction

Challenge Abundant surface, liquid water is not common in our solar system. Have students do research on this topic and write a short report on what they find. In general, large amounts of surface, liquid water are unique to Earth in our solar system. Encourage students to report on the search for water on the Moon and Mars. L3 P

12 ◆ H CHAPTER 1 Water

Figure 7 Water molecules are farther apart in ice than in liquid water. This causes ice to float on water.

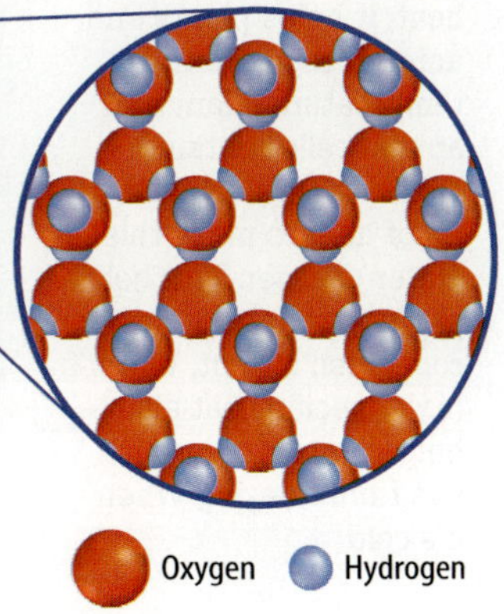

Effects of Bonding The polarity of water molecules makes water great for dissolving other substances, such as sea salts and substances that travel through your body. Polarity also means that ice will float on liquid water. As water freezes, the weak bonds between the molecules form an open arrangement of molecules, as seen in **Figure 7.**

The molecules, therefore, are farther apart when they are frozen than they were as liquid. This causes ice to have a lower density than liquid water, letting it float on water. In large bodies of water, floating ice prevents the water below from freezing. If lakes froze from the bottom up, they would freeze solid every winter, killing all living things inside.

Applying Science

How does water behave in space?

Astronauts onboard space shuttles and space stations have performed experiments to find out how water behaves in the weightlessness that exists in Earth's orbit. Do you think water behaves differently in space than on Earth?

Identifying the Problem

The photograph to the right was taken onboard *Skylab,* a former United States space laboratory. It shows astronaut Joseph Kerwin forming a sphere of water by blowing droplets of water through a straw.

Solving the Problem

1. Why does the water drop remain suspended?
2. Why would water form nearly spherical drops onboard the space shuttle or an orbiting space station?

Visual Learning

Figure 7 Challenge students to draw what they think water molecules look like in solid water, liquid water, and gaseous water. Students' drawing should show molecules more tightly packed in liquid water than in ice. The molecules should be far apart in their drawings of water vapor.

Applying Science

Answers

1. The water drop remains suspended because the space station was orbiting Earth.
2. Drops of water on the space shuttle are spherical because the molecules of water in the drop are attracted to each other because of their polarity. Molecules at the surface of a drop are attracted to molecules inside the drop. On Earth, gravity pulls the drop downward, forming the familiar teardrop shape. On the shuttle, no such force is applied, so the drop is spherical.

LAB DEMONSTRATION

Purpose to demonstrate the effects of different specific heat values

Materials sand; water; 2 thermometers; 2 small, flat dishes

Preparation Place a 1.5-cm layer of sand in one dish and an equal depth of water in the other.

Procedure Place the two flat dishes in sunlight. Record the initial temperatures of the sand and the water. Continue to record both temperatures every 5 minutes for 20 minutes.

Expected Outcome Students will see that sand heats up faster than water.

Assessment

Which of the two substances has the higher specific heat? Water; it does not warm up as fast as sand.

SECTION 1 The Nature of Water H ◆ 13

INTEGRATE Life Science

Specific Heat The high specific heat of water also benefits aquatic organisms. Oceans, lakes, and ponds change temperature slowly. Aquatic organisms do not experience the rapid, short-term temperature variations that often occur in the atmosphere.

Reading Check

Answer the amount of energy that is needed to raise the temperature of 1 kg of a substance 1° C

3 Assess

DAILY INTERVENTION

Check for Understanding

Interpersonal Identify one of water's unique properties and challenge the class to suggest some observed effects. For example, you might identify cohesion and students might respond that water beads on a freshly waxed car, that a glass can be filled above the rim, and that spherical drops of water can float in the space shuttle.

Reteach

Study Chart Draw a chart on the board and work with the class to fill in as many properties of water as they can remember from the chapter. Have students copy the chart and use it as a study aid. LS **Visual-Spatial**

Assessment

Oral Ask students to explain why ice is less dense than liquid water. Weak bonds in ice cause the water molecules to form a structure in which they are farther apart than they are in liquid water.

INTEGRATE Life Science

Specific Heat Because water has a high specific heat, it helps plants and animals regulate their temperature. Plant and animal cells contain mostly water. It takes a lot of heat to make this water warmer. This helps plants and animals stay cool when it's hot. Water's high specific heat also helps protect living tissue from freezing when it's cold.

Specific Heat Have you ever burned your feet running across the sand on a beach? You make a mad dash for the water, where you can cool them off. But think about this: the same hot Sun is shining on the water, too. Why is the water so cool? One reason is water's high specific heat. **Specific heat** is the amount of energy that is needed to raise the temperature of 1 kg of a substance 1°C. The same amount of energy raises the temperature of the sand much more than the temperature of the water. In fact, when you compare water with most other naturally occurring materials, it will increase its temperature the least when heat is added. In other words, water has one of the highest values of specific heat.

Reading Check *What is the definition of specific heat?*

High specific heat means water also will cool off slower when the energy is taken away. Go back to the beach at night and you can find this out yourself. At night the water feels warmer than the sand does.

This property is the reason water often is used as a coolant. Water in a car's radiator cools the engine by carrying away heat without becoming too hot itself. The specific heat of water is just one more of the characteristics that make it so important to life. In the next section, you will learn some of the ways the special properties of water can be put to work.

section 1 review

Summary

Forms of Water

- Water exists in three states on Earth—solid, liquid, and gas.
- Water molecules include two hydrogen atoms and one oxygen atom.

Latent Heat

- When ice melts and when liquid water changes to gas, heat is absorbed.
- When water freezes and when water vapor condenses to a liquid, heat is given off.

A Polar Molecule

- Water is a polar molecule. The oxygen atoms have a partial negative charge, and the hydrogen atoms have a partial positive charge.
- Water has a high specific heat. This means that it takes a lot of heat to increase the temperature of water.

Self Check

1. **List** the three states of water and give examples of where water is found in each state.
2. **Define** the latent heat of fusion. Define the latent heat of vaporization.
3. **Explain** why cold water sinks in warm water and why salt water sinks in freshwater.
4. **Explain** why ice floats on liquid water.
5. **Think Critically** Water molecules have slightly positive and slightly negative ends. How does this property help water dissolve similar substances and substances made of charged atoms?

Applying Skills

6. **Predict** what would happen to the level of water in a bowl if you were to leave it on the counter for several days. Explain any change you predict.

Science Online bookh.msscience.com/self_check_quiz

section 1 review

1. solid, glacier ice; liquid, oceans; gas, water vapor in the atmosphere
2. The amount of heat needed to change water from a solid to a liquid; the amount of heat needed to change water from a liquid to a gas
3. Cold water is more dense than warm water. Salt water is more dense than freshwater.
4. Ice is less dense than liquid water.
5. The charged atoms are attracted to the oppositely charged ends of water molecules. This property allows water to dissolve most ionic compounds easily.
6. Water level would go down as water molecules on the surface gradually evaporate.

LAB

Discovering Latent Heat

Ice melts at 0°C. After you raise the temperature of ice to the melting point, the ice remains at a constant temperature until all of it melts. The same thing happens to water at the boiling point of 100°C. What is happening to the water that explains this?

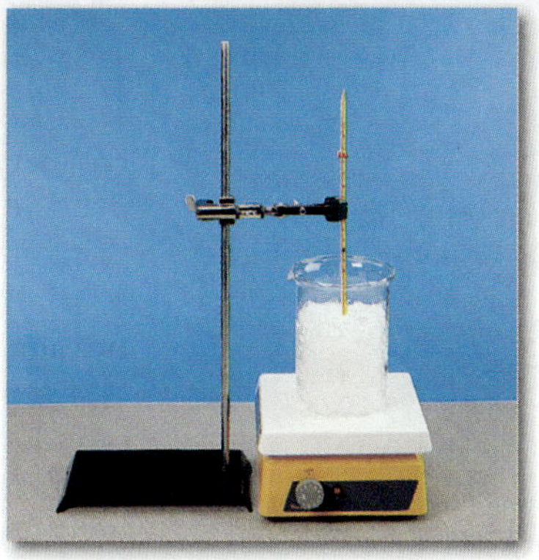

Real-World Question

How does the temperature of water change as it is heated from below the melting point to the boiling point?

Goals

- **Describe** what happens to the temperature of ice and water as thermal energy is added.
- **Explain** what you see on your graph of temperature versus time.

Materials

hot plate
ring stand
400-mL beaker
crushed ice
clamp for thermometer
Celsius thermometer
250-mL graduated cylinder

Safety Precautions

WARNING: *Do not touch hot glassware.*

Procedure

1. Create a data table to take time and temperature measurements. Half fill a 400-mL beaker with crushed ice. Suspend a thermometer in the ice and record its initial temperature in your data table at time = 0 min.
2. Plug in the hot plate and adjust the temperature as directed by your teacher. Record the temperature of the ice every minute in your data table until it is completely melted.
3. Continue recording the temperature of the water every minute until it begins to boil. Note the temperature at which this occurs.
4. Continue recording the temperature of the water for three more minutes.
5. Stop heating the water and let it cool before cleaning up your lab station.
6. **Plot** all your data on a sheet of graph paper with temperature on the vertical axis and time on the horizontal axis.

Conclude and Apply

1. What happened to the temperature while the ice was melting?
2. What happened to the temperature after all the ice was melted but before any water started to boil?
3. What happened to the temperature while the water was boiling?
4. **Hypothesize** what would happen to the temperature of the water vapor after all the water evaporated if thermal energy were added continually.

BENCH TESTED

Real-World Question

Purpose Students will observe temperatures as water changes state. L2 LS **Logical-Mathematical**

Process Skills make and use graphs, measure, observe and infer, measure in SI, compare and contrast, interpret data, form a hypothesis

Time Required 40 minutes

Procedure

Safety Precaution Students should use care with hot items.

Teaching Strategy Students should occasionally stir the ice and water mixture to avoid uneven heating.

Conclude and Apply

1. It remained steady at 0° C.
2. It steadily increased.
3. It remained steady at 100° C.
4. It would increase.

Assessment

Process Ask students what happens to the thermal energy that you transfer to the ice and water mixture when you heat it. The thermal energy is used to break the bonds that maintain the ice's crystal structure.

Communicating Your Data

Students can prepare their data tables and graphs using a computer spreadsheet and graphing program. Have them prepare a poster that presents the graph and explains the concepts involved.

Data Table (sample data)

Measurement	Time (minutes)	Temperature (°C)
Initial	0	−2
1	1	−1
2	2	0
3	3	0
4	4	0

1 Motivate

INTERACTIVE CHALKBOARD PowerPoint® Presentations

Bellringer

Section Focus Transparencies also are available on the Interactive Chalkboard CD-ROM. L2 ELL

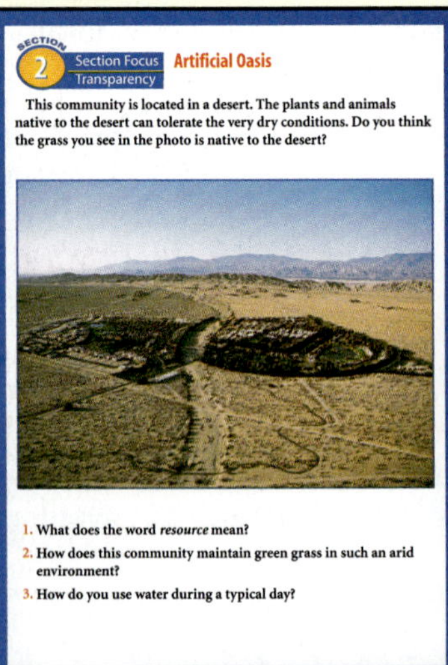
Section 2 Section Focus Transparency **Artificial Oasis**

This community is located in a desert. The plants and animals native to the desert can tolerate the very dry conditions. Do you think the grass you see in the photo is native to the desert?

1. What does the word *resource* mean?
2. How does this community maintain green grass in such an arid environment?
3. How do you use water during a typical day?

Tie to Prior Knowledge

Using Water Have students brainstorm the many ways they use water each day. Write their responses on the board. Then tell students that in this section they will learn why life on Earth could not exist without water.

section 2

Why is water necessary?

as you read

What You'll Learn

- **Explain** that water is essential for life on Earth.
- **Describe** ways that society uses water.
- **Describe** methods for conserving freshwater.

Why It's Important

Less than one percent of Earth's water is available for human use; therefore, it must be used wisely.

Review Vocabulary
agriculture: the science of growing crops and raising farm animals

New Vocabulary
- irrigation
- water conservation

Water and Life

Everyone knows that people can't survive long without water to drink. Even more than food, water is critical to your immediate survival. If you think about all the ways water allows you to live and thrive on Earth, you'll come up with a long list. You drink it, wash in it, and play in it. Water is all around you, and it's all through you. About 70 percent of your body is water. It fills and surrounds the cells of your body, enabling many of your body's processes to occur. Water helps move nutrients throughout your body, control your body temperature, eliminate wastes, digest food, and lubricate joints. When you feel thirsty, your body is telling you that you need more water.

Water not only is important to humans but to all life on Earth. The oceans, streams, rivers, and lakes are full of activity and provide habitats for organisms within and around them.

The properties of water allow plants to get water from soil. Water molecules are attracted to other polar molecules. Together with cohesion, this provides the capillary action that draws water upward through narrow tubes inside plant stems. This is shown in **Figure 8.** Water moves by this method from the ground to the plant leaves, where photosynthesis occurs.

Figure 8 A plant stem works the same way a capillary tube does. Water travels up the tube or plant stem because of capillary action.

Section 2 Resource Manager

Chapter *FAST FILE* Resources
Transparency Activity, p. 45
Directed Reading for Content Mastery, p. 21
MiniLAB, p. 3
Lab Activity, pp. 13–16
Reinforcement, p. 28
Enrichment, p. 31

Life Science Critical Thinking/Problem Solving, p. 23

Performance Assessment in the Science Classroom, p. 61

Water and Society

Take one look at a globe and you'll soon realize how important water is to society. The need for water explains why major cities often are located near large bodies of water. Desert areas have far fewer towns and sparser populations.

Water for Production Industry uses water for many purposes. For example, water is used for processing and cooling during the production of paper, chemicals, steel, and other products. Water also is necessary for transporting manufactured goods. Mining and refining Earth's natural resources call for large amounts of water. **Figure 9** shows how water is used in the United States. In general, communities located near water are better able to attract industry and often have the most productive economies.

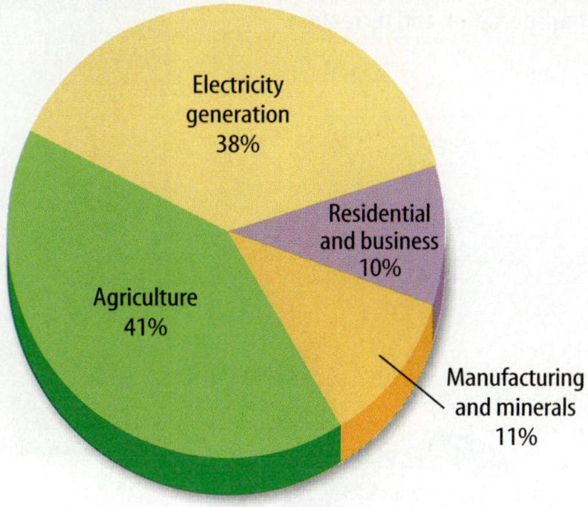

Figure 9 Water has many uses in the United States, as shown in the circle graph.

Agriculture uses about 41 percent of all the water used in the United States, the largest percentage of any sector. Most of this water is used for irrigation of farmland. **Irrigation** means piping in water from elsewhere and using it to grow crops, as shown in **Figure 10.** The water could come from a nearby lake, river, or reservoir or it might be pumped from the ground.

Reading Check *What process do farmers use when they pump water from somewhere else to water their crops?*

Figure 10 Several forms of irrigation are used worldwide. **Describe** *any forms of irrigation used in areas near you.*

2 Teach

Use Science Words

Word Meaning Explain that *adhesion* is the attraction between particles of different substances. It, along with cohesion, is the force that pulls water through the vascular tissues of plants. Have students hypothesize how plants on Earth might look if water did not exhibit this property. Possible hypothesis: Plants would be nonvascular, and thus small. They would absorb and distribute water through their surfaces rather than through a system of roots and tubes.

Discussion

Water for Cities Why are most major cities located near water? Water is necessary for production and residential use. How can some major cities, such as Phoenix, Arizona, exist in the desert? Local water is managed carefully, and some water is directed to the city. Phoenix gets water from a Colorado River reservoir system, in addition to local reservoirs and groundwater. Almost all surface water is captured and used. Phoenix also reclaims water for purposes such as irrigation.

Reading Check

Answer irrigation

Caption Answer

Figure 10 Answers will vary.

Teacher FYI

Water and Plants Water diffuses from the leaves of a plant during transpiration, a part of the water cycle. Plant roots take in water by osmosis, and it is transported upward by adhesion and cohesion. Water is essential for photosynthesis and cellular respiration.

Visual Learning

Figure 9 As students study the graph, point out that irrigation accounts for a huge percentage of the water use in the United States. How could the United States reduce its use of irrigation water and conserve water resources? Accept all reasonable answers, which could include growing crops that have been engineered to be more drought tolerant.

Reading Check

Answer Goods are moved across rivers and oceans by ship.

Mini LAB

Purpose Students will compare the amount of water used for a bath and for a shower. L2 LS **Logical-Mathematical**

Materials bathtub, showerhead, water, meterstick

Teaching Strategies

- Students should measure their bathtubs in centimeters. Explain that the volume can be expressed in liters using the conversion 1,000 cm^3 = 1 L.
- Remind students that the water should not be wasted. Before they begin, they should plan how to use the water when they finish their measurements.

Analysis

1. Answers will vary depending on water pressure and flow rates.
2. A bath will probably use more water.
3. A 10-minute shower would use about twice as much as a 5-minute shower.

Assessment

Content Have students describe how they could change their showering habits to conserve more water. Possible answers: Shut off the water while soaping up and turn it on only to rinse off; take shorter showers. Use **Performance Assessment in the Science Classroom**, p. 89.

Try at Home

Figure 11 Ferries are used for transportation and recreation.

Mini LAB

Predicting Water Use

Procedure

1. Determine the average inside area (width times length) of your **bathtub.**
2. Run a typical bath and measure the depth of the water. Pull the stopper and allow the water to drain.
3. Close the stopper of your bathtub and run the shower for 5 min. Measure the depth of the water at the end of the 5 min.
4. Calculate the volume of water used during a typical bath and during a typical 5-min shower. To do this, multiply the area of the bathtub by the bath depth and then by the shower depth.

Analysis

1. Compare the volumes of water used during a bath and a shower.
2. Determine which uses more water—a bath or a 5-min shower.
3. Infer how much water a 10-min shower would use.

Try at Home

Water for Transportation Although superhighways and airports are important to modern society, water remains valuable for transportation, as shown in **Figure 11.** Passenger liners are popular for vacations. Saving the time and expense of long land trips, ferries move people, cars, and freight across bays, straights, and rivers. Big ships often are the most economical way to move large freight within the country or across the ocean, as shown in **Figure 12.**

Reading Check *Why is water important for transportation?*

Water for Recreation Don't forget the role water plays in recreation. Many people equate water with fun—fishing, swimming, scuba diving, waterskiing, and boating. Boating provides entertainment for many people. Sailors, canoeists, kayakers, powerboaters, and whitewater rafters all spend recreation time on the water.

Water Use

When you consider all the ways that water is used, it is clear that water is a valuable natural resource that must be conserved and protected. Not only is clean water important for use in homes and by society, it is necessary for maintaining the ecological balance in nature. Many species of wildlife live in and around bodies of water.

Bodies of water must be clean to support the animals and plants that live in them. If too much pollution enters rivers, lakes, or ponds, wildlife might be harmed. Oceans also must be kept clean of pollutants such as oil, chemical and radioactive wastes, and litter.

Curriculum Connection

History Many societies throughout history have used irrigation to grow food in arid areas. These societies include the early Romans, the Mesopotamians, and several Native American cultures, such as the Maya of Central America. Have students research the irrigation system of a past society and write a short report about their findings. L2 ELL LS **Linguistic** P

Differentiated Instruction

Behaviorally Disordered Most students enjoy some type of water recreation. Have these students discuss or write about the ways that they enjoy water. Students also might have photos that they would like to share. This topic then can be expanded to address the importance of water to society.

NATIONAL GEOGRAPHIC

VISUALIZING SHIPPING

Figure 12

For thousands of years, people have floated down rivers and sailed across oceans to move goods from place to place. Modern ships may look different from those used in ancient times, but waterways continue to be vital transportation routes, as shown on this page.

▲ OIL TANKERS Carrying more than a million barrels of crude oil, tankers transport cargo from offshore wells, such as this one in the Gulf of Mexico, to refineries, which process the fuel.

▲ BARGES Long, shallow-bottomed barges transport goods along rivers, canals, and other waterways. Here, a barge carrying coal glides down Louisiana's Bayou Lafourche. The bayou branches from the Mississippi River, one of the busiest waterways in the world.

▲ CONTAINER SHIPS Container ships transport cargo that has been packed into large rectangular containers. The containers are lifted aboard by huge cranes and unloaded in the same way when the ship reaches its destination.

◀ AIRCRAFT CARRIERS Some ships transport not only goods but services, as well. The USS *Saratoga*, left, acts as a moving air base for the United States Navy. Steam-powered catapults on the ship can increase a plane's velocity from 0 to 266 km/h in 2 s, making takeoffs and landings possible.

NATIONAL GEOGRAPHIC

Visualizing Shipping

Have students examine the pictures and read the captions. Then ask the following questions.

Huge cranes must be available at docks where container ships onload and offload cargo. What sort of facilities must be available at docking sites for the other ships shown here? Possible answers: Oil tankers need refineries, storage tanks, pipelines, tanker trucks or railroad cars at on- and off-loading locations; barges might need truck or railroad terminals; aircraft carriers would need access to military bases with mechanical and personnel services.

What sort of hazards might be involved with each type of shipping? Possible answers: Oil tanker accidents can result in disastrous fuel spills; barges and container ships can lose their cargoes in severe seas or collide with commercial or recreational vessels; by their nature, aircraft carriers are often associated with international conflicts.

Activity

City Geography Display a map of the world and make a chart on the board with these headings: ocean; lake; river; no water access. Have students inspect the map and record the names of large cities falling into each category. Work as a class to make a bar graph of the data and to discuss its meaning.

Cultural Diversity

Important Rivers Many rivers have crucial roles in the economic lives of the countries through which they flow. These rivers are often the main avenues for transportation as well as important sources of water for irrigation and drinking. Egypt's Nile River, India's Ganges River, Vietnam's Mekong River, and China's Yangtze River are examples. Have groups of students choose a river system to research. Have the groups make a multimedia presentation to the class on the river's importance to a people, both past and present. L2 ELL COOP LEARN LS **Interpersonal**

Make a Model

Irrigation Have students research several irrigation methods used by farmers. Ask them to work in groups of three to build a model of one of the methods. Have them explain how they might use the system shown in their model to irrigate a large garden or field. Then discuss if this method of irrigation conserves or wastes water. LS **Kinesthetic**

Activity

Daily Water Use During one day, have students write down each time they use water from the time they wake up in the morning until they go to sleep at night. Tell them to include what the water is being used for and an estimate of how much is being used. Use the lists as a basis for class discussion on how water is used and wasted. Have students discuss ways that water can be conserved. Then have students repeat the activity, but have them try to conserve as much water as they can. Have a second discussion in class about how successful conservation efforts were for class members. LS **Logical-Mathematical**

Reading Check

Answer They can use overhead sprinklers or slow-drop systems. They also can mulch the ground to retain moisture.

Conserving Water Photographs of Earth taken from space show a watery planet that has blue oceans and scattered continents. But, less than one percent of Earth's water is available for any of the uses described, except for ocean transportation. That's why it is important to conserve the available freshwater on Earth. The careful use and protection of water is called **water conservation.** Water can be saved and kept clean in many ways.

Much of the water used for irrigation is lost to evaporation. Better methods can conserve water. Instead of flooding the fields, farmers can use overhead sprinklers on their crops. Some farmers install tubing that slowly drips water directly above the roots of the plants. Farmers even use computers in the fight for water conservation. Sensors installed in the ground and connected to a computer can signal when crops need to be watered. In residential and commercial areas, mulching the ground around the plants, as shown in **Figure 13,** helps prevent water loss.

Reading Check *What can gardeners and landscapers do to conserve water?*

Industries also can conserve water. Companies can treat and recycle the water that is used in industrial plants. Innovative manufacturing processes that conserve water also might increase plant productivity.

Figure 13 There are several ways to conserve water when watering crops and other plants.

Mulching is a good way to conserve water in residential and commercial areas.

Overhead sprinklers use less water than flooding fields does.

Teacher FYI

Wasting Water According to some estimates, about half the water used in the United States is wasted through evaporation, water system leaks, and other losses.

Differentiated Instruction

Challenge Have students research how water conservation can reduce water pollution. Ask them to evaluate the benefits and costs of various water conservation techniques. They then might want to develop a water conservation plan for their home or community. L3

What can you do? Think about the activities you do on a daily basis that include water. Do you know how much water you use in your home and at school? Some estimates are shown in **Figure 14.** Wouldn't it be helpful to conserve some of this water? It can be done. For starters, you could conserve water in your shower. Turn the water off when you are soaping up, then use it just for rinsing. Do the same while brushing your teeth and washing your hands. See whether your home has a low-flow toilet system. Toilets are available that use only 6 L per flush compared to about 19 L per flush for older toilets.

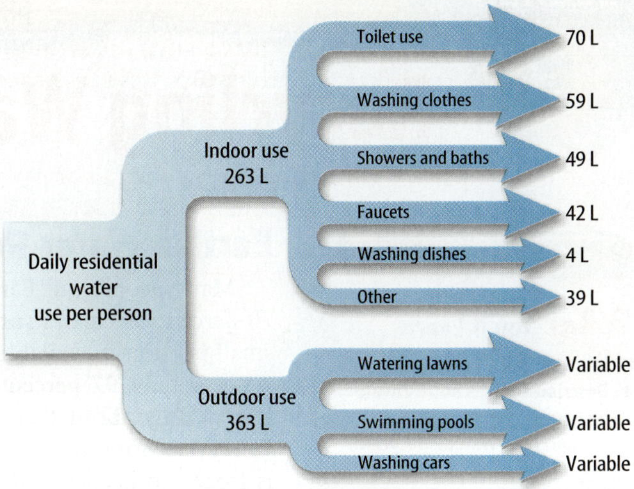

Figure 14 Humans use water every day.
Describe *how you use water in your home.*

Some ornamental plants require less water than others. What if your school administrators used these plants to landscape your school grounds? Over the course of a year or more, small measures such as these can save large amounts of water.

Did you know the water you showered in this morning could have been in Galileo's teapot or in King Tut's reflecting pool? The water on Earth today has been around for millions, even billions, of years. In the next section, you will see how water continuously cycles through the environment.

section 2 review

Summary

Water and Life

- Water is essential for life.
- Your body is about 70 percent water.

Water and Society

- Society uses water for many purposes, including industry, agriculture, transportation, and recreation.

Water Use

- Society and nature depend on abundant, clean water.
- Conserving water helps to ensure a steady supply of clean water.

Self Check

1. **Describe** three ways that water is essential for life on Earth.
2. **List** five ways that water is used for recreation.
3. **Explain** why clean water is important for nature and society.
4. **Describe** some methods that agriculture and industry can use to conserve water.
5. **Think Critically** Why might cities restrict residential water use?

Applying Skills

6. **Concept Map** Make a network-tree concept map that shows the ways water is used in society.

Caption Answer

Figure 14 Answers will vary depending on individual use.

3 Assess

DAILY INTERVENTION

Check for Understanding

Visual-Spatial Ask students to look around the classroom and describe the role water had in the manufacture, transportation, or production of various items.

Reteach

Wasting Water Have each student list three ways they have seen water being wasted during the last 24 hours. Record their answers on the board, and use them as a basis for class discussion. Possible answers: overwatering lawns, faucets or hoses pouring out water when no one is using it; using water to wash down a sidewalk or driveway, running a dishwasher loaded with just a few plates or utensils

Assessment

Performance Have students write a poem or compose a song that covers the importance of water conservation. Use **Performance Assessment in the Science Classroom,** p. 151.

section 2 review

1. moving nutrients, controlling temperature, digesting food, habitat, among others
2. Possible answers: swimming, boating, fishing, scuba diving, water skiing
3. All living things on Earth need water for life processes. Society also uses water for agriculture, manufacturing, and transportation, among other uses.
4. agriculture: overhead sprinklers, drip irrigation, mulch, sensors to signal when crops need water; industry: treat and recycle wastewater, use water conservation measures in manufacturing processes
5. Possible answer: Cities sometimes restrict residential water use when water is in short supply. This often occurs during a long drought.
6. Check students' work. Network tree should include water for drinking, growing crops, industry, transportation, and recreation.

section 3

Recycling Water

as you read

What You'll Learn

- **Identify** Earth's water reservoirs.
- **Describe** sources of freshwater on Earth.
- **Explain** how water is recycled.

Why It's Important

The water cycle makes water available for many processes on Earth.

Review Vocabulary

cycle: a repeating sequence of events

New Vocabulary

- groundwater
- aquifer
- soil water
- surface water

Earth's Water Reservoirs

Many people call Earth the "water planet" because about 70 percent of Earth's surface is covered by water. But, only a small portion is available for human use. Of the world's total water supply, 97 percent is located in the oceans and is salt water. Only about three percent is freshwater, as shown in **Table 1.** More than three-quarters of Earth's freshwater is frozen in glaciers. Of the less than one percent of Earth's total water supply that is available for human use, much of it lies beneath Earth's surface. Next you can think big as you begin your examination of Earth's water reservoirs with the oceans.

Oceans *Water, water everywhere, but not a drop to drink,* so goes the poem. Ocean water is plentiful on Earth, but it's salty and therefore not readily available for human use. You would have to remove the salt if you wanted to use ocean water for drinking, bathing, or irrigating crops. Unfortunately, removing salt from ocean water usually isn't practical, and it is only done in a few arid regions. To find a lot of freshwater, you might want to travel to Earth's frozen poles.

Table 1 Distribution of Earth's Water

Location of Water	Total Supply (%)
Oceans	97.2
Glaciers	2.15
Groundwater	0.62
Freshwater lakes	0.009
Saline lakes and inland seas	0.008
Soil water	0.005
Atmosphere	0.001
Stream channels	0.0001

Ice Glaciers are common in Earth's polar regions. For example, large areas of Greenland and Antarctica are covered by ice. These ice sheets lock up a high percentage of Earth's freshwater. Ice accounts for just more than two percent of the total water on Earth, but that's 77 percent of the planet's freshwater supply. People have proposed using ships to tow large pieces of polar ice to places that need freshwater. It's an expensive proposition, and imagine how difficult it would be to tow a large iceberg. Melting the ice into usable water also would be a difficult task.

1 Motivate

INTERACTIVE CHALKBOARD PowerPoint® Presentations

Bellringer

Section Focus Transparencies also are available on the Interactive Chalkboard CD-ROM. L2 ELL

SECTION 3 Section Focus Transparency **Big Drop**

Angel Falls are the highest waterfalls in the world. Located in Venezuela, the falls plummet 979 meters. They are situated on the Churún River.

1. Is most of the water on Earth freshwater or salt water?
2. What happens to water flowing in a river? How might it end up in your drinking water?
3. How could you try to obtain freshwater in an area where there are no lakes or streams?

Water

Tie to Prior Knowledge

The Water Cycle If billions of people use water each day all over the world, and billions more people have used water throughout history, why hasn't Earth's water supply run out? Use students' responses as the basis for discussion. Then tell students they will learn in this section how Earth recycles the same water over and over again.

Section 3 Resource Manager

Chapter *FAST FILE* Resources

- Transparency Activity, pp. 46, 47–48
- Directed Reading for Content Mastery, pp. 21, 22
- Reinforcement, p. 29
- Enrichment, p. 32
- Lab Worksheet, pp. 7–8

Lab Management and Safety, pp. 61, 62

Reading and Writing Skill Activities, p. 19

Groundwater You might have wondered what happens to rain after it falls. If you leave a bucket outside in a rainstorm, it can fill quickly with water. Where does the water that lands on the ground go? Some of it runs off and flows into streams, and some evaporates. A large amount soaks into the ground. Water that is held underground in layers of rock and sediment is called **groundwater.** The part of the groundwater that is held within openings in the soil is called **soil water.** It keeps plants and crops alive. People who live in houses that get water from wells are drinking groundwater. The water is purified as it slowly permeates through layers of sediment and rock. However, if groundwater becomes polluted, it can be extremely difficult to clean.

Reading Check *Where is groundwater found?*

Aquifers An **aquifer** is a layer of rock or sediment that can yield usable groundwater. Water collects in the open spaces between rock particles. This water flows slowly from one open space to another at rates of a few meters per year. Sometimes aquifers are used to supply water to towns and farms, as shown in **Figure 15.** The water is pumped to Earth's surface through a well. Sometimes Earth's surface dips below the level where groundwater would be. This is where natural lakes and rivers often are located.

INTEGRATE History

Dakota Aquifer Water from the Dakota Aquifer, which exists beneath much of the Great Plains region of the United States, was critical for development of this region. In many places, the water in this aquifer is under pressure, so it can flow from wells without pumping. Because of the Dakota Aquifer, early pioneers had easy access to water. The Dakota continues to supply water to the region today.

Figure 15 Aquifers are found at different depths. Some are near the surface, and others are hundreds of meters below the surface. **Describe** *how water is extracted from aquifers.*

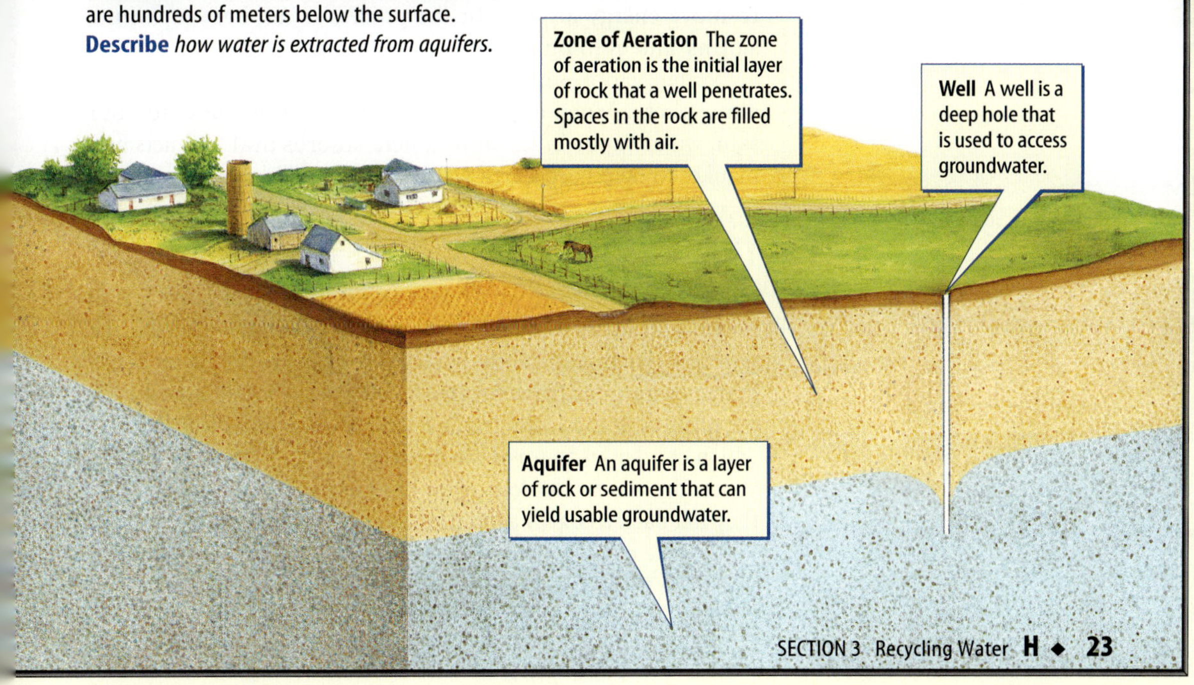

2 Teach

Visual Learning

Table 1 Ask students to use the data in the table to explain why drinking water is such a precious resource on Earth. Students should determine that most of Earth's water is in the oceans, which are salty. The percentage of freshwater that can be used for drinking is a very small percentage of the total.

INTEGRATE History

Dakota Aquifer The Dakota Aquifer is an artesian system through much of its area. This means that overlying rock layers confine the water-bearing rock layers and that the groundwater is under enough pressure to reach the surface if a well is made. The aquifer is an important source of water in the Dakotas, Kansas, Nebraska, and parts of Wyoming and Montana.

Reading Check

Answer underground in layers of rock and sediment

Caption Answer

Figure 15 It is pumped out through wells

Use an Analogy

Porosity Explain to students that sponges can hold water because pores exist within them. Tell them that rock and sediment also can hold water in pores.

Differentiated Instruction

Challenge Have students do research to find out whether plants benefit from groundwater beneath the soil layer. Plants generally do not benefit from this groundwater. Capillary forces in sediment and soil only can pull water a couple of meters above the water table. L3

Teacher FYI

Because of the extensive use of water in the United States and the dependence of many people on water from aquifers, groundwater is being withdrawn faster than it is being replaced. This threatens the long-term survival of many important groundwater supplies.

Reading Check

Answer streams, rivers, ponds, lakes, and reservoirs

Activity

Visiting Water Visit a nearby surface water body with students. Organize them into groups of three or four to list the kinds of land uses (factories, homes, farms, parks, highways, marshes) that are around the body of water. Tell students to research the water quality of the water body and whether it is used as a drinking water supply. If there are water quality problems have students infer, based on nearby land uses, what might be the cause. Have students also report on what is being done to address any pollution problems.

Inquiry Lab

Model the Water Cycle

Question How can processes in the water cycle be modeled?

Possible Materials aluminum pie pan, ice cubes, large glass, warm water

Estimated Time two class sessions: one to plan and gather materials and a second one to perform the experiments

Teaching Strategies

- Remind students that processes in the water cycle include evaporation, transpiration, condensation, precipitation, and runoff. Students might choose to model any or several of these processes.
- You might want to use the following model as a demonstration: fill a glass halfway with warm water, place a pie pan containing ice cubes on top of the glass. Water should evaporate, condense on the bottom of the pie pan, and precipitate back into the glass.

For additional inquiry activities, see *Science Inquiry Labs*.

Figure 16 Your city might have a water-intake plant like this one in Cincinnati, Ohio. Here, water is processed for drinking and many other uses.

Lakes, Rivers, and Ponds You already learned about groundwater. The water at Earth's surface is called **surface water.** It is found in streams, rivers, ponds, lakes, and reservoirs. This is the water you easily can see and use, as shown in **Figure 16.**

Reading Check *What are three sources of surface water?*

Water in the Atmosphere Have you tried to dry your towel on a towel rack and discovered it remained damp hours later? That's because air holds water, too. Earth's atmosphere can consist of anywhere from near zero to about four percent water vapor by volume. Water vapor in Earth's atmosphere has several important roles. Clouds need it to form; therefore, water vapor is the source of rain, sleet, and snow. But the role of water vapor in the atmosphere is even more complex.

Recall that each time water undergoes a change of state it gives off heat or absorbs heat. For example, heat is given off when water vapor condenses to form the water droplets in clouds. But, heat is absorbed when water evaporates to water vapor. That's why you feel cold when you are soaking wet. The atmosphere uses these heating and cooling properties of water to move energy around. In the process it can brew up wind, storms, and even hurricanes.

Water vapor serves another important function in the atmosphere. Much as a sweater keeps warm air next to your skin, water vapor in the atmosphere absorbs heat and acts as a blanket to help keep Earth warm and hospitable to life.

The Water Cycle

Are you still wondering how water from King Tut's reflecting pool might make it into your morning shower? It is because the water on Earth constantly is recycled through the water cycle. Water evaporates from oceans, lakes, rivers, puddles, and even the ground. It then rises into Earth's atmosphere as water vapor, which condenses to form the droplets and ice crystals in clouds. When the water droplets or ice crystals become heavy enough, they fall back to Earth as rain, snow, or sleet. Rainwater runs off the surface back to rivers, lakes, and finally the ocean, as shown in **Figure 17.** Water continuously circulates in this way.

Differentiated Instruction

English-Language Learners When student groups are planning models for the inquiry lab, make English-language learners responsible for listing and gathering materials. Have students write their materials list in both languages. ELL

Curriculum Connection

Geography Have students research and list any large bodies of water located near your school that would play an important part in the evaporation portion of Earth's water cycle. Obtain a map of the area around your school, and allow students to mark the bodies of water they have listed.

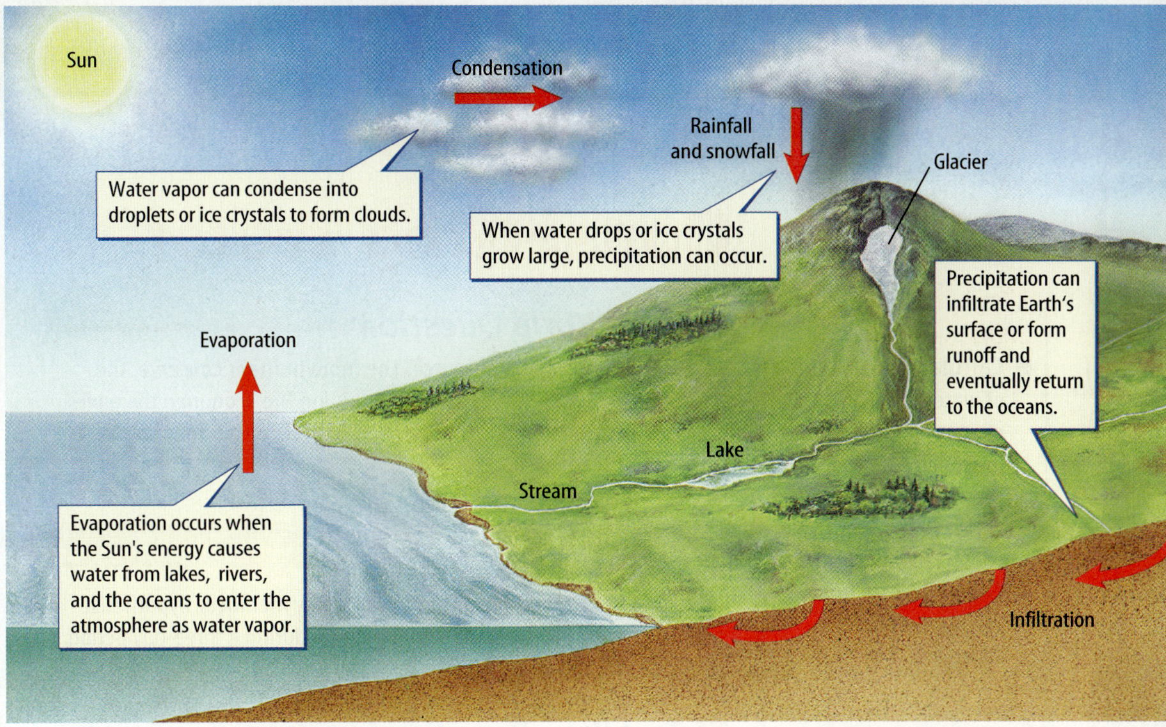

Figure 17 Water constantly cycles from Earth to the atmosphere and back to Earth again.

A Valuable Resource Humans are lucky that water circulates, but this doesn't mean that it doesn't need to be cared for. Now that you have learned how important water is to Earth, you can understand why it is crucial that it be protected. Earth holds many treasures, and water is truly a valuable resource.

section 3 review

Summary

Earth's Water Reservoirs

- About 70 percent of Earth's surface is covered by water.
- Earth's water reservoirs include the oceans, glacier ice, groundwater, surface water, and water in the atmosphere.
- Groundwater is water that is underground in layers of rock and sediment.

The Water Cycle

- Water constantly is cycled from Earth to the atmosphere and back again.
- Water is one of Earth's most valuable resources.

Self Check

1. **Identify** the percentage of Earth's total water supply that is available as surface freshwater.
2. **Explain** how soil water differs from other groundwater.
3. **Define** the term *aquifer*.
4. **Describe** sources of freshwater on Earth.
5. **Think Critically** How could a water molecule in a puddle near your home get to the ocean?

Applying Math

6. **Use Numbers** Look back at **Table 1.** How many times more water is in Earth's atmosphere than in all of Earth's streams? How many times more water is in soil than in all of Earth's streams?

Science Online bookh.msscience.com/self_check_quiz

Visual Learning

Figure 17 Have students trace the path of water in the water cycle. Then have them identify where evaporation and condensation occur. Stress that all parts of the cycle are occurring at the same time.

3 Assess

DAILY INTERVENTION

Check for Understanding

Interpersonal Ask student groups to make tables that include specific examples of water reservoir types. For example, under the heading *Oceans*, students might list *Pacific*, *Atlantic*, *Indian*, *Arctic*.

Reteach

Water Concept Map Have students use the section text to make a concept map that includes the major points about Earth's water reservoirs and the water cycle. LS **Visual-Spatial** P

Assessment

Oral Why is the water cycle essential for the maintenance of life on Earth? It supplies a steady source of freshwater that is essential for both plant and animal life.

section 3 review

1. less than one percent
2. Soil water is trapped within the openings in soil, unlike other groundwater, which occurs in layers of rock or sediment beneath the soil horizons.
3. A layer of rock or sediment that can yield usable groundwater.
4. lakes, rivers, reservoirs, groundwater
5. Possible answer: It could evaporate and become water vapor. Then it could become part of a cloud as it cools and condenses to form liquid water in the upper atmosphere. That cloud could move over the ocean. The molecule could then fall into the ocean as a part of precipitation.
6. 10 times more; 50 times more

BENCH TESTED

Real-World Question

Purpose Students design and perform an experiment to evaluate the water conservation efforts of their families over a three-day period. L2 LS **Logical-Mathematical**

Process Skills measure in SI, design an experiment, form a hypothesis, separate and control variables, observe and infer, make and use tables, interpret data

Time Required 30 minutes to design the experiment; 6 days to observe water use at home; 20 minutes to answer questions

Materials If the calculations are done in school, ask students who have them to bring calculators from home.

Form a Hypothesis

Possible Hypothesis If my family follows my conservation plan, we will conserve 300 L of water over a three-day period.

Test Your Hypothesis

Possible Procedures Measure the amount of water used by the family over a three-day period. Suggest ways the family can reduce water consumption. Observe and measure the amount of water used over the next three-day period. Compare the usage in both periods.

Teaching Strategy Encourage students to discuss their plan with their families ahead of time, so that all members of the family are familiar with the goals of the experiment.

LAB Design Your Own

Conserving Water

Goals

- **Calculate** the amount of water the members of your family use during a three-day period.
- **Design** techniques by which your family can conserve water.
- **Calculate** how much water your family conserves.

Materials

calculator

Real-World Question

Clean water is a valuable resource. The more humans conserve, the fewer problems society will have maintaining the economy, the environment, and general health. How much water can a family conserve during a three-day period?

Form a Hypothesis

Based on the water-use practices of your family members, hypothesize how much water you can conserve as a group during a three-day period.

Test Your Hypothesis

Make a Plan

1. **Consider** the problem, develop a hypothesis, and decide how you will test it. Identify results that will support your hypothesis.
2. **List** the steps you will need to take to test your hypothesis. Be specific. Describe exactly what you will do in each step. List your materials.
3. **Prepare** a data table in your Science Journal to record your observations.
4. **Read** the experiment to make sure all steps are in logical order.
5. **Identify** the variables in your experiment. List all variables that must be kept constant.

Follow Your Plan

1. Make sure your teacher approves your plan before you start.
2. Carry out the experiment as planned.
3. While doing the experiment, record your observations and complete the data table in your Science Journal.

Water Conserved During Three Days

Item	Water Conserved (L)
Toilet	
Washing machine	Answers will vary.
Others	
Total	

Differentiated Instruction

Learning Disabled Have students write simple descriptive reports about ways that their families can conserve water. Allow students to use less detail and a minimum of mathematical calculations in setting up their procedures.

Alternative Inquiry Lab

Conservation To make this Lab an Inquiry Lab, have students develop a water-conservation plan for your school or community.

26 ◆ H CHAPTER 1 Water

Analyze Your Data

1. **Calculate** the average amount of water that your family uses on a typical day.
2. **Determine** how much water your family conserved during the three-day period of your experiment.
3. **Compare** your results with those of others.
4. What had to be kept constant in this experiment?
5. What were the variables in this experiment?

Conclude and Apply

1. Did the results support your hypothesis? Explain.
2. **Describe** what effect water conservation would have on the day-to-day habits of a typical American family.
3. **Infer** whether steps taken at home can contribute significantly to water conservation.

Communicating Your Data

Write a step-by-step brochure about water conservation that can be shared with all students and their families. Include simple suggestions about how to conserve water at home.

Expected Outcome Students will note a decrease in the amount of water used over the second three-day period.

Analyze Your Data

Answers to Questions

1. Answers will vary depending on individual water use.
2. Answers will vary depending on how much water was used initially compared to how much was used after conservation measures were put in place.
3. Results will vary by family.
4. Possible answers: limit experiment to family members; limit measured water to specific uses such as brushing teeth or showering
5. Possible answer: the amount of water used for each type of water use observed

Error Analysis If students find they have not saved much water, there may have been errors in the way they measured or recorded water use, more people than they anticipated may have been using the water, or their conservation methods may have been flawed.

Conclude and Apply

1. Answers will vary depending on the hypothesis and the way the experiment was carried out.
2. They would have to become more aware of water waste. Showers would be shorter and water wouldn't run in sinks when not being used. Accept other reasonable answers.
3. Results should confirm that they could contribute significantly to water conservation.

Assessment

Process Have students compare the water use values in **Figure 14** with values they obtained in their homes. Students should prepare graphs comparing the data sets. Use **Performance Assessment in the Science Classroom,** p. 111.

Communicating Your Data

Have students prepare the brochure using computer desktop publishing software. Students can distribute copies of their brochures to other students in the school.

TIME

Content Background

It is estimated that the world's population will increase by almost two billion by 2025. As the population increases, maintenance of drinkable water supplies and adequate freshwater resources for food production will become increasingly important. Eighteen percent of the land used today for crops is irrigated and produces almost 40% of the world's food. In developing countries 90% of the water used is for irrigation of cropland. Worldwide, two thirds of water use is for food production. Researchers such as Sandra Postel and Peter Gleick estimate that by 2025 an additional 192 mi^3 of water will be required for food production. It is not yet known how this additional water will be provided.

Discussion

Water Use In what ways do we use water, directly and indirectly, every day? Possible answers: Direct uses include bathing and washing; cooking and cleaning; flushing toilets, etc. Indirect uses include water used for production of food, gasoline, steel and aluminum products, etc.

Activity

Freshwater Have students search for pictures of different sources of freshwater and water uses, and then organize the pictures into several categories according to usage. Direct them to construct collages of these images, and to include suggestions for reducing water usage.

TIME SCIENCE AND Society

SCIENCE ISSUES THAT AFFECT YOU!

Not a Drop to Drink

During the next few decades, finding and protecting freshwater supplies will be more important—and more difficult—than ever. Freshwater is needed to keep organisms such as humans alive and well. It also is used in factories and for growing crops.

In some parts of the world, water is especially precious. People might walk from their homes to a community well that might be several miles away. There, they load up on water for cooking, drinking, and bathing. Then they must carry the water back home.

Not a Drop to Drink

This isn't the case in the United States, but the lack of freshwater is a problem in this country, too. In the Southwest, for example, water is scarce. As the population of the region grows, more and more water is used. Growing cities, such as Phoenix, Arizona, are using freshwater faster than it can be replaced by nature.

To grow food, farmers pump more and more water out of underground aquifers to irrigate, or water, their fields. An aquifer is a layer of rock or sediment that can yield usable groundwater. In some places, the aquifers might run dry because of all the water that is being used.

More Crop for the Drop

Cities and farms often compete for the same water. Irrigation uses a lot more water than people use, so research is being done to find ways for farmers to use less water while still increasing the amount of food they can grow.

Still, even where freshwater is available, it can be too polluted to use. Together, preventing pollution and using water more wisely will help Earth's freshwater supplies last a long time.

Postel's Passion

The problems of freshwater use and their solutions are the concern of Sandra Postel. Postel studies how people use water around the world and how governments are dealing with water use. Her book, *Pillar of Sand*, talks about methods to improve irrigation efficiency. She shows how food and water—two basic human needs—are tied together.

Research **Investigate xeriscaping (ZI ruh skay ping)—landscaping homes with native plants that don't require much water. What kinds of local plants can people use? How will this save water?**

Science Online
For more information, visit bookh.msscience.com/time

Research The term *xeriscaping* refers to landscape techniques that use plants adapted to dry, or *xeric*, environments. Decorating landscapes in dry areas with plant species adapted to use less, and retain more, water can lead to significant decreases in water demands. Grasses used in lawns require large amounts of water, while cactus, aloe and other plants are designed for water retention.

Resources for Teachers and Students

"Making Every Drop Count" by Peter H. Gleick, *Scientific American*, February 2001

"Growing More Food With Less Water" by Sandra Postel, *Scientific American*, February 2001

Reviewing Main Ideas

Section 1 The Nature of Water

1. Water exists in three states on Earth—solid, liquid, and gas. Energy is either absorbed or given off when water changes state.
2. Water is a polar molecule.
3. The formation of weak bonds among water molecules causes water to have cohesion.

Section 2 Why is water necessary?

1. Water is essential for life on Earth.
2. Agriculture and industry use freshwater.
3. Water also is used for transportation and recreation.

Section 3 Recycling Water

1. Most of Earth's water is in the oceans. Less than one percent is available as freshwater.
2. Groundwater is held underground in layers of rock and sediment.
3. Water continuously circulates through Earth's water cycle.

Visualizing Main Ideas

Copy and complete the concept map using the following terms: high specific heat, transportation, surface water, ice, production, *and* polar molecules.

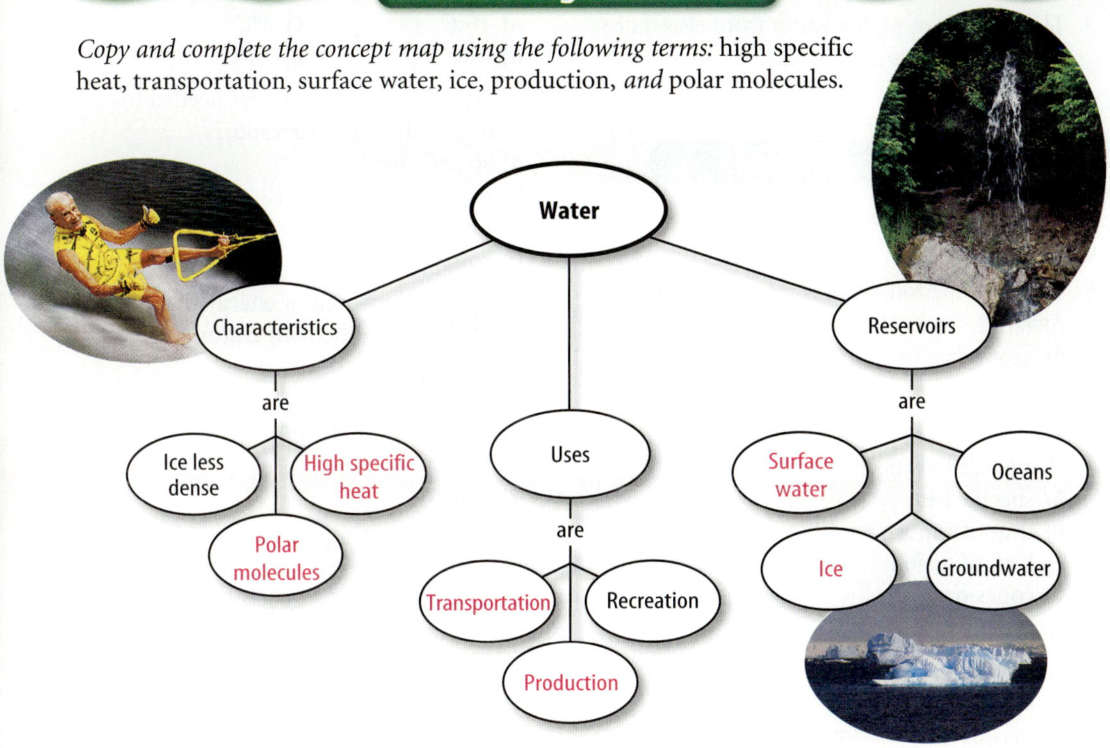

Reviewing Main Ideas

Summary statements can be used by students to review the major concepts of the chapter.

Visualizing Main Ideas

See student page.

Science Online

Visit bookh.msscience.com
/self_check_quiz
/interactive_tutor
/vocabulary_puzzlemaker
/chapter_review
/standardized_test

Assessment Transparency

For additional assessment questions, use the *Assessment Transparency* located in the transparency book.

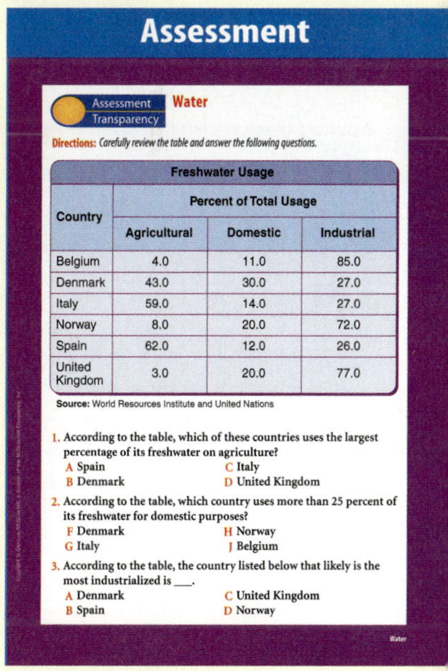

Assessment

Assessment Transparency **Water**

Directions: *Carefully review the table and answer the following questions.*

Freshwater Usage

Country	Percent of Total Usage		
	Agricultural	Domestic	Industrial
Belgium	4.0	11.0	85.0
Denmark	43.0	30.0	27.0
Italy	59.0	14.0	27.0
Norway	8.0	20.0	72.0
Spain	62.0	12.0	26.0
United Kingdom	3.0	20.0	77.0

Source: World Resources Institute and United Nations

1. According to the table, which of these countries uses the largest percentage of its freshwater on agriculture?
   A Spain C Italy
   B Denmark D United Kingdom
2. According to the table, which country uses more than 25 percent of its freshwater for domestic purposes?
   F Denmark H Norway
   G Italy J Belgium
3. According to the table, the country listed below that likely is the most industrialized is ___.
   A Denmark C United Kingdom
   B Spain D Norway

Water

Using Vocabulary

1. density
2. cohesion
3. surface water
4. soil water
5. irrigation

Checking Concepts

6. A
7. C
8. D
9. C
10. B
11. A
12. C
13. D

chapter 1 Review

Using Vocabulary

aquifer p. 23	polar molecule p. 12
cohesion p. 12	soil water p. 23
density p. 11	specific heat p. 14
groundwater p. 23	surface water p. 24
irrigation p. 17	water conservation p. 20

Fill in the blanks with the correct vocabulary word or words.

1. The _________ of a substance means how massive it is for its size.

2. The attraction between water molecules is called _________.

3. Lakes, streams, and rivers are _________.

4. Water that is held between spaces in the soil is _________.

5. The practice of piping water from elsewhere to water plants is _________.

Checking Concepts

Choose the word or phrase that best answers the question.

6. Which of the following is water in layers of rock or sediment?
A) groundwater
B) aquifer
C) well
D) surface water

7. Which is the amount of heat needed to change ice to water?
A) specific heat
B) heat of vaporization
C) heat of fusion
D) cohesion

8. Which of the following is the attraction among water molecules?
A) specific heat
B) density
C) conservation
D) cohesion

Use the graph below to answer question 9.

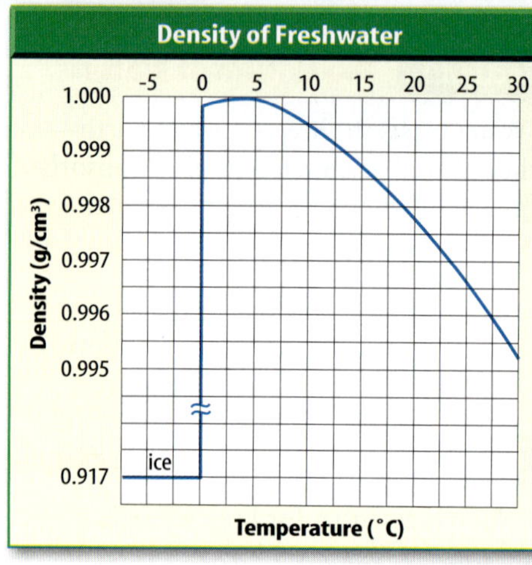

9. Examine the graph above. At which temperature is freshwater most dense?
A) 10°C
B) 0°C
C) 4°C
D) 15°C

10. What is the amount of heat needed to change water to water vapor?
A) specific heat
B) heat of vaporization
C) heat of fusion
D) cohesion

11. What is the amount of energy that is needed to raise the temperature of 1 kg of a substance 1°C?
A) specific heat
B) heat of vaporization
C) heat of fusion
D) density

12. Which is the pumping of water from one place to another for watering crops?
A) groundwater
B) soil water
C) irrigation
D) aquifer

13. Which is a recreational use of water?
A) drinking
B) bathing
C) irrigation
D) kayaking

Science Online bookh.msscience.com/vocabulary_puzzlemaker

Use the ExamView® Pro Testmaker CD-ROM to:

- create multiple versions of tests
- create modified tests with one mouse click for inclusion students
- edit existing questions and add your own questions
- build tests aligned with state standards using built-in State Curriculum Tags
- change English tests to Spanish with one mouse click and vice versa

Thinking Critically

14. Explain When ice melts, the temperature of the water stays constant even as heat is added. Explain why this happens.

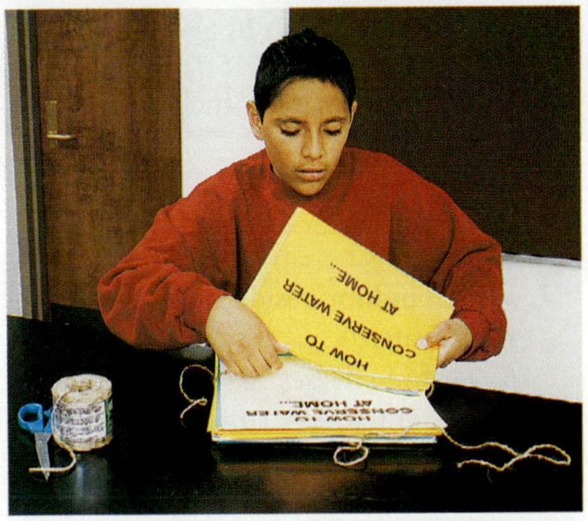

15. Infer Diving into a pool on your stomach can hurt. Which property of water explains why this is so?

16. Recognize Cause and Effect How does the water cycle produce freshwater from salt water?

17. Draw Conclusions After decades of pumping groundwater, some wells go dry. Explain.

18. Recognize Cause and Effect How is the formation of weak bonds among water molecules responsible for cohesion?

19. Form Hypotheses How could a river continue to flow during a drought?

20. Research Information Select a topic that was discussed in this chapter. Do further research about the topic in the library and write a short paper to inform others about your topic. Make sure that your paper has an introduction, body, and conclusion.

21. Communicate Write a newspaper article about the importance of soil water.

Performance Activities

22. Poster Use photographs from old magazines to create a poster illustrating a variety of different uses for water in society. Combine your poster with those of your classmates.

23. Letter Write a letter to the local water company asking for suggestions about how to conserve water in the home. With your family's help, decide whether you will use some of these suggestions in your home.

24. Communicate Use a word processor to produce a brochure that encourages people to conserve water in their homes.

Applying Math

Use the table below to answer questions 25–27.

Process	Amount of Water (km^3)
Evaporation from ocean	350,000
Precipitation on ocean	310,000
Precipitation on land	100,000
Evaporation from land	60,000
Stream and groundwater flow to oceans	?

25. Ocean Water According to the table, how much more water is evaporated from the ocean than falls on the ocean as precipitation?

26. Water on Land How much more water falls on the land than is evaporated from the land?

27. Water Flow Assuming that Earth's water budget is in balance, how much water must flow from land to the oceans in streams or as groundwater each year?

Assessment Resources

Reproducible Masters

Chapter *Fast File* Resources
- Chapter Review, pp. 37–38
- Chapter Tests, pp. 39–42
- Assessment Transparency Activity, p. 49

Glencoe Science Web site
- Chapter Review Test
- Standardized Test Practice

Glencoe Technology
- Assessment Transparency
- Exam*View*® Pro Testmaker
- MindJogger Videoquiz
- Interactive Chalkboard

chapter 1 Review

Thinking Critically

14. The temperature remains constant because the energy is needed to change the state of water from solid to liquid. This energy is called latent heat of fusion.

15. cohesion

16. When water evaporates from the ocean or a saline lake, the salt is left behind. The water vapor might later condense and fall to Earth as freshwater.

17. Water was withdrawn from the aquifer faster than it could be replaced.

18. The weak bonds hold water molecules together.

19. Groundwater could enter the river if the river's channel is below the water table.

20. Encourage students to cite the sources of information that they use for their papers.

21. Student articles should mention that soil water is essential for plant growth.

Performance Activities

22. Posters could include uses such as swimming, boating, fishing, hydroelectric dams, cooling, or transport ing goods. Use **Performance Assessment in the Science Classroom,** p. 145.

23. Letters should be in the proper business letter form. Letter writers should identify themselves as students and clearly ask for information about home water conservation tips. Use **Performance Assessment in the Science Classroom,** p. 139.

Applying Math

National Math Standards

5

24. 40,000 km^3

25. 40,000 km^3

26. 40,000 km^3

FAST FILE

Answer Sheet A practice answer sheet can be found at bookh.msscience.com/answer_sheet.

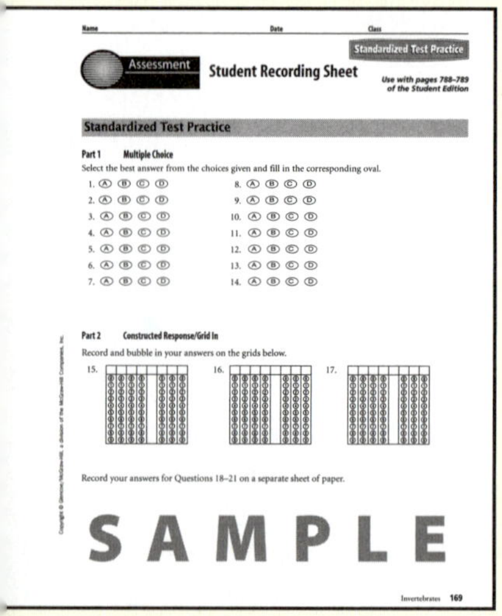
Name Date Class

Assessment **Student Recording Sheet** Standardized Test Practice

Use with pages 788–789 of the Student Edition

Standardized Test Practice

Part 1 Multiple Choice

Select the best answer from the choices given and fill in the corresponding oval.

1.–14. Ⓐ Ⓑ Ⓒ Ⓓ

Part 2 Constructed Response/Grid In

Record and bubble in your answers on the grids below.

15. 16. 17.

Record your answers for Questions 18–21 on a separate sheet of paper.

SAMPLE

Part 1 Multiple Choice

1. B
2. C
3. D
4. B
5. A
6. B
7. B
8. A
9. B
10. C

Part 2 Short Response

11. Water helps organisms regulate temperature, move nutrients, eliminate wastes, digest food, among others. Water also provides habitat for many organisms.
12. Answers will vary. Some possible choices are transportation, industry, irrigation, recreation, and personal consumption.
13. Latent heat of fusion is absorbed or released during state changes between liquid water and ice. Latent heat of vaporization is absorbed or released during state changes between liquid water and water vapor.

chapter 1 Standardized Test Practice

Part 1 Multiple Choice

Record your answers on the answer sheet provided by your teacher or on a sheet of paper.

Use the photo below to answer question 1.

1. Which use of water is shown in the photo above?
   A. water for industry
   B. water for agriculture
   C. water for recreation
   D. water for transportation
2. About what percent of Earth's surface is covered by water?
   A. 40% **C.** 70%
   B. 10% **D.** 90%
3. Which water reservoir contains the most water?
   A. groundwater **C.** lakes
   B. streams **D.** oceans
4. Which elements are in water?
   A. carbon and oxygen
   B. hydrogen and oxygen
   C. nitrogen and oxygen
   D. hydrogen and carbon

Test-Taking Tip

Consider All Choices Be sure you understand the question before you read the answer choices. Make special note of words like NOT or EXCEPT. Read and consider all the answer choices before you mark your answer sheet.

5. What is a layer of rock or sediment that can yield usable groundwater?
   A. aquifer **C.** soil
   B. well **D.** glacier
6. Which type of water includes lakes, streams, and ponds?
   A. groundwater **C.** soil water
   B. surface water **D.** ocean water
7. Which indoor, residential activity uses the most water?
   A. washing dishes **C.** washing clothes
   B. flushing toilets **D.** showers and baths

Use the diagram below to answer questions 8–10.

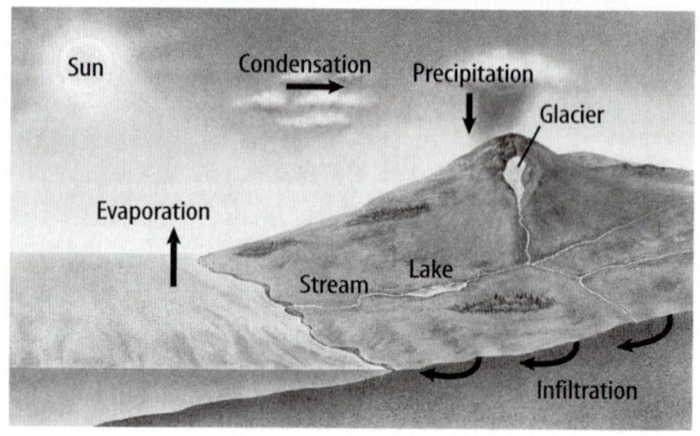

8. During which process does water change from a liquid to a gas?
   A. evaporation **C.** condensation
   B. runoff **D.** infiltration
9. Which process transfers water from the atmosphere to Earth's surface?
   A. evaporation **C.** infiltration
   B. precipitation **D.** runoff
10. During which process does water change from a gas to a liquid?
    A. evaporation **C.** condensation
    B. runoff **D.** infiltration

14. Water exists in the pore spaces in sediment and rock.
15. The temperature rises to the freezing point, where it remains constant until all ice is melted. Temperature then rises to the boiling point, where it remains constant until all water has changed to steam.
16. Ice is less dense than liquid water.
17. Answers will vary but may include bathing, washing, flushing toilets, watering lawns, and washing cars.
18. 4186 J/Kg °C
19. 294 J/Kg °C
20. 3892 Joules
21. water, ice, air, granite rock

Part 2 Short Response/Grid In

Record your answers on the answer sheet provided by your teacher or on a sheet of paper.

11. Why is water essential for life? List some life processes that need water.

12. How is water used in society? List at least three different ways.

13. What's the difference between latent heat of fusion and latent heat of vaporization?

14. Explain how water can exist beneath Earth's surface.

15. Explain how the temperature of water changes as it is heated from below freezing to boiling.

16. Explain why ice floats on liquid water.

17. List the ways that you use water everyday.

Use the table below to answer questions 18–21.

Substance	Specific Heat (J/kg°C)
Water (pure)	4,186
Ice	2,093
Air (dry)	1,005
Granite rock	294

18. What is the specific heat value for pure water?

19. What is the specific heat value of granite rock?

20. How many more joules does it take to heat one kilogram of water 1°C than one kilogram of granite rock 1°C?

21. Order the four substances from the one that would change temperature most slowly to the one that would change temperature most quickly when heat is added.

Part 3 Open Ended

Record your answers on a sheet of paper.

Use the photo below to answer question 22.

22. What is cohesion? Why can the insect shown above dart across the surface of a pond?

23. Explain why it is important to protect the quality of Earth's freshwater supply.

24. Explain how water is important for transportation of products.

25. How can water be used more efficiently during irrigation?

26. Where is most freshwater located on Earth? Why is it difficult for people to use this water?

27. How does the water cycle supply freshwater to Earth's land surface?

28. Why does water rise through a plant's stem to the leaves?

29. How does water exist in Earth's atmosphere?

30. Make a drawing of a water molecule. Label each atom in the molecule.

31. Expain why the water molecule is a polar molecule. Which parts of the molecule have a slight positive charge? Which part has a slight negative charge?

28. An attraction between water molecules and other polar molecules, and cohesion between water molecules create the forces required to move water through a plant stem.

29. Water exists mostly in the gas phase. Some water exists as droplets and ice crystals in clouds.

30. The drawing should include one large atom of oxygen and two smaller atoms of hydrogen. The two hydrogen atoms should be independently bonded to the oxygen atom. See **Figure 6** for reference.

31. The positive and negative charges on the water molecule are unequally distributed. The side of the molecule toward the hydrogen atoms is positive. The oxygen side is negative.

Rubrics

For more help evaluating open-ended assessment questions, see the rubric on p. 10T.

Part 3 Open Ended

22. Cohesion is the attraction between polar molecules. The cohesive forces between water molecules allow the insect to dart across the pond's surface.

23. Humans and other organisms depend on clean freshwater.

24. Ships transport products to markets. Water transportation is an efficient and economical way to move goods.

25. More efficient irrigation methods can be used or watering can be done at times of day when the rate of evaporation is low.

26. Most freshwater is stored in glaciers. Most of this water is in polar regions.

27. Salt water evaporates. Water vapor condenses and falls to Earth as precipitation.

Section/Objectives	Standards		Labs/Features
	National	State/Local	
Chapter Opener	See pp. 9T–10T for a Key to Standards.		**Launch Lab:** How do lake nutrients mix, p. 35 **Foldables,** p. 35
Section 1 Streams 2 sessions 1 block 1. **Discuss** the distribution of Earth's water. 2. **Compare and contrast** stream types. 3. **Describe** how streams erode and deposit sediment.	National Content Standards: UCP.1, UCP.2, UCP.3, UCP.5, A.1, A.2, C.1, C.4, C.5, D.1, F3		**MiniLAB:** Modeling Stream Flow, p. 39 **Science Online,** p. 42
Section 2 Lakes and Reservoirs 3 sessions 1.5 blocks 4. **Explain** how lakes form. 5. **Discuss** organisms found in lakes. 6. **Discuss** how lakes change through time.	National Content Standards: UCP.1, UCP.2, UCP.3, UCP.5, A.1, A.2, C.1, C.4, C.5, D.1		**Science Online,** p. 45 **Integrate Life Science,** p. 47 **Applying Math:** Calculating with Sedimentation Rates, p. 47 **Visualizing Eutrophication,** p. 48 **Lab:** Lake Nutrients, p. 50
Section 3 Wetlands 1 session .5 block 7. **Describe** a wetland. 8. **Compare and contrast** different types of wetlands. 9. **Identify** some organisms that live in wetlands.	National Content Standards: UCP.1, UCP.2, UCP.3, UCP.5, A.1, A.2, C.1, C.4		
Section 4 Pollution of Freshwater 4 sessions 2 blocks 10. **Identify** sources of water pollution. 11. **Explain** the difference between point source and nonpoint source pollution. 12. **Describe** laws that regulate water pollution.	National Content Standards: UCP.1, UCP.2, UCP.3, UCP.5, A.1, A.2, F2		**MiniLAB:** Evaluating Dilution, p. 55 **Integrate History,** p. 56 **Lab:** Adopt a Stream, p. 58 **History of the Great Lakes,** p. 60

Glencoe Exclusive! TeacherWorks™ All-In-One Planner and Resource Center

Lab Materials	Reproducible Resources	Section Assessment	Technology
Launch Lab: graph paper, colored pencils	**Chapter *FAST FILE* Resources** Foldables Worksheet, p. 13 Directed Reading Overview, p. 15 Note-taking Worksheets, pp. 31–34	Glencoe's Assessment Advantage	TeacherWorks includes: • Interactive Teacher Edition • Lesson Planner with calendar • Access to all program blacklines • Correlations to standards • Web links Video Lab
MiniLAB: large flat pan or stream table, sand, wedge, flexible tubing, water, bucket	**Chapter *FAST FILE* Resources** Transparency Activity, p. 44 MiniLAB, p. 3 Lab Activity, pp. 9–10 Enrichment, p. 27 Reinforcement, p. 23 Directed Reading, p. 16	**Portfolio** Science Journal, p. 38 **Performance** MiniLAB, p. 39 Applying Math, p. 43 **Content** Section Review, p. 43	Section Focus Transparency Guided Reading Audio Program Interactive Chalkboard CD-ROM
Lab: felt-tipped marker, 4 small baby-food jars with lids, 100-mL graduated cylinder, distilled water, 0.4% nitrate solution, 0.4% phosphate solution, nitrate-phosphate solution, algae cultures, dropper	**Chapter *FAST FILE* Resources** Transparency Activity, p. 45 Lab Activity, pp. 11–12 Enrichment, p. 28 Reinforcement, p. 24 Directed Reading, p. 16 Transparency Activity, pp. 49–50 Lab Worksheet, pp. 5–6	**Portfolio** Cultural Diversity, p. 46 **Performance** Applying Math, p. 47 Applying Skills, p. 49 **Content** Section Review, p. 49	Section Focus Transparency Teaching Transparency Guided Reading Audio Program Interactive Chalkboard CD-ROM
Need materials? Contact Science Kit at 1-800-828-7777 or www.sciencekit.com on the Internet.	**Chapter *FAST FILE* Resources** Transparency Activity, p. 46 Enrichment, p. 29 Reinforcement, p. 25 Directed Reading, p. 17	**Performance** Applying Skills, p. 53 **Content** Section Review, p. 53	Section Focus Transparency Guided Reading Audio Program Interactive Chalkboard CD-ROM
MiniLAB: 4 drinking glasses, tap water, red food coloring **Lab:** bucket, rope, alcohol thermometer, string, plastic sample bottles, beakers (100-mL), graduated cylinder (500-mL), pH paper, dissolved oxygen kit, alkalinity kit, nitrate test kit	**Chapter *FAST FILE* Resources** Transparency Activity, p. 47 MiniLAB, p. 4 Enrichment, p. 30 Reinforcement, p. 26 Directed Reading, pp. 17, 18 Lab Worksheet, pp. 7–8 **Lab Management and Safety,** p. 69	**Portfolio** Assessment, pp. 57, 59 **Performance** MiniLAB, p. 57 Applying Math, p. 57 **Content** Section Review, p. 57	Section Focus Transparency Virtual Labs CD-ROM Guided Reading Audio Program Interactive Chalkboard CD-ROM

End of Chapter Assessment

Glencoe's Assessment Advantage

Blackline Masters	Technology	Professional Series
Chapter *FAST FILE* Resources Chapter Review, pp. 37–38 Chapter Tests, pp. 39–42 **Standardized Test Practice,** pp. 11–14	MindJogger Videoquiz Virtual Labs CD-ROM Exam*View*® Pro Testmaker TeacherWorks CD-ROM Interactive Chalkboard CD-ROM	**Performance Assessment in the Science Classroom (PASC)**

chapter 2 Freshwater at Earth's Surface

Transparencies

Section Focus

SECTION 1 Section Focus Transparency **Gift of the Nile**

There is a saying that goes, "Egyptian civilization is the gift of the Nile." In addition to supplying water, seasonal floods deposit sediments that make the Nile's floodplains excellent for growing crops. In the 1950s and 1960s, the Egyptian government built the Aswan High Dam, which is shown below, to provide a source of power and to help regulate the seasonal floods.

1. Why might ancient civilizations have flourished along the Nile?
2. In the photo, which way does the river flow? Why did you choose that direction?
3. How does the Aswan High Dam help Egypt? What difficulties arise from damming rivers?

L2

SECTION 2 Section Focus Transparency **Hold Back the Lake**

Navigating massive freighters between Lake Ontario and Lake Erie might be difficult since Lake Erie is about 99 meters higher than Lake Ontario! A series of eight locks in the Welland Canal raises or lowers commercial vessels between these Great Lakes. One of these locks is shown below.

1. How are lakes and streams similar? How are they different?
2. How do people use lakes?
3. What are some ways that lakes are formed?

L2

SECTION 3 Section Focus Transparency **Islands in the Marsh**

This photo shows the mud and reed island homes of the Ma'dan (Marsh Arabs) in the marshes of southern Iraq. These marshes, with an area of over 15,000 km², have supported diverse human populations for over 5,000 years.

1. What difficulties would you encounter living in an area like the one in the photo?
2. In general, what kind of plants can you see in this photo?
3. Can you name some other marshy or swampy areas?

L2

This is a representation of key blackline masters available in the Teacher Classroom Resources. See Resource Manager boxes within the chapter for additional information.

Assessment

Assessment Transparency **Freshwater at Earth's Surface**

Directions: *Carefully review the table and answer the following questions.*

Precipitation in Flower Stream 1990-1999				
Year	Ca Calcium (mg/L)	K Potassium (mg/L)	NH_4 Ammonia (mg/L)	NO_3 Nitrates (mg/L)
1990	0.14	0.009	0.19	0.62
1991	0.18	0.057	0.55	0.08
1992	0.37	0.026	0.29	0.94
1993	0.45	0.023	0.29	0.86
1994	0.63	0.033	0.38	1.11
1995	0.69	0.027	0.30	0.88
1996	1.48	0.040	0.20	0.73
1997	0.49	0.035	0.24	0.89
1998	0.77	0.025	0.28	0.96
1999	0.65	0.031	0.34	1.13

1. According to the table, which element or compound had the greatest concentration increase from 1993 to 1999?
   A Ca B K C NH_4 D NO_3
2. In which year did nitrates have the lowest concentration?
   F 1990 G 1991 H 1995 J 1999
3. According to the table, the element or compound that had the highest concentration at a given point during the nine-year period was ____.
   A Ca B NO_3 C NH_4 D K

L2

Teaching

SECTION 2 Teaching Transparency **Lake Life**

L2

Key to Teaching Strategies

The following designations will help you decide which activities are appropriate for your students.

- **L1** Level 1 activities should be appropriate for students with learning difficulties.
- **L2** Level 2 activities should be within the ability range of all students.
- **L3** Level 3 activities are designed for above-average students.
- **ELL** ELL activities should be within the ability range of English-Language Learners.
- **COOP LEARN** Cooperative Learning activities are designed for small group work.
- **LS** Multiple Learning Styles logos, as described on page 6T, are used throughout to indicate strategies that address different learning styles.
- **P** These strategies represent student products that can be placed into a best-work portfolio.
- **PBL** Problem-Based Learning activities apply real-world situations to learning.

Hands-on Activities

Student Text Lab Worksheet

Name ____ Date ____ Class ____

Activity **Lake Nutrients**

Hands-On Activities

Lab Preview

Directions: *Answer these questions before you begin the Activity.*

1. Why are you instructed to wear goggles during this activity?
2. What is the significance of nitrates and phosphates in lake ecosystems?

Plants and animals that die settle to the bottom of a lake where they decay and release phosphates and nitrates to the water. The amount of algae that is present in lake water depends on the amount of nutrients in the water.

What You'll Investigate

How does the presence of nutrients in lake water affect the amount of algae that can survive?

Materials

felt-tip marker
small baby food jars with caps (4)
100-mL graduated cylinder
distilled water (200 mL)
0.4% nitrate solution (200 mL)
0.4% phosphate solution (200 mL)
nitrate-phosphate solution (200 mL)
(mix 100 mL of 0.4% nitrate solution with 100 mL of 0.4% phosphate solution)
algae cultures
dropper

Goals

- **Predict** which solution will show the most algal growth.
- **Predict** how nutrients will affect algal growth in the lake.

Safety Precautions

Procedure

1. Label each jar with your name or lab group.
2. Label and fill each jar with the following:
   - 0.4% nitrate solution
   - 0.4% phosphate solution
   - nitrate/phosphate solution
   - distilled water
3. Prepare the algae culture according to the directions and place eight drops of the algae solution in each of the four different jars.
4. With the lid on, shake each jar gently and place each on a windowsill.
5. Remove the lid and check the bottles for algal growth each day for five days. Replace the lid after each observation. Record your observations in the data table on the next page.

L2

Laboratory Activities

Name ____ Date ____ Class ____

LAB 1 Laboratory Activity **Water Desalination**

Hands-On Activities

Freshwater is essential to all life forms. But what about a situation in which you do not have freshwater available? Life rafts on boats are equipped with an apparatus that can be used to distill freshwater from salt water. Desert safety survival rules provide another means to distill water.

Strategy

You will desalinate water by using a simple distillation process.
You will discuss how this process could be used in an emergency situation.

Materials

2 coat hangers, or bendable wire
salt
water
cereal bowl
pen (felt-tip)
pan (larger than the circumference of the bag)
plastic bag (clear)
sunlamp or bright sunshine

Procedure

1. Bend the coat hangers into a frame. See Figure 1.
2. Mix the salt into water in the cereal bowl. Mark the water level on the inside with the pen.
3. Place the cereal bowl in the pan and place the wire frame over it.
4. Pull the plastic bag over the frame until it touches the pan. Record the appearance of the water.
5. Set the apparatus in direct sun or under a sunlamp.
6. Allow the apparatus to stand undisturbed. Observe and record your observations after 10 min and again after 30 min in the Data and Observations table.

Figure 1

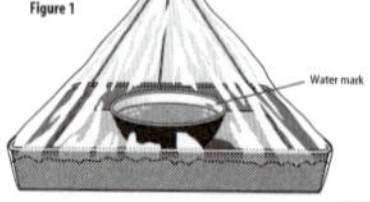

L2

Meeting Different Ability Levels

Content Outline

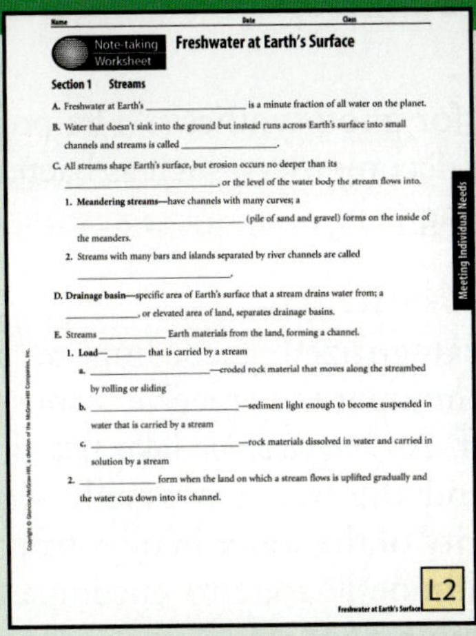

Reinforcement

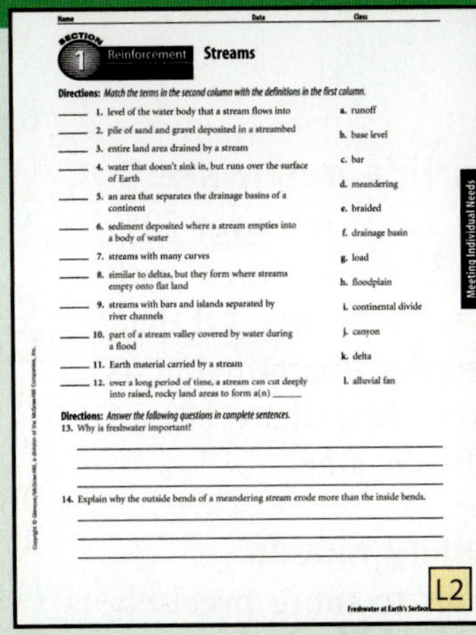

Enrichment

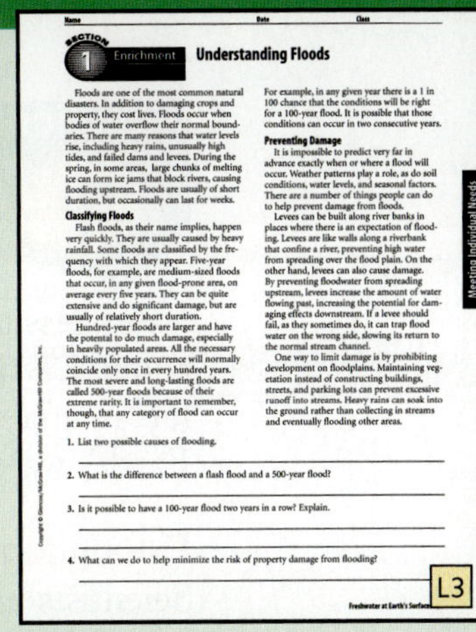

Directed Reading (English/Spanish)

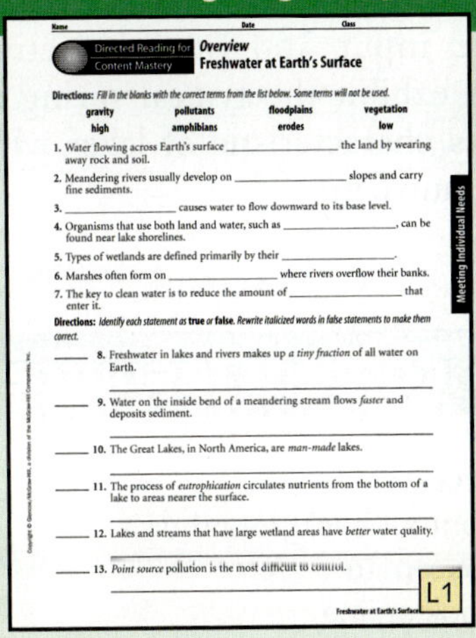

Study Guide

Reading Essentials

Assessment

Test Practice Workbook

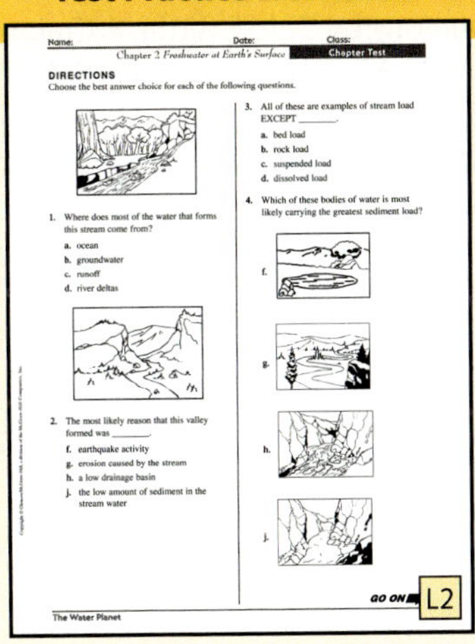

Chapter Review

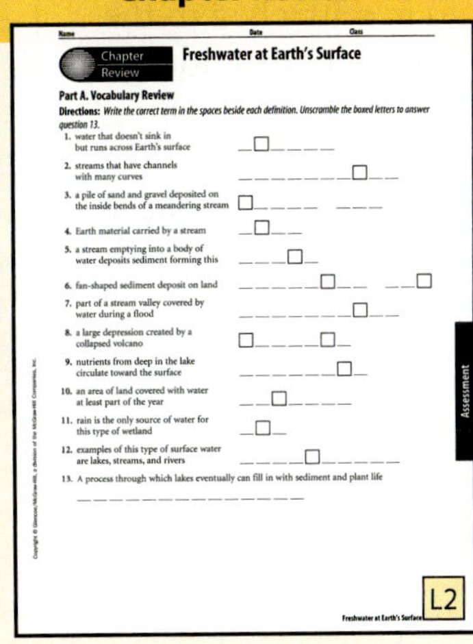

Chapter Tests

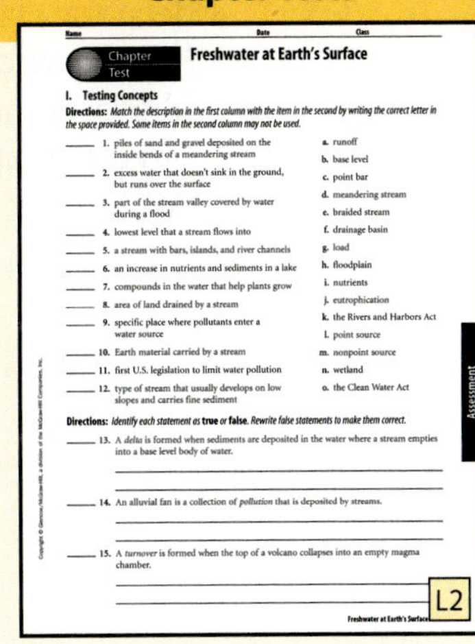

Science Content Background

section 1 Streams

Where is Earth's freshwater?

Despite the fact that just over 70% of Earth's surface is covered with water, less than 3% exists as freshwater. Of that, about 2% remains realistically unattainable as ice in glaciers. This leaves less than 1% of Earth's water as available freshwater.

Floodplains and Controlling Floods

Scientists are using satellites to more precisely map floodplain areas. Many people do not realize that their property is on a 50- or 100-year floodplain. In many cases it is a lack of accurate records that accounts for this. By analyzing satellite images, scientists are able to identify areas that have been affected by large floods. When such an area is located, detailed images are studied and scientists map the area carefully. The information is then logged into a database that is available for public use.

section 2 Lakes and Reservoirs

Life in Lakes

The near-shore zone of a lake, which is shallow enough to allow aquatic plants to root and for light to penetrate and be used for photosynthesis, is called the *littoral zone.* The littoral zone has the greatest diversity of organisms. The *limnetic zone* is the area farther from shore where the water is too deep for plants to be able to live attached to the bottom and capture enough light to undergo photosynthesis. In this area, surface-dwelling phytoplankton are eaten by zooplankton; fish are secondary consumers. The *profundal zone* is found near the lake bottom. Here, not enough light is available for photosynthesis. The populations consist of decomposers, such as bacteria, and bottom feeding fish.

Sediments and Nutrients

Lakes can be categorized according to the amount of nutrient input they receive from sediment and runoff. An *oligotrophic* lake has little nutrient input and the water is usually clear. The extreme clarity of the water in oligotrophic lakes allows the limnetic zone to encompass a wider area, continuing out into deeper water. A lake that receives a high amount of nutrient input is called *eutrophic.* The water is cloudy as a result of the input and organic activity. Eutrophic lakes exhibit the typical spring and fall algal blooms characteristic of lakes with a high amount of nutrients.

chapter content resources

Internet Resources

For additional content background, visit **bookh.msscience.com** to:

- access your book online
- find references to related articles in popular science magazines
- access Web links with related content background
- access current events with science journal topics

Print Resources

Process Geomorphology by Dale F. Ritter, R. Craig Kochel, and Jerry, R. Miller, McGraw-Hill Higher Education, 2002

Limnology: Lake and River Ecosystems by R.G. Wetzel, Academic Press, 2001

section 3 Wetlands

Types of Wetlands

Wetlands can consist of saltwater as well as freshwater. Wetlands that exist in coastal areas, often referred to as salt marshes or tidal marshes, can occur in estuaries, areas where rivers empty into oceans. These areas provide a habitat and nursery ground for many economically important marine species of fish, crab, shrimp, and oysters. Similar to inland wetlands, they also serve as breeding and nesting areas for many species of birds.

Mangrove swamps are another example of coastal wetlands. Mangrove swamps, located in tropical regions worldwide, are host to 80 different species of mangrove trees and shrubs. Mangrove swamps provide a unique habitat for many marine organisms including fish, stingrays, shrimp, lobsters, snails, crabs, seabirds, and even crocodiles.

Stanley Breeden/DRK Photo

section 4 Pollution of Freshwater

Pollution Sources

Acid rain can affect the quality of freshwater in lakes and streams and can have far-reaching ecological consequences. Acid rain originates with the input into the atmosphere of sulfur dioxide from coal- or oil-burning power plants or nitrous oxide, which is released by car and industry emissions. These substances react with water vapor to form sulfuric acid and nitric acid, respectively. The acids enter freshwater lakes and streams through precipitation. If no chemical buffering occurs, the acids accumulate, eventually bringing the pH of a freshwater system to dangerously low levels. If the pH becomes too low, organisms that inhabit these lakes can no longer survive. Some lakes in the northeastern United States are lifeless as a result of acid rain. Acid rain can affect the quality of drinking water as well. The increased acidity of groundwater can dissolve pipes, releasing high amounts of lead into the water supply. In some areas of the northeastern United States, drinking water must be treated with a chemical buffer that decreases acidity before the water can be consumed by the public.

Teacher to Teacher

Steve Federman
Loveland Middle School
Loveland, Ohio

"To demonstrate to my students how little freshwater there is at Earth's surface, I use a gallon bucket of water to represent all the water on Earth and ask students to estimate how much of that gallon would represent freshwater. I pour out $\frac{1}{4}$ cup of water which represents (3%) of all Earth's water while the remaining amount (97%) is saltwater. Next I ask students to predict how much of the $\frac{1}{4}$ cup is available for us to use at Earth's surface, then with an eyedropper demonstrate that it is one drop. The rest is ice, groundwater, in biomass, or in the atmosphere."

Steve Federman

chapter 2

Chapter Vocabulary

runoff, p. 36
meandering stream, p. 37
point bar, p. 37
braided stream, p. 38
drainage basin, p. 38
load, p. 39
stream discharge, p. 41
floodplain, p. 42
nutrient, p. 47
eutrophication, p. 47
turnover, p. 49
wetlands, p. 51
point source pollution, p. 54
nonpoint source pollution, p. 55

Science Journal This exercise develops students' descriptive writing skills. Their responses should reveal prior knowledge about rivers.

INTERACTIVE CHALKBOARD with Image Bank

PowerPoint® Presentations

This CD-ROM is an editable Microsoft® PowerPoint® presentation that includes:

- a pre-made presentation for every chapter
- interactive graphics
- animations
- audio clips
- image bank
- all new section and chapter questions
- Standardized Test Practice
- transparencies
- pre-lab questions for all labs
- Foldables directions
- links to bookh.msscience.com

chapter 2

Freshwater at Earth's Surface

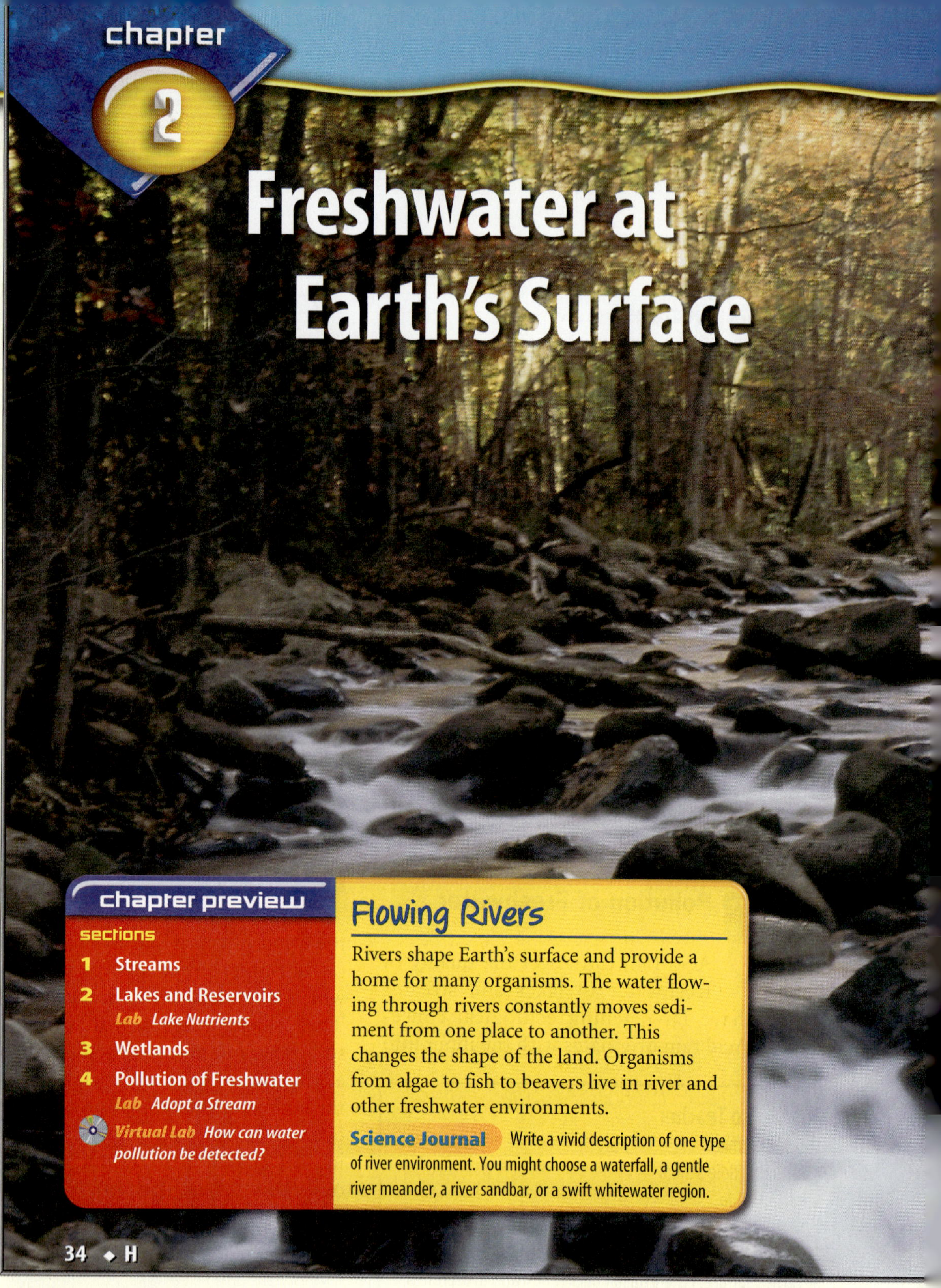

chapter preview

sections

1. Streams
2. Lakes and Reservoirs
   Lab *Lake Nutrients*
3. Wetlands
4. Pollution of Freshwater
   Lab *Adopt a Stream*
   Virtual Lab *How can water pollution be detected?*

Flowing Rivers

Rivers shape Earth's surface and provide a home for many organisms. The water flowing through rivers constantly moves sediment from one place to another. This changes the shape of the land. Organisms from algae to fish to beavers live in river and other freshwater environments.

Science Journal Write a vivid description of one type of river environment. You might choose a waterfall, a gentle river meander, a river sandbar, or a swift whitewater region.

34 ◆ H

Theme Connection

Systems and Interactions The chapter describes the many types of bodies, such as streams and lakes, that form Earth's surface water system. Students also learn about human interactions with surface waters and how those interactions can pollute the water.

About the Photo

Flowing Rivers The photo shows the Roaring Fork River in Great Smoky Mountains National Park, Tennessee.

Start-Up Activities

Launch LAB

How do lake nutrients mix?

The temperature of lake water affects the distribution of nutrients. During summer, warm surface water can be low in nutrients. During autumn, surface water cools, and nutrient-rich deep water mixes throughout the lake.

1. Label a sheet of graph paper with depth on the vertical axis and temperature on the horizontal axis.
2. Plot the data points for June and November with a different colored pencil. Connect each set of data points with a smooth curve or line.
3. **Think Critically** Explain how seasonal changes in a lake's temperature profile allow nutrients to mix throughout the lake.

Lake Temperature

	June	Nov.
Depth (m)	T (°C)	T (°C)
0	18	7
5	16	7
10	13	7
15	11	7

FOLDABLES™ Study Organizer

Freshwater at Earth's Surface Make the following Foldable about lakes, streams, and wetlands.

STEP 1 **Fold** a sheet of paper from side to side. Make the front edge about 1.25 cm shorter than the back edge.

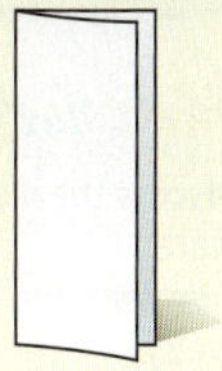

STEP 2 **Turn** lengthwise and **fold** into thirds.

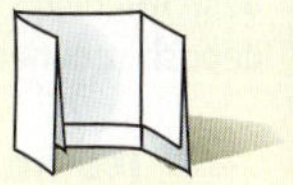

STEP 3 **Unfold and cut** only the top layer along both folds to make three tabs.

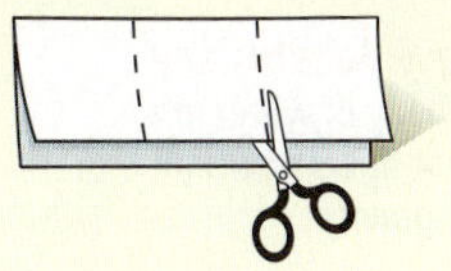

STEP 4 **Label** each tab as shown.

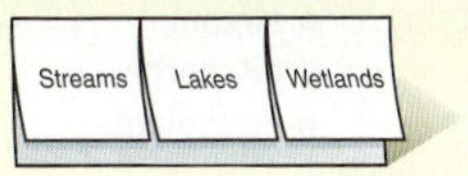

Read and Write Before you read the chapter, describe each source of freshwater on the front of the tabs. As you read the chapter, write information about each under the tabs.

Science Online — Preview this chapter's content and activities at bookh.msscience.com

Launch LAB

Purpose Students make a graph that shows a lake's temperature profile during June and November. L2 LS **Logical-Mathematical**

Preparation Remind students that because warm water is less dense than cool water, warm water tends to float above cool water.

Materials graph paper, colored pencils

Teaching Strategy Suggest that students prepare their graphs with a depth of 0 m at the top of the graph and a depth of 15 m at the bottom of the graph.

Think Critically

During June the lake is warm near the surface and cool near the bottom. During November the lake's temperature is uniform throughout its depth. When the lake's temperature is uniform, all of the water has the same density and mixing can occur.

Assessment

Oral Ask students to explain how the lake's temperature affects nutrient distribution. In June many of the nutrients will remain at the bottom of the lake because cool water will not rise to mix with warm surface water. In November the water has a uniform temperature, so the nutrients will be distributed throughout the lake. Use **Performance Assessment in the Science Classroom,** p. 89.

FOLDABLES™ Study Organizer — Dinah Zike Study Fold

Student preparation materials for this Foldable are available in the Chapter *FAST FILE* Resources.

section 1 Streams

1 Motivate

Bellringer

INTERACTIVE CHALKBOARD PowerPoint® Presentations

Section Focus Transparencies also are available on the Interactive Chalkboard CD-ROM. L2 ELL

Section 1 Section Focus Transparency **Gift of the Nile**

There is a saying that goes, "Egyptian civilization is the gift of the Nile." In addition to supplying water, seasonal floods deposit sediments that make the Nile's floodplains excellent for growing crops. In the 1950s and 1960s, the Egyptian government built the Aswan High Dam, which is shown below, to provide a source of power and to help regulate the seasonal floods.

1. Why might ancient civilizations have flourished along the Nile?
2. In the photo, which way does the river flow? Why did you choose that direction?
3. How does the Aswan High Dam help Egypt? What difficulties arise from damming rivers?

Freshwater at Earth's Surface

Tie to Prior Knowledge

Stream Observations Ask students to recall times when they observed a stream. Have them discuss its size, the movement of water, and land uses nearby. Tell students that in this section they will learn about different types of streams and how they change.

---

as you read

What **You'll Learn**

- **Discuss** the distribution of Earth's water.
- **Compare and contrast** stream types.
- **Describe** how streams erode and deposit sediment.

Why **It's Important**

Streams change Earth's surface and can cause flooding.

Review Vocabulary
stream: a body of water in a channel that flows downhill because of gravity

New Vocabulary
- runoff
- meandering stream
- point bar
- braided stream
- drainage basin
- load
- stream discharge
- floodplain

Where is Earth's freshwater?

Water is essential for life. Plants use water, along with carbon dioxide, to make food with the aid of sunlight. Animals use water to move nutrients and wastes through their bodies. Humans can survive for weeks without food, but for only a few days without freshwater. How much of the water on Earth is freshwater? As **Figure 1** shows, freshwater at Earth's surface is a minute fraction of all the water on the planet. Learning about freshwater on Earth and the factors that can affect this precious resource helps protect the life that depends on it.

Runoff

Think about the last time you were outside when it suddenly started raining. As you ran for cover, what did you notice about where all the water was going? Some of it was sinking into the ground, but much of it was flowing over the surface of Earth. The water that doesn't sink in, but instead runs across Earth's surface, is **runoff.**

As water falls on Earth's surface, runoff moves from higher to lower areas. Over time, the water enters small channels and eventually flows into streams. The slope of the land, how much vegetation is on it, and how much it rains help determine how much water runs off into stream channels.

Figure 1 If 1 L of water represents all of Earth's water, a drop represents all the water in lakes and rivers.

Section 1 Resource Manager

Chapter *FAST FILE* Resources
- Transparency Activity, p. 44
- Directed Reading for Content Mastery pp. 15, 16
- MiniLAB, p. 3
- Note-taking Worksheets, pp. 31–34
- Enrichment, p. 27
- Lab Activity, pp. 9–10
- Reinforcement, p. 23

Home and Community Involvement, p. 39
Cultural Diversity, p. 33

36 ◆ H CHAPTER 2 Freshwater at Earth's Surface

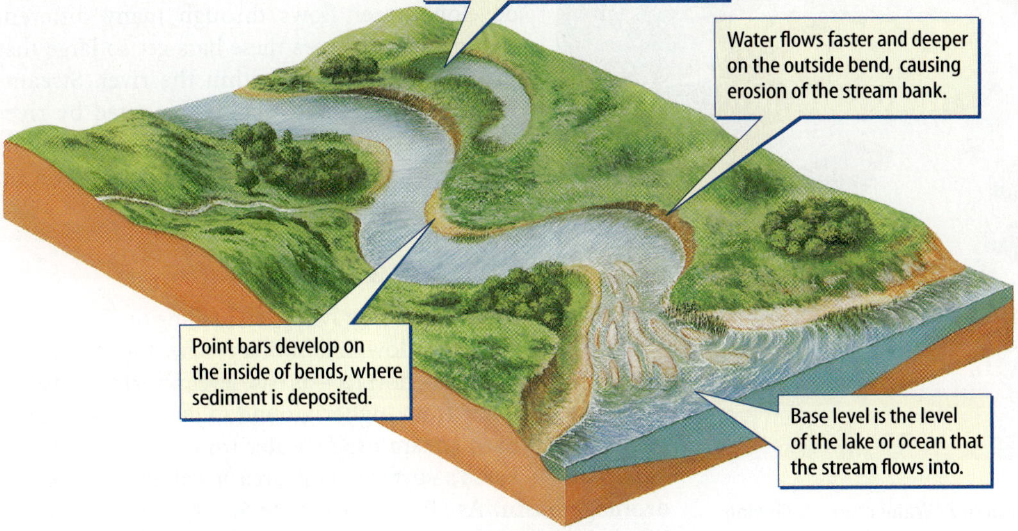

Figure 2 Meandering streams erode channel banks and deposit sediment in point bars.

Types of Streams

Do all streams and rivers look the same to you? Some are narrow, and others are wide. Some streams have a lot of turns and twists, and others flow straight. However, all streams shape Earth's surface. Water flowing over Earth's surface erodes the land by wearing away rock and moving weathered rock and soil. However, a stream can erode Earth's surface no deeper than its base level, the level of the water body that the stream flows into.

Meandering Streams Streams that have channels with many curves, as shown in **Figure 2,** are called **meandering streams.** Rivers that meander usually develop on low slopes and carry a lot of fine sediment, such as silt and clay. The water in meandering streams erodes the land in some places and deposits sediment in others. For example, water on the outside of a bend, or meander, is deeper and flows faster, so it erodes the bank of the stream. Water on the inside of a meander flows slower, so sediment gets deposited there. Soon a **point bar,** a pile of sand and gravel that is deposited on the inside of a meander, forms.

Reading Check *How do point bars form?*

Meandering streams are always changing. Erosion on the outside of a meander and deposition of sediment on the inside cause the stream to move back and forth across its valley over time. Meanders also slowly migrate down the valley.

2 Teach

Visual Learning

Figure 1 Where is most of Earth's water found? It is in oceans.

Make a Model

Meandering Streams Have pairs of students make models of a meandering stream channel in clay or soil. Instruct them to show areas of erosion and deposition at the proper places along the banks of the stream. Erosion should be on the outside of a bend and deposition of sediment should be on the inside of a bend. L2 COOP LEARN

LS **Kinesthetic**

Reading Check

Answer Water on the inside of a meander flows more slowly, causing the water to drop sediments. This forms a point bar.

Differentiated Instruction

Physically Challenged Partner these students with other students to make the stream model. Students who have difficulty manipulating materials can verbally contribute ideas for the design of the model. These students also might give a short oral presentation about the model to the class. L2

Use Science Words

Word Usage Have students use the term *drainage basin* in a sentence that describes something on the **Figure 4** map. Possible answer: The Mississippi River drainage basin is east of the continental divide. L2

Activity

Mapping Divides Provide students with a map of the United States that shows all major river systems. Have them trace where they think divides are located. Encourage students to work in groups and to compare their work with that of others. Then provide a map of the United States that shows the actual location of major divides. Ask students to evaluate how close their divide locations are to the actual locations. L2 COOP LEARN
LS **Visual-Spatial**

Visual Learning

Figure 4 Do you live east or west of the continental divide? Answers will vary depending on the location of your community.

Caption Answer

Figure 4 Precipitation falling west of the Continental Divide drains to the west, while precipitation falling east of the Continental Divide drains to the east.

Figure 3 Water channels flowing around bars in the Platte River resemble a braid.

Braided Streams Eroded sediment, such as sand and gravel, can be deposited in the middle of a stream channel, forming bars. As the bars develop, water flows through many different channels. Sometimes these bars get so large that they become islands within the river. Streams with many bars and islands separated by river channels are called **braided streams.** One well-known braided stream is the Platte River in Nebraska, shown in **Figure 3.**

Drainage Basins

Because of gravity, water falling on Earth's surface flows downhill until it reaches the lowest level that it can flow to—its base level. Streams transport water from higher ground to an ocean or a lake. Each stream drains water from a specific area of Earth's surface. This area is called the stream's **drainage basin.** As shown in **Figure 4,** large rivers, like the Mississippi, can have huge drainage basins that include many smaller streams.

Divides An elevated area of land called a divide separates the drainage basin of one stream from the drainage basin of another. Divides can be as small as a ridge or as large as a mountain range. An area that separates drainage basins of an entire continent is called a continental divide, shown in **Figure 4.**

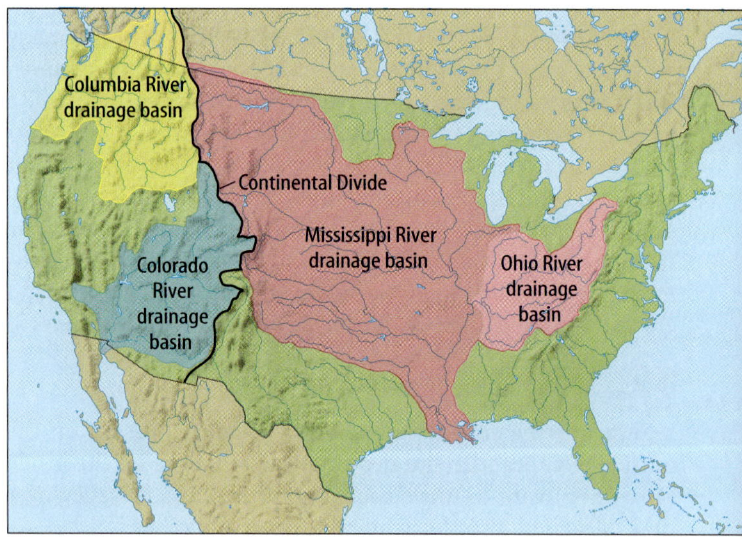

Figure 4 Some of the major drainage basins in the United States are shown here.
Explain *why the highest ridge of the Rocky Mountains is called the Continental Divide.*

Science Journal

River Poems and Songs Have students write a poem or song in their Science Journals about experiences they have had with streams or rivers. Set up a poetry corner to display the poems and songs. Have interested students share their work orally in class. L2 LS **Linguistic** P

Teacher FYI

Nile River The longest river in the world is Africa's Nile River at 6,695 km (4,158 mi.). The drainage basin of the Nile covers about one-tenth the area of the continent of Africa, or 3,349,000 km^2 (1,293,000 mi^2.).

38 ◆ H CHAPTER 2 Freshwater at Earth's Surface

Figure 5 A stream carries material from one place to another.
Identify *which type of load cannot be seen in the water.*

Stream Erosion

As you have learned, streams erode Earth's surface. Streams remove rock and sediment from some places and transport it to other places. Streams erode in two important ways. The force of the flowing water in a stream can be strong enough to move loose sediment, such as silt, sand, and gravel. The faster the water flows, the more sediment a stream moves. The sediment that is moved by the water allows streams to erode in another way. When the grains of sediment hit rock on the stream bottom, small pieces can be broken off of the rock and carried away.

INTEGRATE Physics

Load Sediment that is carried by a stream is called stream **load.** The three different types of stream load, as shown in **Figure 5,** are bed load, suspended load, and dissolved load. Sediment that moves along the bed of a stream by rolling, bouncing, or sliding is the bed load. These are particles that are too heavy to be carried by the flowing water. When sediment is light enough to become suspended in the water, it is carried by the stream as suspended load. Suspended load makes some streams appear to be muddy. Some rock materials become dissolved in water and are carried in solution by the stream as dissolved load. Much of the dissolved load is carried into streams by groundwater.

Reading Check *What type of stream load rolls, bounces, and slides along a stream's bottom?*

Mini LAB

Modeling Stream Flow

Procedure

1. Obtain a **large, flat pan or stream table.**
2. Fill one end of the pan with **sand,** tilt the pan at a 10° angle, and form a small channel through the sand.
3. Using **flexible tubing,** direct a flow of **water** into the channel in the sand.
4. Using a **bucket,** catch the water at the bottom of the pan.

Analysis

1. Observe what happened to the sand as the water flowed in the pan.
2. Did you model a meandering or a braided stream?

Caption Answer

Figure 5 dissolved load

Mini LAB

Purpose Students model a stream channel. L2 LS **Kinesthetic**

Materials flat pan or stream table; wedge; sand; flexible tubing; water; bucket

Teaching Strategies

- Students can direct water into the tubing either by attaching it to a faucet or by attaching it to a funnel and pouring water into the funnel.
- Have students experiment with increasing and decreasing the speed of the water.

Troubleshooting Caution students that if the water flows too quickly, the sand will be washed away without showing the desired effect.

Analysis

1. Sand was eroded from some places and deposited in others.
2. Answers will vary.

Assessment

Oral Have students discuss the processes that move material through natural stream channels. Students should discuss movement along the bed, suspension of material, and dissolution of material. Use **Performance Assessment in the Science Classroom,** p. 89.

Reading Check

Answer bed load

Differentiated Instruction

Challenge Motivated students might enjoy extending the MiniLAB by filming the way the flow of water changed the stream channel and then using the short film in a presentation. Students might try speeding up the film or using stop action to better illustrate the process. L3 LS **Visual-Spatial**

Discussion

Canyons Ask students why canyons sometimes are called rejuvenated river valleys. The term *rejuvenated* means "to become youthful once again." If a stream experiences uplift, it will begin to erode its channel. This will form the steep-sided walls of a canyon.

Reading Check

Answer by the downcutting of a river as the land on which it is located is gradually uplifted

Use an Analogy

Sugar Sediment Tell students that the deposition of sediment by streams is analogous to what happens when stirred sugar in a cold glass of water falls to the bottom of the glass. When the water is moving quickly in a stream or in a glass because of stirring, the particles that have not dissolved are suspended in the water. When the speed of the water slows down in the stream or in the glass of water, the water can no longer "hold up" the particles. The particles drop to the bottom of the stream or the bottom of the glass.

Caption Answer

Figure 7 Flood water flows faster. It can pick up and move the sediment in the bars.

Figure 6 This canyon formed when the Big Horn River in Montana cut downward as the land rose.

Canyon Formation Sometimes the land on which a stream is located is uplifted gradually. This uplifting changes the distance from the stream to its base level. The streambed becomes steeper, and the stream cuts down into its channel. Even if the stream is wide with many meanders, the gradual change in elevation can cause it to cut down while it continues to follow its original channel. Over time, the stream can carve a steep canyon along its course, as shown in **Figure 6.**

Reading Check *How does a canyon form?*

Stream Deposition

You have been learning about how a stream erodes its channel, but do you know why the load carried by a stream eventually is deposited somewhere, such as in the bars shown in **Figure 7?** Think about the last time you worked until you were physically exhausted. When you started the job, you had lots of energy and worked quickly. As the job continued, you had less energy and became tired. It became harder to move things after your energy supply was reduced. You can think about water in the same way. When water has a lot of energy, as it does when a stream has large volumes of fast-moving water, it carries and moves more sediment. As water loses energy, it drops some of its load. Heavier pieces of sediment drop out first, then the lighter pieces are deposited. Where does this occur? It occurs where the speed of the water decreases.

Figure 7 Whether stream sediment is carried or deposited depends upon the speed of flowing water. Bars form where the speed of water slows, and stream load is deposited.
Explain *why these bars might change during a flood.*

Teacher FYI

Mississippi River Over thousands of years, the Mississippi River has added thousands of square kilometers of land to the area that is now the state of Louisiana. New Orleans, which is more than 100 kilometers from the mouth of the Mississippi today, was sea bottom 5,000 years ago.

Deltas form where streams such as the Dungeness River in Washington State deposit sediment into a body of water.

Alluvial fans form where a stream deposits sediment onto a flat area of land.

Figure 8 Deltas and alluvial fans form where a stream deposits sediment.

Deltas and Alluvial Fans Water in a stream loses energy and might deposit sediment where the streambed becomes flatter, or where the stream channel widens. However, a stream has the least amount of energy when it reaches its base level. As a stream empties into a body of water such as a lake, gulf, or bay, the remaining sediment is deposited, forming a delta such as the one shown in **Figure 8.** The deposited sediment eventually raises the bottom of the water body, and the stream forms new channels through the deposit. Eventually, the delta grows and new land is created.

When a fast-moving mountain stream flows onto a flat plain or valley, a fan-shaped deposit of sediment, called an alluvial fan often forms. Alluvium is rock and sediment that is deposited by streams. Alluvial fans are similar to deltas, except that alluvial fans form on land, whereas a delta forms in the water.

Floods

The volume of water flowing through a stream in a given amount of time is called **stream discharge.** For example, if 200 m^3 of water flow past a point along the stream channel in 1 s, the discharge is 200 m^3/s. Heavy rain or melting ice and snow can lead to an increase in discharge. As discharge increases, the stream's speed also increases and it can carry more sediment. This can lead to erosion of the stream channel.

IDENTIFYING Misconceptions

Floods Most students are aware of the negative aspects of floods, such as destruction of property. Explain that floods are a part of a stream's natural cycle that helps maintain floodplain ecosystems. When flooding occurs, the stream or river deposits sediment and nutrients on the adjacent floodplain. This helps the survival of floodplain plants and the animals that depend on them. It also creates fertile soil in which to plant crops.

Fun Fact

During record-breaking floods in the Midwest in 1993, the Mississippi River flowing past St. Louis carried six times its normal discharge.

Activity

Pros and Cons of Dams Dams are used for flood control and to produce electricity, but some people object to them. Have students construct posters that show the pros and cons of dams. Possible answers: Pros: hydroelectric power, recreation, flood control; cons: destroys and floods land behind dam, displaces people and wildlife, disrupts downstream ecosystems L3

Curriculum Connection

Geography Have students research the Midwest floods of 1993. Instruct them to work in pairs to draw maps of the United States that show the extent of the flooded area. Have students write short reports that explain why the flood occurred, how it spread, and how it disrupted communication and the movement of people and goods through that area of the country. L2

Discussion

Living on a Floodplain Having learned about the risks and benefits of living on a floodplain, would you build a home or business in such an area? Accept all reasonable answers as a basis for discussion. Some students will think the dangers of flooding are too great. Others will think there are advantages, such as fertile soil for farmers, that can be enjoyed, and that levees and dams keep people safe most of the time.

Activity

Floodplain Observations If your community is located on or near a floodplain, arrange a class trip to the river that flows through it. Have students observe and photograph or sketch any structures (such as levees or dams) that are part of a flood control system, and determine which parts of the developed area nearby would flood if the system failed. Observe the flatness of the landscape, and visit any natural areas of the floodplain that remain. Have students observe and photograph or sketch plants and animals in the area and contrast this natural area of floodplain with developed areas nearby. Invite students to make a class bulletin board that tells the story of the floodplain with photographs, sketches, quotes, and stories. L2
LS **Visual-Spatial and Linguistic**

Figure 9 Floodplains are flat regions of land often covered by water during floods. Floodplains act as natural flood control for downstream areas.

Floodplains Sometimes a stream's discharge increases so much that the river channel cannot hold the increased volume of water. Water flows onto the stream's floodplain. The **floodplain** of a stream is the region of land next to the stream channel that is covered by water during a flood, as shown in **Figure 9.** When the water overflows its banks, it spreads out over the floodplain and the speed of the water flow decreases. Sediment is deposited on the floodplain. Floodplains help reduce stream discharge downstream, reduce erosion, and collect new, fertile soil from stream deposits.

> **Science Online**
> **Topic: Flood Control**
> Visit bookh.msscience.com for Web links to information about flood control.
>
> **Activity** Using photos and diagrams, prepare a public information brochure about one type of flood control. Describe the risks and benefits associated with this flood-control method in your brochure.

Controlling Floods Floodplains have some features of ideal building sites. Flat, level ground that is full of rich soil is a perfect place to build a community. Unfortunately, as many people have experienced, living on a floodplain can have serious consequences—the floodplain often is covered with water.

One way to control a flood is to build artificial levees, which are mounds of soil placed on the sides of a stream channel. The levees help keep the water within the channel, protecting the property and lives of people who live nearby. However, building levees sometimes can affect downstream regions. Farther downstream, where levees might not be built, more severe flooding could occur because more water is flowing in the stream channel. The water also flows faster and causes more erosion. During times of heavy stream discharge, levees might not remain intact. If a levee breaks in one area, the floodwater becomes trapped on the floodplain and cannot easily return to the stream channel. The area can remain flooded for weeks.

Differentiated Instruction

English-Language Learners Team English language learners with native speakers. Invite the teams of students to create an English/second language dictionary that includes important terms in this section. Retain the dictionaries for classroom reference. L2

Changing Stream Flow Dams have been used for flood control for many years. Flood-control dams store the additional stream discharge and let it out a little at a time, thus preventing the flood. However, dams are expensive and can be difficult to build. They also create large water areas that once were land, and they block the movement of river organisms.

Stream channels can be altered to hold more water. Dredging the channel makes it deeper and straighter, allowing the water to move faster and perhaps preventing it from rising to flood stage. This flood-control method, shown in **Figure 10,** is called channelization. However, some problems are associated with this method. As more water moves faster, erosion of the stream banks and channel increases. Often, trees that shade the stream and keep the water cool must be removed to channelize the river. Cooler water has more oxygen and often has a wider variety of organisms in it than warm water does. In addition, increasing the stream discharge in one part of a stream can make downstream flooding worse.

Perhaps the best solution to flood problems is to use the land wisely. Identify areas of the floodplain that are likely to be flooded and then use the area appropriately. For example, in agricultural areas, farmers can wait until after spring rains to plant some crops. Many communities carefully control development, permitting buildings only on areas above the floodplain.

Figure 10 Channelizing a stream involves removing Earth materials to straighten and deepen the stream channel.
Explain *why this might be a temporary solution to flooding.*

section 1 review

Summary

Types of Streams

- Meandering streams have channels with many curves.
- Bars separate channels in braided streams.

Stream Erosion

- Streams pick up and move sediment.
- A stream's load includes bed load, suspended load, and dissolved load.

Stream Deposition

- Streams deposit sediment where water speed decreases.
- Floods occur when streams overflow their banks.

Self Check

1. **Describe** how much of Earth's water is found in streams and lakes.
2. **Explain** why streams deposit sediment on the inside of a meander.
3. **Explain** how divides separate different drainage basins.
4. **Think Critically** A community near a river that floods frequently builds levees and straightens the river channel. How might this affect downstream regions?

Applying Math

5. **Solve One-Step Equations** The slope of a stream equals the change in elevation divided by the distance over which it flows. If a stream drops 75 m over 3 km, what is the stream's average slope?

Science Online bookh.msscience.com/self_check_quiz

Caption Answer

Figure 10 Over time, the stream's channel will return to a more natural condition.

3 Assess

DAILY INTERVENTION

Check for Understanding

Auditory-Musical Lead students to create an impromptu song about a cobble in a river. Include the cobble's source, movement, and destination in the song. L2

Reteach

Straight Channels Draw two stream channels on the chalkboard—one with a straight channel and one with several meanders. The stream channels should extend over the same linear distance on the chalkboard. Ask students which stream channel is longer and why water can move more quickly through a straight channel. Water moving through the straight channel could cover the distance without traveling as far because a straight line is the shortest distance between two points. L1
LS **Visual-Spatial**

Assessment

Content Ask students to explain in their own words why channelization could cause a change in the type of organisms in a river. Trees that shade the river often are removed. This causes the water in the river to be warmer and limits the variety of organisms that can live there. In addition, some organisms prefer gravelly stream bottoms and turbulent water. Dredging often changes the bottom substrate and removes riffles. Use **Performance Assessment in the Science Classroom**, p. 89.

section 1 review

1. 0.01 percent
2. The water's speed slows, causing it to lose energy and drop sediment.
3. Divides have higher elevation than surrounding land, so drainage systems cannot cross them.
4. Water that would normally spill onto the floodplain in this community was kept in the river channel. This increased discharge was carried further downstream, where it finally spilled out and flooded a different area.
5. 75 m ÷ 3 km = 25 m/km

1 Motivate

INTERACTIVE CHALKBOARD
PowerPoint® Presentations

Bellringer

Section Focus Transparencies also are available on the Interactive Chalkboard CD-ROM. L2 ELL

Section 2 Section Focus Transparency **Hold Back the Lake**

Navigating massive freighters between Lake Ontario and Lake Erie might be difficult since Lake Erie is about 99 meters higher than Lake Ontario! A series of eight locks in the Welland Canal raises or lowers commercial vessels between these Great Lakes. One of these locks is shown below.

1. How are lakes and streams similar? How are they different?
2. How do people use lakes?
3. What are some ways that lakes are formed?

Freshwater at Earth's Surface

Tie to Prior Knowledge

Lake Observatations Ask students to recall a visit to a lake or a photograph of a lake they have seen. Have them describe the plants or animals they saw around and in the lake. Then ask students how a lake is different from a stream or river. Possible answers: Its shape is usually not linear like a stream; it doesn't flow like a stream. Tell students that in this section they will learn about lakes.

section 2 Lakes and Reservoirs

as you read

What You'll Learn

- **Explain** how lakes form.
- **Describe** organisms found in lakes.
- **Summarize** how lakes change through time.

Why It's Important

Natural and human-made lakes are an important source of drinking water and recreation.

Review Vocabulary
density: the mass of a substance divided by its volume

New Vocabulary
- nutrient
- eutrophication
- turnover

Formation of Lakes

Do you remember the last time you were outside after a heavy rain? You probably had to step around a lot of puddles. Puddles form when water collects in a depression in the ground. Lakes and ponds form in a similar way.

Lakes and ponds are bodies of relatively still, or standing, water. Although no definite size distinguishes a lake from a pond, lakes generally are larger and deeper than ponds. Water enters lakes from streams and rivers, by rain or snow falling on or near the depression, and from groundwater. **Figure 11** shows that lakes can form in many different ways. You will see how this happens by first looking at natural and then human-made lakes.

Reading Check *How does water enter lakes?*

Natural Lakes Many North American lakes, including the Great Lakes, formed when glaciers scoured out depressions that filled with rainwater, groundwater, or water from melting ice. Lakes also formed when sediment that was deposited by glaciers dammed the flow of streams.

Figure 11 Lakes form when natural and human-made structures capture freshwater.

The Finger Lakes in New York state formed when glaciers scoured depressions in Earth's crust.

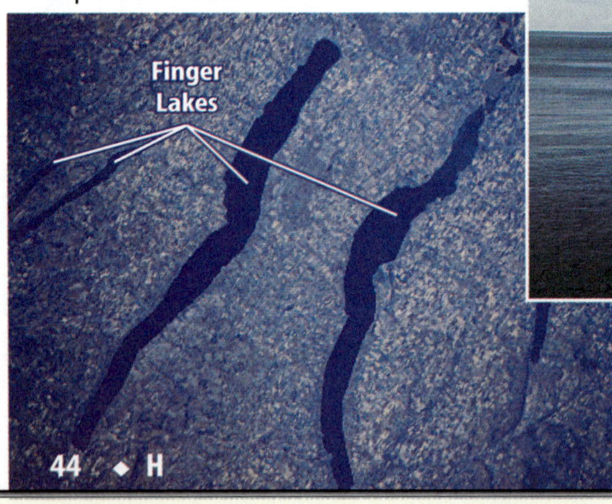

Some lakes form when Earth's surface is stretched. Lake Baikal in Russia is among the world's deepest lakes.

Section 2 Resource Manager

Chapter *FAST FILE* Resources
Transparency Activity, pp. 45, 49–50
Directed Reading for Content Mastery, p. 16
Enrichment, p. 28
Reinforcement, p. 24
Lab Worksheet, pp. 5–6
Lab Activity, pp. 11–12

Life Science Critical Thinking/Problem Solving, p. 13
Mathematics Skill Activities, p. 9
Lab Management and Safety, p. 38

Crustal Movements Movement of Earth's crust can create depressions that fill with water. These lakes often form along faults—surfaces along which rocks break and move. Rock lying between two faults can sink to form a depression. If Earth's crust continues to move along the faults, the lake will deepen. Lake Baikal in Russia, shown in **Figure 11,** formed because blocks of rock moved along faults.

Lakes also can form after a volcanic eruption. Magma from beneath Earth's surface can spill out of a volcano. The top of the volcano then might collapse into the partially emptied magma chamber and form a large depression called a caldera. Rainwater and runoff later fill the caldera.

Reading Check *How do calderas form?*

Human-Made Lakes When a dam is built within a stream, a human-made lake forms. Often, the lake that is formed behind the dam is used to supply drinking water for nearby cities and for irrigation of farmland. When water in a natural or artificially created lake is used for human consumption, the lake is called a reservoir.

Lake Mead, on the Nevada-Arizona border, was created in 1936 when the Hoover Dam was built across the Black Canyon, blocking the flow of the Colorado River. As shown in **Figure 11,** Hoover Dam is one of the largest human-made dams in the world. Like many human-made lakes, Lake Mead is used for flood control, water supply, and recreation. Lake Mead also is used to generate power. Water moving through Hoover Dam produces enough electricity to serve 1.3 million people per year.

Science Online

Topic: Hoover Dam
Visit bookh.msscience.com for Web links to information about Hoover Dam and Lake Mead.

Activity Create a call-out poster illustrating important details about Hoover Dam and Lake Mead.

This lake in Equador formed when the top of a volcano collapsed and filled with rainwater.

Hoover Dam blocks the flow of the Colorado River to form Lake Mead.

2 Teach

Reading Check

Answer Water can enter a lake from streams and rivers, by rain or snow falling directly into the lake, and from groundwater seepage.

IDENTIFYING Misconceptions

Lake Clarity Students might think that fresh, clear water in a lake is best for the life and health of the lake ecosystem. Explain that very clear lake water probably lacks nutrients, meaning that it cannot support a wide variety of organisms.

Reading Check

Answer When a volcano erupts and expels magma, the top of the volcano sometimes collapses into the magma chamber to form a caldera.

Discussion

Providing Irrigation Why might a dam and reservoir be useful in a dry environment? The reservoir could store water for drinking and irrigation.

Visual Learning

Figure 11 What function does Lake Mead perform that the other lakes shown here do not? Water flows through a dam and generates power.

LAB DEMONSTRATION

Purpose to illustrate the formation of an additional type of natural lake, a kettle lake

Materials plastic tub, sand, large block of ice

Preparation Make the ice block in a margarine tub.

Procedure Fill the plastic tub with sand. Bury the ice block in the sand and allow the ice to melt. Observe the depression that remains.

Expected Outcome After the ice melts, a depression will be left in the sand. A similar phenomenon occurs at the margins of melting glaciers. Ice blocks are buried in sediment, melt, and leave a depression that fills with water. Such lakes are called kettle lakes.

Assessment

Ask students to infer the size of most kettle lakes. They generally are small.

SECTION 2 Lakes and Reservoirs H ◆ 45

Discussion

Lake Depth If you see a lake that is choked with plants, is the lake probably deep or shallow? If plants can grow in the water, the lake bottom is fairly close to the surface, so the lake is shallow.

Activity

Lake Zones Organize the class into several groups, each of which will research one of the zones of life found in lakes as seen in **Figure 12.** Have each group present what it finds to the class in a short oral report. Encourage students to use visual aids. L2 COOP LEARN

LS **Interpersonal**

Fun Fact

The Great Lakes are the world's largest group of freshwater lakes, and they contain 18 percent of the fresh surface water on Earth.

Quick Demo

Life in Lakes

Materials samples of typical lake vegetation

Estimated Time 10 minutes

Procedure Show students samples of vegetation from a nearby lake. Describe the vegetation and explain how it exists in the lake.

Life in Lakes

Do you remember the last time you went to a lake? Perhaps you saw one on a nature program at school or at home. Although lakes usually look peaceful, they are full of life. Insects skim across the surface as birds swoop after them. A frog jumps into the water from the shore. A large swirl tells you a fish has caught its dinner.

Many types of organisms can be found in different parts of lakes, as shown in **Figure 12.** Near a shallow, gently sloping shoreline, light penetrates to the lake bottom, allowing many types of rooted aquatic plants to grow. As plants use sunlight to make food, they produce oxygen that other organisms need. Plants also provide many places for small organisms to hide from hungry predators. Organisms that use land and water, such as amphibians, also are found here.

If you take a boat ride to the middle of the lake, you'll notice fewer insects as the boat moves toward open water. You won't see any frogs. Plenty of organisms still live here, but as **Figure 12** shows, most are too small for you to see. Because sunlight cannot penetrate to the bottom of a deep lake, rooted plants cannot survive there, so frogs and small fish have nowhere to hide.

Wormlike organisms, bacteria that feed on dead plants and animals, and animal wastes that fall from higher layers of the lake are found on the lake bottom. Some fish species, such as lake trout, are found in deep waters of large lakes where they feed on the organisms found there.

Figure 12 The depth to which sunlight penetrates into lake water helps determine the types of organisms found in various lake zones.

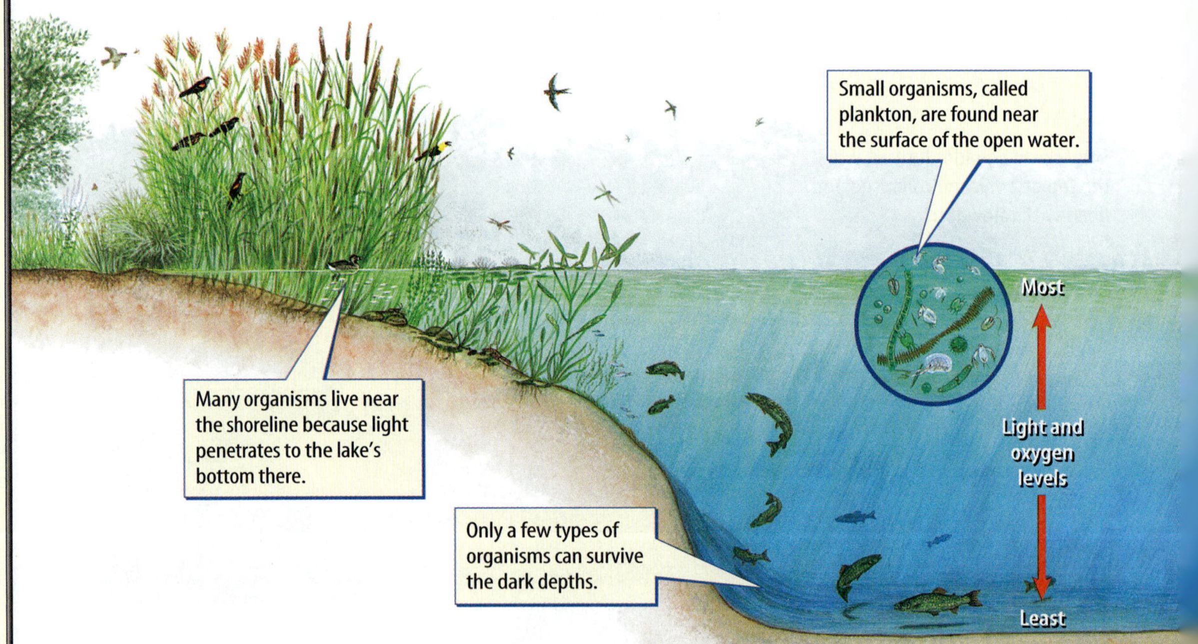

Cultural Diversity

Rice from the Lakes In Wisconsin and Minnesota, Native American tribes such as the Ojibwa and the Menominee have traditionally harvested the wild rice that grows in and around many of the region's lakes. It was always important to thank the spirit of the lake for the gift of the rice, in part, by leaving rice grains in their waters during harvest. Have students research to find out what wild rice is. Have them also find out and write brief reports on the importance of wild rice, the lakes where it grows, and the culture of these tribes. Wild rice is actually a type of grass and is unrelated to rice. L3 P

Sediment and Nutrients

Is a lake forever a lake? When streams and rivers empty into lakes, they deposit sediment on the lake bottom. A lake might fill more quickly if **nutrients**—compounds such as nitrates and phosphates that are used by plants, algae, and some plankton to help them grow—are in the water.

Over time, the deposited sediment decreases the depth of the lake so sunlight can penetrate more of the lake bottom. More plants grow, so more organisms can hide in them and use the oxygen they produce. More life means more death. As organisms die, they sink and decay, adding more material to the lake bottom and releasing more nutrients into the water. This cycle, shown in **Figure 13,** continues until the lake becomes eutrophic. **Eutrophication** (yew troh fuh KAY shun) is an increase in nutrients and organisms that is a normal part of a lake's life. Eventually, the lake will become dry land.

INTEGRATE Life Science

Plankton Two types of plankton are found in lakes—phytoplankton and zooplankton. Which of these makes food from sunlight and which doesn't? Write your answer in your Science Journal.

Applying Math **Solve One-Step Equations**

BAY DEPTH A small bay in a large lake is filling at a rate of 0.1 cm per year. If the water in the bay averages 100 cm deep, how long will it take for the bay to become land?

Solution

1. *This is what you know:*
   - infill rate: $r = 0.1$ cm/y
   - average depth of bay: $d = 100$ cm
2. *This is what you need to find out:* time: t
3. *This is the procedure you need to use:*
   - Put the known values into the equation.
     $$time = \frac{depth}{infill\ rate}$$
   - Solve the equation for *time:*
     $$time = \frac{100\ \text{cm}}{0.1\ \text{cm/y}} = 1{,}000\ \text{y}$$
4. *Check your answer:* Multiply the answer by the infill rate: $1{,}000\ \text{y} \times 0.1\ \text{cm/y} = 100\ \text{cm}$. The answer is correct.

Practice Problems

1. A different bay in the same lake is filling at a rate of 0.3 cm/y. This bay averages 150 cm deep. How long will it take to fill?
2. A boat-launch area is filling at a rate of 1 cm/y. It has an average depth of 200 cm. At least 50 cm are needed to safely launch the boats. How long will boats be able to use the area?

Science Online — For more practice, visit bookh.msscience.com/math_practice

INTEGRATE Life Science

Plankton Phytoplankton are tiny plant-like organisms that use sunlight to make food through photosynthesis. Zooplankton are tiny animal-like organisms that feed on phytoplankton.

Use Science Words

Word Origin Have students look up the origin of the word *eutrophic* and explain how this word relates to the life cycle of a lake. It is from the Greek word *eutrophos,* which means "well-nourished." Eutrophication refers to an increase in nutrients and organisms.

Applying Math

National Math Standards

Correlation to Mathematics Objectives

1, 2, 9

Teaching Strategy

Review how to cancel units before students do the practice problems.

Answers to Practice Problems

1. 150 cm ÷ 0.3 cm/y = 500 years
2. (200 cm − 50 cm) ÷ 1 cm/y = 150 years

Science Journal

Eutrophication Have students write a brief report in their Science Journals about how building a dam could lead to increased eutrophication in a lake. When a dam is constructed and forms a reservoir, sediment can collect behind the dam and quickly fill in the reservoir. L2 LS **Linguistic**

NATIONAL GEOGRAPHIC

Visualizing Eutrophication

Have students examine the pictures and read the captions. Then ask the following questions.

What two ingredients contribute to lake eutrophication? sediments, which decrease lake depth, and nutrients, which encourage plant growth

How might eutrophication-causing nutrients be introduced into a lake by humans? Possible answers: fertilizer from nearby farms or yards can be washed into the lake by rainwater; sewage runoff from boats, septic tanks or sewage treatment plants

Activity

Eutrophication Posters Have students make posters showing a lake in each of the various stages of eutrophication. Be sure the posters show the roles of nutrients, plant life, and oxygen in the eutrophication process. L2
LS **Visual-Spatial**

NATIONAL GEOGRAPHIC **VISUALIZING EUTROPHICATION**

Figure 13

Eutrophication is a natural process that ultimately turns a lake into land. Under natural conditions, this process normally takes hundreds, even thousands, of years. The stages by which a lake can become dry land are shown here.

A With little sediment and few nutrients flowing into this lake, its clear waters remain deep and relatively free of living organisms.

B Sediment accumulating in a lake can decrease its depth, making light available to more of the lake bottom. Nutrients encourage plant growth.

C Plants and algae gradually cover the lake's entire surface. Dead plant matter accumulates and decays on the bottom, depleting the oxygen content of the water.

D Eventually the lake fills in and becomes land.

Differentiated Instruction

Challenge Have students make hypothetical graphs showing the trends in dissolved oxygen, nutrients, and plants growth for the stages of lake eutrophication depicted in their posters. Have them explain their graphs to the class. L3
LS **Logical-Mathematical**

Teacher FYI

Human Activities About one-third of the 100,000 medium to large lakes and about 85% of the large lakes near large urban centers in the United States are experiencing accelerated eutrophication because of human activities.

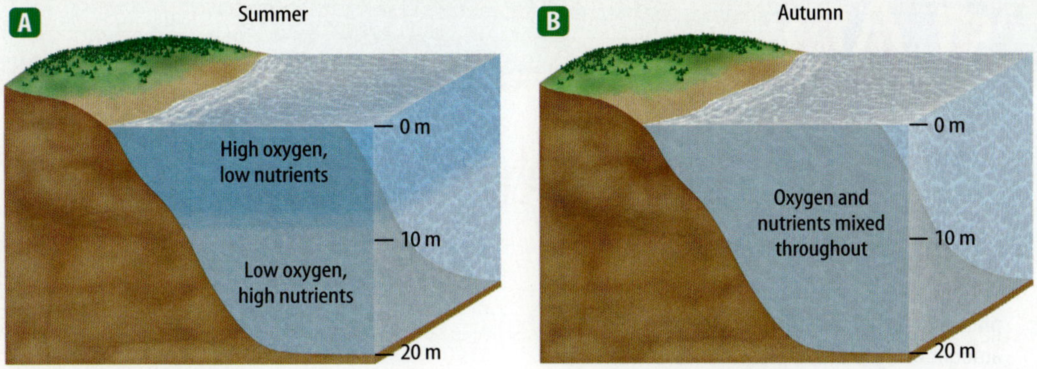

Figure 14 **A** Little mixing of bottom water with surface water occurs in the summer. **B** Cool water sinking in the fall mixes the entire water column, circulating oxygen and nutrients throughout the lake.

Turnover When organisms die, they sink to the lake bottom, decay, and nutrients are released into the water. In deep lakes of the northern United States, these nutrients are concentrated at the bottom of the lake during summer and are not available to many organisms. The water in the lake's upper layer warms during summer and becomes less dense than the layers below. Because of the different densities, little mixing occurs between layers. When summer ends, the surface water becomes cooler and denser. It sinks and mixes with nutrient-rich water. This mixing of lake water, as shown in **Figure 14,** is called **turnover.** Turnover causes nutrients from deep in the lake to move upward toward the surface. This process circulates nutrients at the lake bottom to the shallow areas and surface water, where organisms use them to grow and reproduce.

section 2 review

Summary

Formation of Lakes

- Lakes form in many ways.
- Some lakes form naturally, and some are made by humans.

Life in Lakes

- Different types of organisms live in different lake environments.

Sediment and Nutrients

- Most lakes eventually will become eutrophic, which means that they will become shallower and contain more nutrients and more life.
- The water in many large lakes mixes during autumn. This process of mixing is called turnover.

Self Check

1. **Describe** two ways that lakes can form naturally.
2. **Explain** how a lake can become dry land.
3. **Identify** the types of organisms that live near a lake's shore.
4. **Infer** why rooted plants don't live on the bottom of a deep lake.
5. **Think Critically** Lake turnover occurs during fall. Could turnover also occur during the spring? Explain why or why not.

Applying Skills

6. **Communicate** In your Science Journal, write a one-page summary that describes life in a lake.

Science Online bookh.msscience.com/self_check_quiz

3 Assess

DAILY INTERVENTION

Check for Understanding

Visual-Spatial Challenge students to create drawings that illustrate various ways that lakes can form. L2

Reteach

Concept Map Ask students to draw a concept map that includes all of the major terms and concepts covered in this section of the chapter. L1 LS **Visual-Spatial**

Assessment

Oral Have students explain why turnover occurs in many lakes.

section 2 review

1. Glaciers scrape out depressions that fill with water; faults in Earth's crust create depressions that fill with water.
2. Over time, as sediments are deposited, sunlight penetrates more of the lake bottom. More plants and organisms grow and die. Eventually, the lake is filled.
3. rooted plants and small animals, such as turtles and frogs
4. Light cannot penetrate, so they would not be able to make food through photosynthesis.
5. Yes; as the surface water warms to 4° C, it reaches its maximum density and the water sinks to the lake bottom, thus mixing the lake.
6. Student summaries should describe life in all three lake zones.

BENCH TESTED

Real-World Question

Purpose Students observe the effect of nutrients on algae growth. L2 LS **Visual-Spatial**

Process Skills observe and infer, communicate, recognize cause and effect, separate and control variables, interpret data

Time Required day 1: 40 minutes; days 2–6: 5 to 10 minutes each day for observations

Materials *Euglena* or other algae cultures can be obtained from biological supply companies. Many plant stores sell liquid fertilizers that can be used to make the nutrient solutions.

Safety Precautions Caution students to wash their hands after handling materials.

Procedure

Teaching Strategy Have students use magnifying lenses to observe the algae growth closely.

Conclude and Apply

1. Nutrient solutions showed algae growth, with the solutions turning dark green. Distilled water shows little or no change.
2. The nitrate and phosphate solution produced the most algae growth.
3. It showed little or no algae growth.
4. The presence of phosphates and nitrates in lakes causes algae growth.

LAB

Lake Nutrients

When plants and animals die, they settle to the bottom of a lake where they decay and release phosphates and nitrates to the water. The amount of algae that is present in lake water depends on the amount of nutrients in the water.

Real-World Question

How does the presence of nutrients in lake water affect the amount of algae that can survive?

Goals

- **Predict** which solution will have the most algal growth.
- **Predict** how nutrients affect algal growth in a lake.

Materials

felt-tipped marker
small baby-food jars with caps (4)
100-mL graduated cylinder
distilled water (200 mL)
0.4% nitrate solution (200 mL)
0.4% phosphate solution (200 mL)
0.4% nitrate-phosphate solution (200 mL)
algae cultures
dropper

Safety Precautions

Procedure

1. Label each jar with your name or lab group.
2. Label and fill each jar with the following:
   - 0.4% nitrate solution
   - 0.4% phosphate solution
   - 0.4% nitrate-phosphate solution
   - distilled water
3. Prepare the algae culture according to the directions and place eight drops of the algae solution in each of the four different jars.
4. With the lid on, shake each jar gently and place each on a windowsill.
5. Remove the lids and check the jars for algal growth each day for five days. Replace the lids after each observation. Record your observations in your Science Journal.

Conclude and Apply

1. **Describe** any changes, such as color or clarity, that you notice in each solution.
2. Based on your observations, which solution or solutions were best for algal growth?
3. What observations can you make about the solution containing distilled water?
4. Relate your observations to the nutrient makeup and the algae present in lakes.

Communicating Your Data

Compare your conclusions with those of other students in your class. **For more help, refer to the** **Science Skill Handbook.**

Assessment

Performance Have students reduce the amount of nutrients in each jar by mixing 100 mL of each solution with 100 mL of distilled water. Ask students to observe the jars and contrast algae growth with that during the original exercise. Use **Performance Assessment in the Science Classroom,** p. 97.

Communicating Your Data

Invite students to use a computer word processing program to write and print out reports.

section 3

Wetlands

Types of Wetlands

When you visit a lake, pond, or stream, do you always know where the land ends and the water begins? At many places, you might be able to walk right up to the water's edge. But what if you step in a muddy area? Is this water or is it land? **Wetlands** are areas of land that are covered with water during some part of the year. Biologists recognize wetlands by the types of plants that grow there.

Reading Check *How do biologists recognize wetlands?*

Wetlands can have different types of vegetation, depending on where they form and how much water is in them during the year. Types of wetlands include swamps, marshes, and bogs and are defined primarily by their vegetation.

Swamps Swamps, like the one in **Figure 15,** are wetlands where the most common types of plants are trees and shrubs. Swamps need a steady supply of water in order for wetland trees to grow. They often are found in low-lying areas near rivers, where water is slow moving.

A swamp is sometimes named for the type of trees that inhabit the area. For example, cypress swamps are named after the bald cypress trees that grow there. Oak swamps are common in the northeastern United States where that tree is found.

as you read

***What* You'll Learn**

- **Describe** a wetland.
- **Compare and contrast** different types of wetlands.
- **Identify** some organisms that live in wetlands.

***Why* It's Important**

Wetlands reduce flooding, purify water, and provide habitat for many types of organisms.

Review Vocabulary
nutrients: substances, such as nitrates or phosphates, that are necessary for the growth of algae and plants

New Vocabulary
- wetlands

Figure 15 Many swamps, such as this cypress swamp at Caddo Lake State Park, Texas, are found in the southeastern and south central United States.

Section 3 Resource Manager

Chapter *FAST FILE* Resources

Transparency Activity, p. 46

Directed Reading for Content Mastery, p. 17

Enrichment, p. 29

Reinforcement, p. 25

Life Science Critical Thinking/Problem Solving, p. 23

1 Motivate

INTERACTIVE CHALKBOARD PowerPoint® Presentations

Bellringer

Section Focus Transparencies also are available on the Interactive Chalkboard CD-ROM. L2 ELL

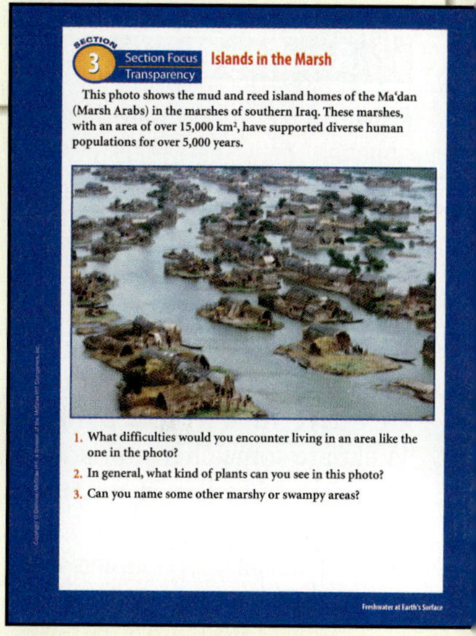

SECTION 3 Section Focus Transparency **Islands in the Marsh**

This photo shows the mud and reed island homes of the Ma'dan (Marsh Arabs) in the marshes of southern Iraq. These marshes, with an area of over 15,000 km^2, have supported diverse human populations for over 5,000 years.

1. What difficulties would you encounter living in an area like the one in the photo?
2. In general, what kind of plants can you see in this photo?
3. Can you name some other marshy or swampy areas?

Freshwater at Earth's Surface

Tie to Prior Knowledge

Wetlands Ask students if they know what a wetland is. If not, ask them to predict the definition based on the word itself. An area of land covered with water during some part of the year. Then have students identify wetlands in your area.

Reading Check

Answer by the types of plants that grow there

2 Teach

Visual Learning

Figure 15 Do we live in the part of the United States with the most swamps? Answers will vary, depending on your community's location.

Caption Answer

Figure 16 Marshes contain grass-like plants. Swamps contain mostly trees and shrubs.

Reading Check

Answer A bog receives water from precipitation only, not from overland sources. Precipitation is low in nutrient content.

Discussion

Wetland Animals All organisms need water to survive, but not all thrive in wetlands habitats. What are some characteristics of organisms that live in wetlands? Possible answers: Ducks have webbed feet for swimming. Beavers and others have oily skin to repel water. Some animals can hold their breath for a long period of time. Roots of trees are near the surface so they can get oxygen.

IDENTIFYING Misconceptions

Wetland Habitat Most students think of wetlands as large areas near a river or lake. Explain that wetlands are also found in small, isolated pockets not associated with lakes or rivers. These wetlands provide needed habitat for frogs, salamanders, and toads.

Marshes A marsh is a type of wetland that doesn't have many trees or shrubs. Sometimes marshes are called wet meadows because they contain mostly grasses and other soft-stemmed plants called rushes and sedges. Marshes, such as the one in **Figure 16,** often form on floodplains where rivers overflow their banks. They also form at the edges of lakes where the shoreline slopes gently. Ponds that are filling with sediment also can become marshes.

Figure 16 Marshes are characterized by grasslike plants.
Explain *how marshes are different from swamps.*

Bogs Most bogs formed in depressions that were created by glaciers. Rain is the only source of freshwater in a bog. Because no streams enter a bog to bring in nutrients, plants have developed unique ways to get them. You know that some insects eat plants, but have you ever seen a plant eat an insect? Some plants, such as pitcher plants and the Venus's-flytrap shown in **Figure 17,** get their nutrients from insects.

Reading Check *Why don't bogs have many nutrients?*

Water doesn't drain from a bog, so dead material builds up. Dead plants sink to the bottom and might eventually form a substance called peat. Most of the peat that is used by gardeners comes from a plant called sphagnum moss. This plant, when dried, can hold as much as 20 times its weight in water. Native Americans used dried sphagnum peat as diapers.

Figure 17 Bogs have a low nutrient content. The Venus's-flytrap obtains its nutrients from insects.

Wetlands Animals

Many different kinds of animals depend on wetlands. Because wetlands are found along the edge between water and land, they provide a habitat for water and land animals. Wetlands are important for migrating birds. They provide cover and resting areas between long flights. They also provide plenty of small insects, fish, and plants to feed on.

Animals such as mink and muskrat live in wetlands. One animal, the beaver, makes its own wetlands by damming up a stream if no wetlands are around. Wetlands also are home to a variety of reptiles, such as alligators, turtles, and snakes. Amphibians, such as salamanders and frogs, also live in wetlands. Many of these animals need the shallow water of wetlands to nest and reproduce. Newly hatched fish depend on wetlands. Young fish find plenty of small organisms to eat, and they can hide from predators among the wetland plants.

Teacher FYI

Losing Wetlands About 475 km^2 (183 mi^2) of inland wetlands are lost in the United States each year as they are filled in, dredged, or drained.

Differentiated Instruction

Challenge Have students research how the city of Arcata, California, uses a coastal wetland as part of its sewage treatment system. Arcata uses a coastal marsh system to provide secondary treatment for the city's wastewater. The area also provides a wildlife habitat and an attractive natural area on Humboldt Bay. L3

Importance of Wetlands

Wetlands once were thought of as wastelands—marshy, buggy, and boggy sites that were better off filled with soil. Today, wetlands are recognized as a valuable resource. Wetlands act as a natural sponge, soaking up excess water from rain, melting snow, and floods and then slowly releasing it. Wetlands near lakes and rivers help protect shorelines and stream banks against erosion because plant roots help hold soil in place and slow the speed of moving water.

Figure 18 Wetlands at this sewage-treatment facility are used to remove nutrients before water flows back into the environment.

Trapping Sediment Wetlands, such as those found on floodplains and deltas, are especially effective at trapping sediment because their vegetation slows the water's flow. Trapped sediment cannot enter the lake or stream, so the water remains clear. Fish, their eggs, and plant life all thrive in clear water.

Removing Nutrients Wetland plants also use nutrients for growth, slowing the rate of eutrophication. Lakes and streams that have large wetland areas generally have better water quality because wetlands remove nutrients from the water. Human wastewater, called sewage, contains high amounts of nutrients. In most cities, the wastewater is treated at a sewage-treatment facility to remove pollutants. The water then is released into a river or other water body. Wetlands are so effective at removing nutrients that they sometimes are created at sewage-treatment facilities to treat sewage, as shown in **Figure 18.**

section 3 review

Summary

Types of Wetlands

- Swamps, marshes, and bogs are types of wetlands.
- Each type of wetlands has different vegetation.

Wetlands Animals

- Many animals depend on wetlands. These animals include mammals, birds, reptiles, amphibians, and fish.

Importance of Wetlands

- Wetlands reduce the impact of floods, prevent shoreline erosion, and keep water clean.

Self Check

1. **Define** the term *wetlands*.
2. **Explain** how wetland areas near lakes and streams improve water quality.
3. **Explain** how wetlands reduce erosion.
4. **Describe** how migratory birds use wetlands.
5. **Think Critically** How might a marsh become a swamp?

Applying Skills

6. **Compare and Contrast** Use a three-circle Venn diagram to compare and contrast swamps, marshes, and bogs.

Science Online bookh.msscience.com/self_check_quiz

3 Assess

DAILY INTERVENTION

Check for Understanding

Linguistic Ask students to write an information brochure about the importance of wetlands. L2

Reteach

Types of Wetlands With the assistance of students, draw a chart on the board that lists the different types of wetlands covered in this section, along with the characteristics of each. Include categories such as "Types of Plants" and "Types of Animals." L2 LS **Visual-Spatial**

Assessment

Oral Have students write original questions and answers about wetlands based on the material in the section and then quiz one another aloud. Use **Performance Assessment in the Science Classroom,** p. 91.

section 3 review

1. Wetlands are areas of land that are covered with water at least part of the year.
2. Wetland areas can absorb excess nutrients and reduce erosion, thus reducing eutrophication and sediment load in lakes and streams.
3. Wetlands soak up extra water from rain, melting snow, and floods. Wetland plants hold soil in place and slow the flow of water, thus reducing erosion.
4. Wetlands provide food, resting areas, and cover for birds along their migration routes.
5. Swamps need a steady supply of slow-moving water, while marshes are located in floodplains or along lake shores. If water flow changes, a marsh could become a swamp.
6. All are wetlands, but vegetation differs. Swamps contain trees and shrubs. Marshes contain mostly grasses. Bogs are low in nutrients and have specialized plants such as the Venus's-flytrap. Bog bottoms often are lined with peat.

section 4

Pollution of Freshwater

1 Motivate

Bellringer

Section Focus Transparencies also are available on the Interactive Chalkboard CD-ROM. L2 ELL

Section 4 Section Focus Transparency **No, Seriously—the River Caught Fire**

A low point in water quality was reached when the Cuyahoga River caught fire in 1969. Pollution was a factor even in the early 1800s, when raw sewage was dumped directly into the river, but it was later industrial pollution that led to this fire. In the years since 1969, the quality of the Cuyahoga River's water has greatly improved.

1. In this fire, is the water actually burning? Explain.
2. In 1969, the Cuyahoga River was a "dead river." What does that mean?
3. How might oil or industrial chemicals get into a river?

Tie to Prior Knowledge

Water Pollution Ask students how they can tell if a river or lake is polluted. Possible answers: debris on the surface, scum, discolored water, bad smell Have students brainstorm possible sources of water pollution. Possible answers: nearby factories, sewage, runoff from nearby land

Caption Answer

Figure 19 The polluted discharge could be stopped at the source or treated before it flows into the water.

as you read

What **You'll Learn**

- **Identify** sources of water pollution.
- **Explain** the difference between point source and nonpoint source pollution.
- **Describe** laws that regulate water pollution.

Why **It's Important**

Clean water is necessary for a healthy environment.

Review Vocabulary
pollutant: a substance that contaminates the environment

New Vocabulary
- point source pollution
- nonpoint source pollution

Pollution Sources

Imagine that you are walking on a footbridge over a river near your home or school. You stop to look at the flowing water, but you see more than water. Perhaps you see a multicolored film of something on top of the water, or maybe the water has a strange odor. A pollutant is a substance that contaminates the environment. Water pollution occurs when substances are added to water that lower its quality. In this section, you'll learn where water pollution comes from. You'll also learn about some of the sources of the pollution that enters freshwater.

Point Source Pollutants How do pollutants get into Earth's freshwater? Are they from one source or are they from many? Pollution that enters water from a specific location, such as shown in **Figure 19,** is called **point source pollution.** When a ditch or pipe discharges water, you can see exactly where the water is coming from. Point sources are fairly easy to control because the location of the pollution source is known.

Figure 19 Point sources of water pollution are easy to see. **Identify** *ways that these sources of pollution might be controlled.*

Section 4 Resource Manager

Chapter *FAST FILE* Resources
- Transparency Activity, p. 47
- Directed Reading for Content Mastery, pp. 17, 18
- MiniLAB, p. 4
- Enrichment, p. 30
- Reinforcement, p. 26
- Lab Worksheet, pp. 7–8

Mathematics Skill Activities, p. 25
Reading and Writing Skill Activities, p. 29
Lab Management and Safety, p. 69

Figure 20 Runoff containing pesticides and herbicides is one type of nonpoint source pollution.

Nonpoint Source Pollutants Most water pollution doesn't enter a body of water from sources you can identify. When pollution comes from a wide area such as lawns, construction sites, and roads, it is called **nonpoint source pollution.** Pollutants can be delivered to a body of water by runoff from yards, parking lots, and farm fields, as shown in **Figure 20.** Nonpoint sources also include pollutants in rain or snow. Nonpoint sources are much harder to control because it is hard to tell exactly where the pollution comes from. For example, when a stream gets muddy, do you know exactly where the sediment came from?

Reading Check *Why are nonpoint sources hard to control?*

Reducing Water Pollution

The key to clean water is to reduce the amount of pollutants that enter it. Controlling point source pollutants might seem easy—reduce the amount of pollutants at the source. However, what if the pollution comes from manufacturing products that you use, like stereos, TVs, and food packaging? You might not want to give these up, or pay more for them at the store. Controlling nonpoint source pollution is more difficult. For example, fertilizers, herbicides, and insecticides are sprayed on crops and gardens over large land areas to encourage plant growth and reduce damage from weeds and insects. However, these substances can be carried into Earth's surface freshwater. Reducing these pollutants requires cooperation and careful use of chemicals.

Controlling pollution is not easy. It requires difficult decisions about the types and amounts of pollutants that should be allowed in water. Laws that limit pollutants are written carefully to protect the water resource and the economy.

Mini LAB

Evaluating Dilution

Procedure

1. Set four empty **drinking glasses** on a counter.
2. Fill the first glass with **tap water** and add three drops of **red food coloring.**
3. Pour half of the water from the first glass into the second glass. Add tap water to fill the second glass.
4. Pour half of the water in the second glass into the third glass. Add tap water to fill the third glass.
5. Pour half of the water in the third glass into the fourth glass. Add tap water to fill the fourth glass.

Analysis

1. What does the water in the fourth glass look like?
2. How does dilution with clean water affect polluted water?

Try at Home

2 Teach

Reading Check

Answer It is difficult to determine the exact source of the pollution.

Mini LAB

Purpose Students investigate the effectiveness of diluting polluted water. L2 LS **Visual-Spatial**

Materials 4 drinking glasses, food coloring, water

Teaching Strategy Students should stir the water after adding the red food coloring.

Analysis

1. a light shade of red (pink)
2. In some cases, diluting might decrease the concentration of pollutants to safe levels, but the pollutants are not removed.

Assessment

Performance Have small groups of students design systems for filtering pollutants in water. Use **Performance Assessment in the Science Classroom,** p. 95.

Try at Home

Virtual Labs

Pollution *How can water pollution be detected?*

Active Reading

Synthesis Journal In this strategy, students reflect on a project, a paper, or a performance in light of their own experiences and plan for personal application. Have each student divide a sheet of paper into three sections. Have students record *What I did, What I learned,* and *How I can use it.* Have students write a Synthesis Journal entry related to the MiniLAB on this page.

Differentiated Instruction

Challenge Have students create a bulletin-board display that illustrates various sources of freshwater pollution. The sources should be classified as point source or nonpoint source and the processes by which the pollutant enters freshwater should be explained. L3

SECTION 4 Pollution of Freshwater H ◆ 55

INTEGRATE History

Research Public opinion was influenced by the publication of Rachel Carson's *Silent Spring* in 1962. This book warned of the dangers of various pesticides. A vocal environmentalist movement developed. In response to this political pressure, the EPA was created on December 2, 1970. William Ruckelshaus, sometimes called "Mr. Clean," was confirmed as its first Administrator a day earlier.

Reading Check

Answer the Rivers and Harbors Act of 1899

Inquiry Lab

Reducing Water Pollution

Question How can water pollution be reduced at home and at school?

Possible Materials stopwatch or watch with second hand, vessel for measuring volume of water, calculator

Estimated Time two to three weeks

Teaching Strategies

- Brainstorm water pollution sources at home and at school.
- Encourage students to develop and evaluate methods for reducing water pollution during their daily activities.
- Have students quantify their results and predictions using graphs and tables.
- Students also should develop a summary report listing their recommendations. L3

For additional inquiry activities, see *Science Inquiry Labs*.

Legislation As governments came to realize the importance of protecting Earth's freshwater, protective laws were enacted. The Rivers and Harbors Act of 1899 was the first attempt to use legislation to help regulate water pollution. This law required that when any work was done in water that ships use for navigation, like dredging the bottom, the wildlife resources of the water must be considered.

Reading Check *What legislation was the first attempt to help regulate water pollution?*

By the late 1960s, water pollution became so bad in the United States that many beaches along the east and west coasts were unhealthy for swimming. Drinking-water supplies were contaminated. Then, in 1969, greasy debris floating on a large river in Cleveland, Ohio, caught on fire. This was a wake-up call for many people. In 1972, the U.S. Congress responded by passing the Federal Water Pollution Control Act. Amendments were added to this act in 1977 and 1987, and it was renamed the Clean Water Act. These laws put limits on the types of pollution that can be discharged into streams and lakes.

In 1970, Congress created the Environmental Protection Agency (EPA), to enforce water pollution limits. The Clean Water Act amendments also provided federal funds for communities and industries to limit their water pollution sources. Today, 55 percent of streams and rivers and 46 percent of lakes that have been tested have good water quality. Still, as **Figure 21** shows, much improvement is needed.

The EPA, America's Clean Water Foundation, and the International Water Association are working together to promote water quality testing. October 18, 2003, was selected as the first World Water Monitoring Day.

INTEGRATE History

Formation of the EPA
The U.S. EPA was created in response to an outpouring of public opinion. Do research to learn about the events that led to the formation of the EPA and the confirmation of its first Administrator. Report what you learn to the class.

Figure 21 The circle graphs below show that the majority of streams and lakes that have been tested have good water quality. However, improvement is still needed.

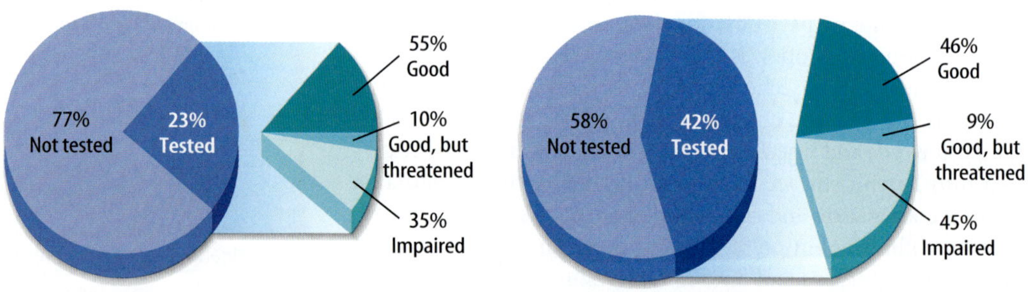

Approximately 55 percent of the tested streams and rivers have good water quality.

Of the 42 percent of lakes that have been tested, 45 percent still are impaired by pollution.

56 ◆ H CHAPTER 2 Freshwater at Earth's Surface

Curriculum Connection

History Have students trace the history of a major water pollution law, such as the Clean Water Act or the Safe Drinking Water Act. Organize the class into two panels for a pro-and-con debate about the usefulness of the laws and the types of changes that might need to be made to them. L2 LS **Interpersonal**

Visual Learning

Figure 21 What is *good, but threatened?* These bodies of water still have pollutant levels that are good, but pollution has increased and further increases might cause them to be impaired.

Figure 22 Piling yard wastes makes compost, which can be used in your garden instead of purchasing commercial fertilizer.

How can you help? You can do many things to reduce water pollution. Pay attention to how much water you use. When you use water, it must be treated before it can be returned to a stream. Just turning off the faucet when you brush your teeth can save more than 19 L of water per day. Keep your yard and driveway free of pet wastes, oil, and other debris. When it rains, these pollutants run off to streams and lakes.

Learn alternate methods to care for your yard and garden, like composting grass clippings as shown in **Figure 22.** Composting yard wastes reduces the amount of fertilizer and other chemicals needed for your yard. Properly dispose of any hazardous substances, such as used oil, antifreeze, and paint. Learn about programs in your community for safe disposal of hazardous waste.

section 4 review

Summary

Pollution Sources

- Substances added to water that reduce its quality cause water pollution.
- Sources of pollution include point sources and nonpoint sources.

Reducing Water Pollution

- Decisions about controlling pollution are difficult to make and are sometimes controversial.
- Many laws have been made to restrict the amount of pollution that can enter freshwater.
- You can help reduce water pollution by not using more water than is necessary.

Self Check

1. **Describe** some sources of water pollution.
2. **Describe** some laws meant to reduce water pollution.
3. **Explain** the purpose of the EPA.
4. **List** three ways you can reduce water pollution.
5. **Think Critically** Compare and contrast point source pollution and nonpoint source pollution.

Applying Math

6. **Use Statistics** An industry released wastewater to a stream every day for a week. The concentrations of a pollutant in the wastewater for each day are the following: 2 mg/L, 1 mg/L, 3 mg/L, 1 mg/L, 4 mg/L, 1 mg/L, 2 mg/L. Calculate the average daily concentration of the pollutant.

Science Online bookh.msscience.com/self_check_quiz

Teacher FYI

Agricultural Pollutants According to the United States EPA, agriculture is the leading source of pollution for rivers and streams. Pollutants include excess sediment, pathogens, nutrients, and pesticides.

3 Assess

DAILY INTERVENTION

Check for Understanding

Interpersonal Work together as a class to develop a table listing causes, effects, and possible remedies for water pollution. L2

Reteach

Pollution Sources Show students various illustrations from books or magazines. Ask students to call out "point source" or "nonpoint source" as you show each visual. Possible illustrations: point: outfall pipe from factory or wastewater treatment plant; nonpoint: farm field, golf course, construction site, runoff from a driveway or road L1

Assessment

Portfolio Suggest students make posters showing ways they can help reduce water pollution. Display the posters in a hallway. Use **Performance Assessment in the Science Classroom,** p. 145. P

section 4 review

1. factories, farms, yards, sewage treatment facilities, among others
2. Laws include the Clean Water Act and the Rivers and Harbors Act.
3. It enforces United States laws that control or prevent pollution. It also does research on the effects of pollution on health, proposes laws to limit pollution, and writes regulations to carry out antipollution laws passed by Congress.
4. Possible answers: Use less water; keep the area around your house free of pet waste, debris, and oil; compost to reduce use of fertilizers; dispose of hazardous household substances properly.
5. Both are pollution sources. The location of point source pollution can be identified and it is therefore easier to control; the location of nonpoint source pollution cannot be identified.
6. 2.0 mg/L

BENCH TESTED

Real-World Question

Purpose Students monitor water quality data through time to predict the health of a water system. L2

Process Skills plan, collect data, record observations, make and use graphs, draw conclusions, predict

Time Required several months

Materials Materials for this lab are available from scientific suppliers.

Safety Precautions When in the field, students should be supervised at all times. Do not allow any students to enter the water. Follow instructions for using chemicals carefully.

Procedure

Teaching Strategies

- Some states have programs that support stream monitoring activities.
- The GLOBE program also provides detailed support and collaboration. Visit their Web site for more details.

LAB

Adopt a Stream

Goals

- **Analyze** data about a local stream.
- **Measure** water temperature, dissolved oxygen, pH, alkalinity, and nitrate concentration.
- **Evaluate** the effects of point and nonpoint source pollution on your stream.

Materials

bucket with rope attached to handle
alcohol thermometer
string
plastic sample bottles
100-mL beakers
500-mL graduated cylinder
pH paper or pH meter
dissolved oxygen kit
alkalinity kit
nitrate test kit

Safety Precautions

WARNING: *Be careful near streams. Do not enter the water. Use testing chemicals with care. Wear gloves and follow your teacher's directions carefully.*

Real-World Question

The quality of the water in a stream determines whether organisms can live in it. How can you monitor the quality of a stream through time?

Procedure

1. With your teacher's help, select a location along a local stream to monitor. You should use this same location each time you collect a water sample. Perform all measurements at the stream site just after collecting your sample.
2. Following your teacher's directions, collect a water sample from the stream each week. All samples should be collected on the same day of the week and at the same time of day. Use a bucket with a rope attached to the handle to collect the water samples.
3. Measure the temperature of your water sample. Tie a piece of string to an alcohol thermometer and lower the thermometer into your water sample. Wait 5 min. and record the temperature. Repeat your measurement two more times.

Differentiated Instruction

Physically Challenged If you have physically challenged students who cannot participate in the field studies, return some bottled water samples to the classroom. Ask these students to supervise the analysis of these returned samples. L2

Alternative Inquiry Lab

Water Quality To extend the activity, have students develop a plan to assess the quality of local groundwater. Evaluate student plans for feasibility and safety. Encourage students to conduct their research.

4. Fill a bottle with your sample water. Make sure that the bottle is completely full. Use your dissolved oxygen kit to determine the amount of dissolved oxygen in your sample. Follow your teacher's directions carefully. Repeat the procedure two more times. Record your data.
5. Half-fill a 100-mL beaker with your sample water. Use pH paper or a pH meter to measure the pH of your sample. Repeat this procedure two more times. Record your data.
6. Half-fill a clean 100-mL beaker with your sample water. Use an alkalinity kit to determine the alkalinity of your sample. Follow your teacher's instructions carefully. Repeat this procedure two more times. Record your data.
7. Use your nitrate test kit to determine the amount of nitrate in your water sample. Follow the kit directions and your teacher's directions carefully. Repeat the procedure two more times. Record your data.
8. Calculate averages for each data category. Eliminate values that seem unreasonable. Repeat procedures as necessary.

Stream Nitrate Concentrations

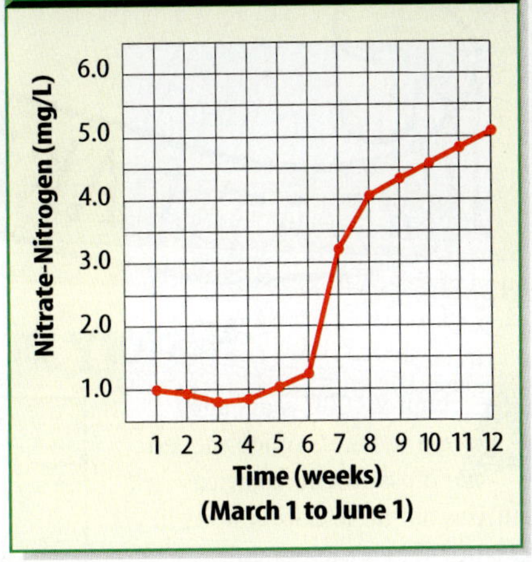

Analyze Your Data

1. **Describe** any trends that you observe in your data through time. Do the data vary with time of year or with the amount of water flowing in the stream?
2. **Graph** your data to illustrate any changes through time.

Conclude and Apply

1. **Evaluate** the quality of the water in your stream. Is your stream healthy? Does the health of your stream vary through time?
2. **Evaluate** the effects of pollution on your stream. Do you think that any point or nonpoint sources of pollution are affecting your stream? Explain.

Communicating Your Data

Create a poster or other visual display that illustrates any effects of point or nonpoint source pollution on your stream.

Analyze Your Data

Expected Outcome Students will monitor the water quality in a stream through time. They will observe changes associated with the season and flow conditions. Students will evaluate the quality of the stream and identify possible sources of pollution.

Answers to Questions

1. Data should vary seasonally. For example, students might notice higher levels of nitrate nitrogen during spring when fertilizers are being applied.
2. Student graphs should show some interpretable trends. After graphing, ask students to examine all of the data, including notes taken at the stream site, before making conclusions.

Error Analysis It is important to replicate all measurements taken at the site. Continue taking data and evaluating procedures until consistent data are obtained. Disregard any unreasonable data.

Conclude and Apply

1. Most streams in the United States are healthy. However, a significant percentage is not. Some streams might have acceptable water quality only during some portions of the year.
2. Point and nonpoint sources likely will affect most streams. Ask students to consider the type of pollution and then correlate with potential source types.

Assessment

Portfolio Have students include their graphs and a summary of their conclusions in their portfolios. P

TIME

Content Background

Glaciers advanced and retreated many times through the Pleistocene Epoch of geologic time (1.8 million years ago to 10,000 years ago). The last great ice sheet to advance through the Great Lakes region was the Laurentide Ice Sheet. Lobes of ice advanced and scoured the lake basins. After the glaciers retreated and the immense weight of the ice was removed, the lakes filled and the land began to rise. The region is still rising slowly today.

Discussion

Why are the Great Lakes such an important waterway? Possible answer: The Great Lakes are the major waterway between Canada and the U.S. for shipping and trade. The waterways also are used to ship goods from all over the world to America's heartland, the Midwest.

Historical Significance

The Erie Canal was an important step toward the settlement and development of much of the United State's heartland. It greatly shortened the amount of time needed to travel to the Great Lakes from the east. Trade flourished in the region, and communities sprung up along its course. Dewitt Clinton, who became governor of New York in 1917, was largely responsible for the construction of the canal. Some thought the idea of digging across New York so ridiculous that the project often was referred to as "Clinton's Folly."

TIME **SCIENCE AND HISTORY** SCIENCE CAN CHANGE THE COURSE OF HISTORY!

GREAT LAKES AT A GLANCE

History of the Great Lakes

About 25,000 years ago, giant sheets of ice moved across North America scouring out large depressions in the land. When the climate warmed about 12,000 years ago, the glacier's edge slowly retreated to the north. As it moved, the depressions filled with water that melted from the glaciers. Streams that ran into the depressions also helped fill them up. Today, the five Great Lakes hold 18 percent of the surface freshwater on Earth. Only Lake Baikal in Siberia contains more freshwater.

The border between the United States and Canada runs through the middle of four of the Great Lakes. So, the United States and Canada share the job of keeping the lakes clean and safe for shipping. The two nations work together to build and maintain the locks and canals that link the Great Lakes to the St. Lawrence River and the Atlantic Ocean. Ships from all over the world use this inland waterway to sail into the heartland of North America. Iron ore, coal, finished steel, and grain are the most important items shipped in and among the Great Lakes.

An important part of the canal system, the Erie Canal connects the Hudson River near Troy, New York, to the Niagara River near Buffalo. The Erie Canal was completed in 1825, in spite of political opposition to the project.

Oral Presentation **Pick one of the Great Lakes and prepare a presentation about how one of the following subjects affects the future of the lake—pollution, erosion, cormorants, zebra mussels, or changing water levels.**

Science Online
For more information, visit bookh.msscience.com/time

Oral Presentation Have students consider the effects of these problems on lake ecosystems and on humans. Of particular significance for those living along the lakeshore is coastal erosion. In some regions, land is falling into the lakes at alarming rates.

Resources for Teachers and Students

The Great Lakes An Environmental Atlas and Resource Book, prepared by the U.S. EPA and Government of Canada, 1995

Erie Canal: New York's Gift to the Nation, co-published by the New York State Archives Partnership and Cobblestone Publishing Company

Reviewing Main Ideas

Section 1 Streams

1. Streams shape Earth's surface by eroding and depositing sediment.
2. Floodplains are covered by water during floods. Severe flooding can kill people and damage property.

Section 2 Lakes and Reservoirs

1. Lake environments support a wide variety of life.
2. Through time, the addition of nutrients can cause lakes to become eutrophic.
3. Turnover during fall or spring mixes nutrients throughout a lake.

Section 3 Wetlands

1. Wetlands include swamps, marshes, and bogs.
2. Wetlands are classified according to their vegetation.
3. Wetlands reduce erosion and flooding, and remove excess nutrients from water.

Section 4 Pollution of Freshwater

1. Water pollution comes from point sources and nonpoint sources.
2. Point sources are specific locations where pollution is released.
3. Nonpoint sources are spread across a wide area.

Visualizing Main Ideas

Copy and complete the following concept map.

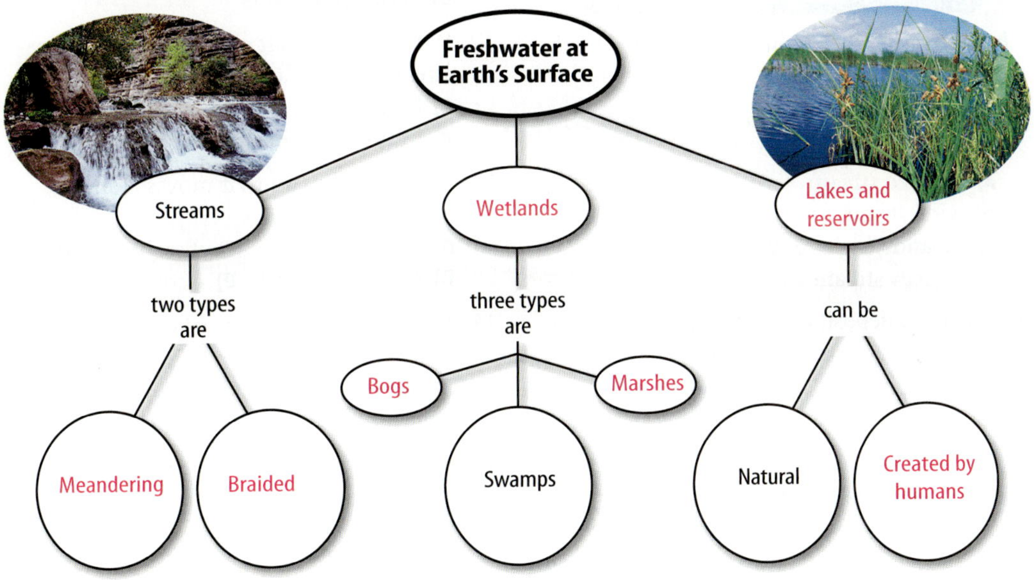

chapter 2 Study Guide

Reviewing Main Ideas

Summary statements can be used by students to review the major concepts of the chapter.

Visualizing Main Ideas

See student page.

Science Online

Visit bookh.msscience.com
/self_check_quiz
/interactive_tutor
/vocabulary_puzzlemaker
/chapter_review
/standardized_test

Assessment Transparency

For additional assessment questions, use the *Assessment Transparency* located in the transparency book.

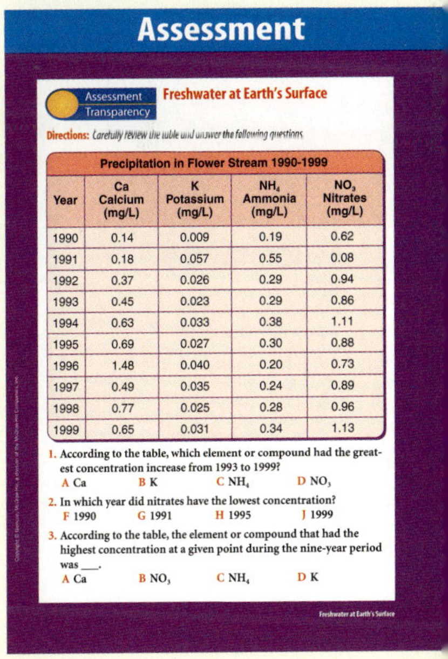

Assessment

Assessment Transparency — **Freshwater at Earth's Surface**

Directions: *Carefully review the table and answer the following questions.*

Precipitation in Flower Stream 1990-1999				
Year	Ca Calcium (mg/L)	K Potassium (mg/L)	NH_4 Ammonia (mg/L)	NO_3 Nitrates (mg/L)
1990	0.14	0.009	0.19	0.62
1991	0.18	0.057	0.55	0.08
1992	0.37	0.026	0.29	0.94
1993	0.45	0.023	0.29	0.86
1994	0.63	0.033	0.38	1.11
1995	0.69	0.027	0.30	0.88
1996	1.48	0.040	0.20	0.73
1997	0.49	0.035	0.24	0.89
1998	0.77	0.025	0.28	0.96
1999	0.65	0.031	0.34	1.13

1. According to the table, which element or compound had the greatest concentration increase from 1993 to 1999?
   A Ca B K C NH_4 D NO_3
2. In which year did nitrates have the lowest concentration?
   F 1990 G 1991 H 1995 J 1999
3. According to the table, the element or compound that had the highest concentration at a given point during the nine-year period was ____.
   A Ca B NO_3 C NH_4 D K

Freshwater at Earth's Surface

Using Vocabulary

1. Meandering streams have channels that form a number of wide bends. Braided streams have bars that form in their channels.
2. When surface water cools, lake water mixes and nutrients from deeper water become available to plants.
3. Stream load is the sediment and dissolved material carried by a stream.
4. Point bars form on the inside of a stream meander because water flows more slowly there.
5. Point sources can be identified easily. Nonpoint sources are not easy to identify because the pollution comes from a large region.

Checking Concepts

6. C	11. B
7. A	12. A
8. D	13. A
9. B	14. C
10. B	15. B

chapter 2 Review

Using Vocabulary

braided stream p. 38	nutrient p. 47
drainage basin p. 38	point bar p. 37
eutrophication p. 47	point source pollution p. 54
floodplain p. 42	runoff p. 36
load p. 39	stream discharge p. 41
meandering stream p. 37	turnover p. 49
nonpoint source pollution p. 55	wetlands p. 51

Use what you know about the vocabulary words in this chapter to answer the following questions.

1. What is the difference between a meandering stream and a braided stream?
2. How does lake turnover make nutrients available to plants?
3. What is stream load?
4. How do point bars form?
5. How do point sources of pollution differ from nonpoint sources of pollution?

Checking Concepts

Choose the word or phrase that best answers the question.

6. Which type of stream has wide bends in its channel?
   A) braided stream
   B) wide stream
   C) meandering stream
   D) narrow stream
7. What is a deposit of sediment that forms on the inside of a stream meander?
   A) point bar **C)** floodplain
   B) runoff **D)** turnover
8. Which type of wetlands contains trees and shrubs and often forms on a floodplain?
   A) marsh **C)** stream
   B) bog **D)** swamp

Use the photo below to answer question 9.

9. Which feature is shown in the photo above?
   A) delta **C)** point bar
   B) alluvial fan **D)** meander
10. What is the entire area of land that is drained by a stream?
    A) base level **C)** floodplain
    B) drainage basin **D)** meander
11. What are pollutants that enter a stream from a wide area?
    A) point source pollutants
    B) nonpoint source pollutants
    C) chemical pollutants
    D) varied pollutants
12. What is the lowest elevation to which a stream can erode?
    A) base level **C)** floodplain
    B) drainage basin **D)** meander
13. What is sediment that moves along the bed of a stream?
    A) bed load **C)** dissolved load
    B) suspended load **D)** nutrient load
14. What part of a stream valley is covered by water during a flood?
    A) bar **C)** floodplain
    B) meander **D)** braid
15. Which of the following causes lake water to mix?
    A) delta **C)** eutrophication
    B) turnover **D)** aquatic plants

Science Online bookh.msscience.com/vocabulary_puzzlemaker

Use the ExamView® Pro Testmaker CD-ROM to:
- create multiple versions of tests
- create modified tests with one mouse click for inclusion students
- edit existing questions and add your own questions
- build tests aligned with state standards using built-in State Curriculum Tags
- change English tests to Spanish with one mouse click and vice versa

Thinking Critically

16. **Infer** why whitewater rapids are common in mountain streams.

17. **Explain** how you might determine whether water pollution comes from a point source or a nonpoint source.

18. **Predict** how a stream channel might change if the stream is dammed to create a reservoir.

19. **Compare and contrast** deltas and alluvial fans.

20. **Recognize Cause and Effect** A clear stream near your home becomes muddy after a heavy rain. Explain.

21. **Draw Conclusions** Are the bars in a braided stream permanent features? Explain.

22. **Recognize Cause and Effect** Why might thick, green mats of algae form on the surface of a lake?

23. **Classify** Obtain several photographs of stream valleys. Classify each stream.

Use the table below to answer question 24.

Depth of Great Lakes

Great Lake	Maximum Depth (m)
Superior	405
Michigan	281
Ontario	244
Huron	229
Erie	64

24. **Make and Use Graphs** Make a bar graph that compares the depths of the Great Lakes. Remember to include a title and axis labels on your graph.

25. **Use Scientific Explanations** Summarize how a stream moves its bedload.

Performance Activities

26. **Poster** Use photographs from old magazines to create a poster that shows the types of plants and animals that use wetland habitats. Combine your poster with those of your classmates to create a classroom display.

27. **Oral Presentation** Develop a land use plan for an imaginary floodplain and illustrate your plan by making a map. Give a persuasive speech to convince others that your plan is a good one.

Applying Math

28. **River Discharge** The Amazon River in South America has a discharge of about 180,000 m^3/s. The Mississippi River has a discharge of about 17,500 m^3/s. How many times higher is the Amazon River's discharge?

29. **Power from Dams** Hydroelectric dams in the United States produce about 3 percent of the country's electricity. What percentage of U.S. electricity is produced by all other sources combined?

Use the graph below to answer question 30.

30.

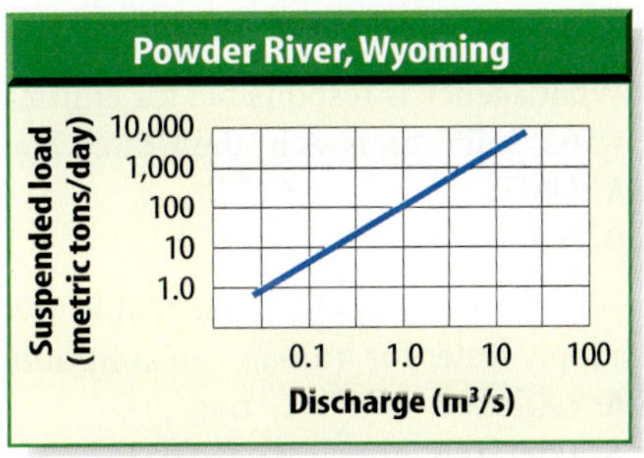

a. **Suspended Load** According to the graph, does the amount of suspended load in this river increase or decrease as discharge increases? How do you know?

b. **Suspended Load and Discharge** Hypothesize about why the relationship shown in the graph exists.

Assessment Resources

Reproducible Masters

Chapter *Fast File* Resources
- Chapter Review, pp. 37–38
- Chapter Tests, pp. 39–42
- Assessment Transparency Activity, p. 51

Glencoe Science Web site
- Chapter Review Test
- Standardized Test Practice

Glencoe Technology
- Assessment Transparency
- Exam*View*® Pro Testmaker
- MindJogger Videoquiz
- Interactive Chalkboard

chapter 2 Review

Thinking Critically

16. Mountain streams have steep gradients.
17. Compare the pollutant type with potential sources.
18. Sediment might accumulate upstream from the reservoir. Downstream areas might have erosion.
19. Both form when flowing water slows. Deltas form where a stream enters an ocean or lake. Alluvial fans form where a mountain stream flows onto flatter land.
20. Runoff carried suspended load into the stream.
21. No, the bars change rapidly during floods.
22. This would occur if the lake contained high amounts of nutrients.
23. Students should note overall stream shape and channel features.
24. Make certain that students remember axis labels and a title.
25. by bouncing, rolling, and sliding

Performance Activities

26. Posters should show a wide variety of plants and animals that live in swamps, marshes, and bogs. Use **PASC,** p. 145.
27. Have students create maps showing how the land on their floodplain will be used. Use **PASC,** p. 123.

Applying Math

National Math Standards

1, 5, 9

28. 10.3
29. 97%
30. a. It increases. The graph shows a positive, linear relationship.
    b. Discharge increases afer heavy rains. The runoff carries sediment into the stream.

FAST FILE

Answer Sheet A practice answer sheet can be found at bookh.msscience.com/answer_sheet.

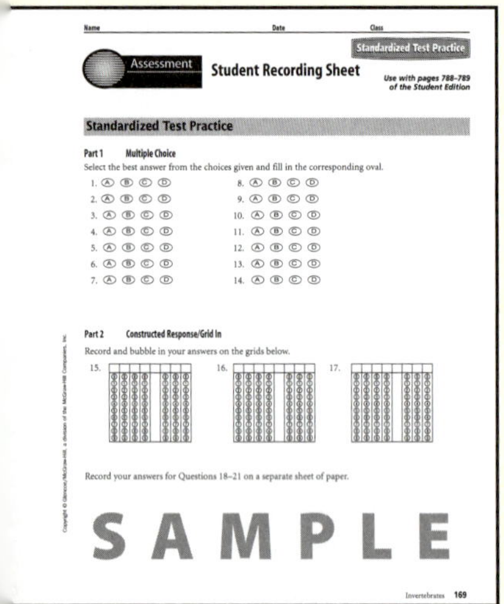

Part 1 Multiple Choice

1. B
2. A
3. C
4. B
5. D
6. C
7. A

chapter 2 Standardized Test Practice

Part 1 Multiple Choice

Record your answers on the answer sheet provided by your teacher or on a sheet of paper.

Use the photo below to answer question 1.

1. Which type of stream is shown in the photo above?
   - **A.** braided stream
   - **B.** meandering stream
   - **C.** straight stream
   - **D.** wide stream
2. What forms when a stream cuts down as the land is uplifted?
   - **A.** a canyon
   - **B.** a delta
   - **C.** a cut bank
   - **D.** an alluvial fan
3. What agency is responsible for enforcing water pollution laws in the United States?
   - **A.** USGS
   - **B.** NASA
   - **C.** EPA
   - **D.** DOE
4. Which term refers to a lake that is used to supply water for human consumption?
   - **A.** caldera
   - **B.** reservoir
   - **C.** bog
   - **D.** marsh

Test-Taking Tip

Skills and Practice Remember that test-taking skills can improve with practice. If possible, take at least one practice test and familiarize yourself with the test format and instructions.

5. How are wetlands classified?
   - **A.** by their shape
   - **B.** by their size
   - **C.** by the type of animals that live in them
   - **D.** by the type of plants that live in them

Use the graph below to answer questions 6 and 7.

Daily Oxygen Concentration in Colorado River

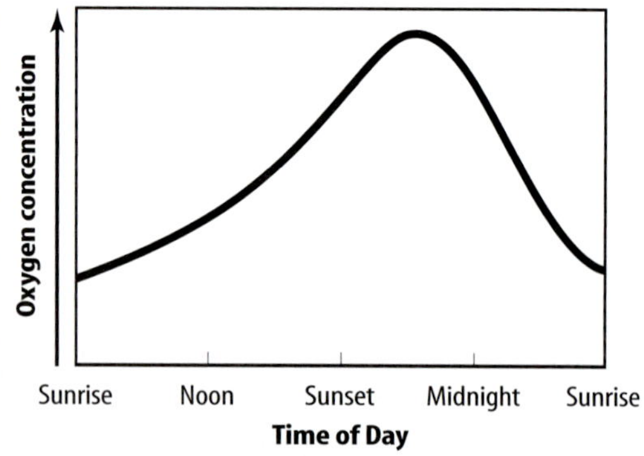

6. According to the graph, at which time of day was the oxygen concentration highest?
   - **A.** between sunrise and noon
   - **B.** between noon and sunset
   - **C.** between sunset and midnight
   - **D.** between midnight and sunrise
7. Which is a reasonable explanation for why the oxygen concentration in a stream might vary this way?
   - **A.** When the Sun rises, plants begin to produce oxygen and the concentration increases until just after sunset.
   - **B.** When the Sun sets, plants use more oxygen and the concentration declines.
   - **C.** When the Sun rises, animals become more active and use more oxygen.
   - **D.** When the Sun sets, animals become more active and use more oxygen.

Part 2 Short Response/Grid In

8. A point source is pollution from one specific location. Possible examples: factory, sewage treatment plant, home septic tank
9. The base level is the level of the stream or lake that the water flows into.
10. Sunlight penetrates to the lake's bottom.
11. by catching insects
12. grasses, rushes, and sedges
13. by constructing Hoover Dam
14. mean = 27.7; median = 28.8
15. 29.8 − 22.1 = 7.7
16. March 4
17. Snow melt or rain washed sediment into the lake.

64 ◆ H STANDARDIZED TEST PRACTICE

Part 2 Short Response/Grid In

Record your answers on the answer sheet provided by your teacher or on a sheet of paper.

8. What is a point source of pollution? List two examples.

9. Explain why the water in a stream cannot flow below the stream's base level.

10. Why do rooted plants only grow in the shallow parts of a lake?

11. How does a Venus's-flytrap obtain nutrients?

12. Describe the vegetation in a marsh.

13. How was Lake Mead created?

Crater Lake in Oregon is the deepest and one of the clearest lakes in the United States. Use the data about Crater Lake to answer questions 14–17.

Clarity Depth in Crater Lake

Date of Measurement	Clarity Depth (meters)
March 1	28.5
March 2	28.8
March 3	29.2
March 4	29.8
March 5	22.1

14. Calculate the mean and determine the median of these data.

15. What is the range of these data?

16. Clarity depths are determined by lowering a disk into the lake until it can no longer be seen. On which day was the water clearest?

17. Hypothesize about why the water clarity might have changed between March 4th and March 5th.

Part 3 Open Ended

Record your answers on a sheet of paper.

Use the graph below to answer questions 18–20.

Discharge of Neuse River near Clayton, NC

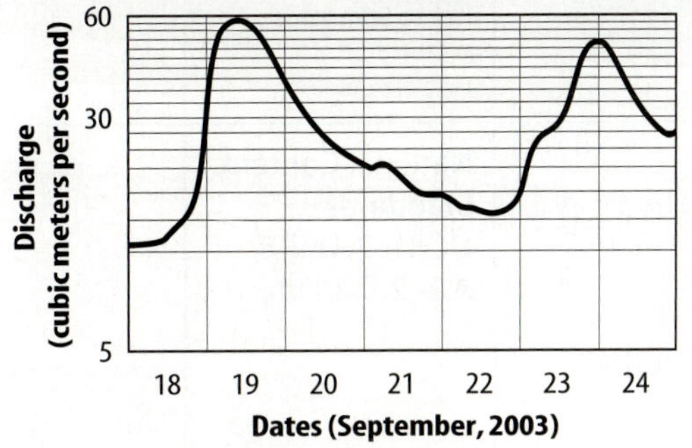

18. On which day was the discharge highest? On which day was it lowest?

19. On September 17, 2003, and September 18, 2003, Hurricane Isabel made landfall in North Carolina. How did this hurricane affect discharge in the Neuse River? Explain why this occurred.

20. On September 22nd and September 23rd, a weather front passed over North Carolina. How did this frontal passage affect discharge in the Neuse River? Explain.

21. Describe the process of eutrophication in lakes.

22. What is lake turnover? Why does it occur?

23. Write a detailed explanation about how deltas form.

24. How are dams used for flood control?

25. What are wetlands? Why are they important?

Science Online bookh.msscience.com/standardized_test

Part 3 Open Ended

18. September 19, September 18

19. Discharge increased significantly. The rain from hurricane Isabel caused the change.

20. It also caused discharge to increase. Because of ground saturation from Isabel, the region was especially vulnerable to additional flooding.

21. Sediment gradually makes the lake shallower and nutrients increase biological productivity. Eventually, the lake fills.

22. the mixing of nutrient-rich deep water throughout the lake

23. Rivers enter a lake or ocean and sediment is deposited to form a delta.

24. Dams hold some flood water in a reservoir. The water then can be released more gradually.

25. areas that are covered by water for some portion of the year; they provide habitat for organisms, prevent erosion, and help keep water clean

Rubrics

The following rubrics are sample scoring devices for short response and open-ended questions.

Short Response

Points	Description
2	The student demonstrates a thorough understanding of the science of the task. The response may contain minor flaws that do not detract from the demonstration of a thorough understanding.
1	The student has provided a response that is only partially correct.
0	The student has provided a completely incorrect solution or no response at all.

Open Ended

Points	Description
4	The student demonstrates a thorough understanding of the science of the task. The response may contain minor flaws that do not detract from the demonstration of a thorough understanding.
3	The student demonstrates an understanding of the science of the task. The response is essentially correct and demonstrates an essential but less than thorough understanding of the science.
2	The student demonstrates only a partial understanding of the science of the task. Although the student may have used the correct approach to a solution or may have provided a correct solution, the work lacks an essential understanding of the underlying science concepts.
1	The student demonstrates a very limited understanding of the science of the task. The response is incomplete and exhibits many flaws.
0	The student provides a completely incorrect solution or no response at all.

Section/Objectives	Standards		Labs/Features
	National	State/Local	
Chapter Opener	See pp. 9T–10T for a Key to Standards.		**Launch Lab:** Groundwater Infiltration, p. 67 **Foldables,** p. 67
Section 1 Groundwater 4 sessions 2 blocks 1. **Explain** the importance of groundwater. 2. **Describe** how groundwater moves. 3. **Compare and contrast** the formation of springs, artesian wells, and geysers.	National Content Standards: UCP.1–3, UCP.5, A.1–2, B.2, D.1		**MiniLAB:** Measuring Porosity, p. 69 **Science Online,** p. 70 **Integrate Chemistry,** p. 71 **Science Online,** p. 73 **Lab:** Artesian Wells, p. 75
Section 2 Groundwater Pollution and Overuse 3 sessions 1.5 blocks 4. **Identify** the sources of groundwater pollution. 5. **Explain** the ways groundwater is cleaned. 6. **Explain** what happens when too much water is pumped from an aquifer.	National Content Standards: UCP.1–3, UCP.5, A.1–2, D.1, F.2, F.4		**Visualizing Sources of Groundwater Pollution,** p. 77 **Integrate Career,** p. 79 **MiniLAB:** Modeling Groundwater Pollution, p. 80 **Science Online,** p. 81 **Applying Science:** Can stormwater be cleaned and reused for irrigation?, p. 81
Section 3 Caves and Other Groundwater Features 4 sessions 2 blocks 7. **Recognize** how caves change through time. 8. **Explain** how cave features form. 9. **Explain** what causes sinkholes and disappearing streams.	National Content Standards: UCP.1–3, UCP.5, A.1–2, D.1–2		**Science Online,** p. 87 **Lab:** Pollution in Motion, p. 90 **Science Stats:** Caves, p. 92

Glencoe Exclusive! TeacherWorks™ All-In-One Planner and Resource Center

Lab Materials	Reproducible Resources	Section Assessment	Technology
Launch Lab: different brands of paper towels (2), large beakers (2), water, graduated cylinder, magnifying glass	**Chapter *FAST FILE* Resources** Foldables Worksheet, p. 17 Note-taking Worksheets, pp. 33–35 Directed Reading Overview, p. 19	Glencoe's Assessment Advantage	TeacherWorks includes: • Interactive Teacher Edition • Lesson Planner with calendar • Access to all program blacklines • Correlations to standards • Web links
MiniLAB: 100 mL gravel, 250-mL beakers (2), 100 mL sand, 100 mL graduated cylinder, water **Lab:** tall, soft-plastic bottle with cap; nail or awl; hammer; protractor; sink with water; plastic straw; scissors; metric ruler	**Chapter *FAST FILE* Resources** Transparency Activity, p. 44 MiniLAB, p. 3 Enrichment, p. 30 Reinforcement, p. 27 Directed Reading, p. 20 Lab Worksheet, pp. 5–6 Lab Activity, pp. 9–12 **Mathematics Skill Activities,** p. 1	**Portfolio** Challenge, p. 70 **Performance** MiniLAB, p. 69 Applying Math, p. 74 **Content** Section Review, p. 74	Section Focus Transparency Virtual Labs CD-ROM Guided Reading Audio Program Interactive Chalkboard CD-ROM
MiniLAB: rubber band; spoon, coffee filter; drinking glass; powdered coffee, hot chocolate, or other powdered drink; water *Need materials?* Contact Science Kit at 1-800-828-7777 or www.sciencekit.com on the Internet.	**Chapter *FAST FILE* Resources** Transparency Activity, p. 45 MiniLAB, p. 4 Enrichment, p. 31 Reinforcement, p. 28 Directed Reading, p. 20 Lab Activity, pp. 13–16 **Home and Community Involvement,** p. 23 **Reading and Writing Skill Activities,** p. 23 **Life Science Critical Thinking/Problem Solving,** p. 6	**Portfolio** Curriculum Connection, p. 83 **Performance** MiniLAB, p. 80 Applying Science, p. 81 Applying Math, p. 84 **Content** Section Review, p. 84	Section Focus Transparency Guided Reading Audio Program Interactive Chalkboard CD-ROM
Lab: clear-plastic box, aquarium gravel, water, pump from a bottle or hair spray or liquid soap, 30 mL salt water, food coloring, 5-cm × 5-cm piece of cloth, 1,000-mL beaker, small rubber band, 50-mL graduated cylinder	**Chapter *FAST FILE* Resources** Transparency Activity, p. 46 Enrichment, p. 32 Reinforcement, p. 29 Directed Reading, pp. 21, 22 Lab Worksheet, pp. 7–8 Transparency Activity, pp. 47–48 **Performance Assessment in the Science Classroom,** p. 44	**Portfolio** Assessment, p. 89 **Performance** Applying Math, p. 89 **Content** Section Review, p. 89	Section Focus Transparency Teaching Transparency Guided Reading Audio Program Interactive Chalkboard CD-ROM Video Lab

End of Chapter Assessment

Glencoe's Assessment Advantage

Blackline Masters	Technology	Professional Series
Chapter *FAST FILE* Resources Chapter Review, pp. 37–38 Chapter Tests, pp. 39–42 **Standardized Test Practice,** pp. 15–18	MindJogger Videoquiz Virtual Labs CD-ROM Exam*View*® Pro Testmaker TeacherWorks CD-ROM Interactive Chalkboard CD-ROM	**Performance Assessment in the Science Classroom (PASC)**

Transparencies

Section Focus

Section 1 Section Focus Transparency **Islands in a Desert Sea**

Oases are fertile areas in the desert. An oasis is usually formed when underground water gets close to the surface. When this happens, springs and wells turn a normally arid place into an area where plants can thrive.

1. How is this oasis like an island?
2. How might people and animals use the resources at an oasis?
3. In order for water under the ground to move, what must be true of the rock and soil?

L2

Section 2 Section Focus Transparency **Land Slick**

What happens when oil or gasoline is spilled on the ground? These products aren't easily broken down, and when it rains, they can wash off the roads and into the soil. The rings in the puddle below are petroleum products floating in rainwater.

1. What do you think happens to gas and oil after they are washed off the roads?
2. What other dangerous substances might seep through the soil?

L2

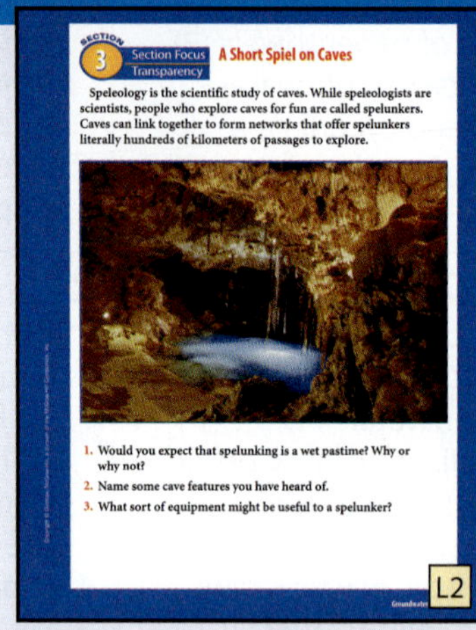
Section 3 Section Focus Transparency **A Short Spiel on Caves**

Speleology is the scientific study of caves. While speleologists are scientists, people who explore caves for fun are called spelunkers. Caves can link together to form networks that offer spelunkers literally hundreds of kilometers of passages to explore.

1. Would you expect that spelunking is a wet pastime? Why or why not?
2. Name some cave features you have heard of.
3. What sort of equipment might be useful to a spelunker?

L2

This is a representation of key blackline masters available in the Teacher Classroom Resources. See Resource Manager boxes within the chapter for additional information.

Assessment

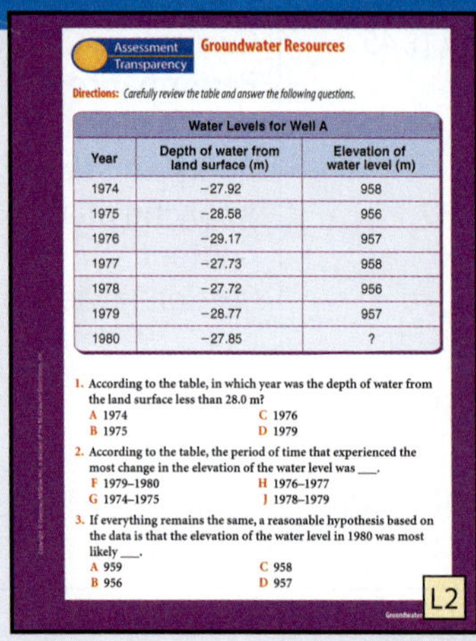
Assessment Transparency **Groundwater Resources**

Directions: *Carefully review the table and answer the following questions.*

Water Levels for Well A		
Year	Depth of water from land surface (m)	Elevation of water level (m)
1974	−27.92	958
1975	−28.58	956
1976	−29.17	957
1977	−27.73	958
1978	−27.72	956
1979	−28.77	957
1980	−27.85	?

1. According to the table, in which year was the depth of water from the land surface less than 28.0 m?
   A 1974 C 1976
   B 1975 D 1979
2. According to the table, the period of time that experienced the most change in the elevation of the water level was ___.
   F 1979–1980 H 1976–1977
   G 1974–1975 J 1978–1979
3. If everything remains the same, a reasonable hypothesis based on the data is that the elevation of the water level in 1980 was most likely ___.
   A 959 C 958
   B 956 D 957

L2

Teaching

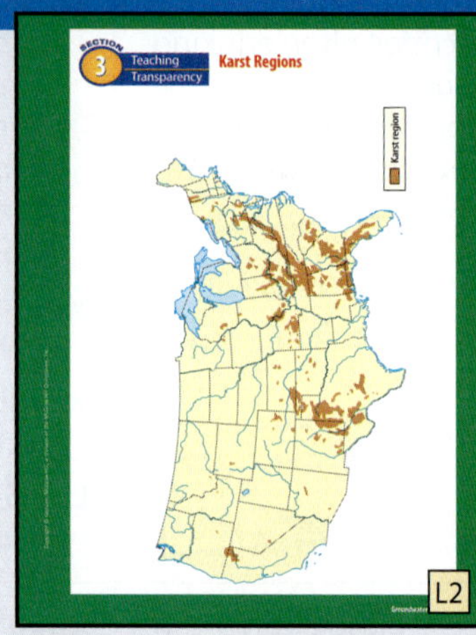
Section 3 Teaching Transparency **Karst Regions**

L2

Key to Teaching Strategies

The following designations will help you decide which activities are appropriate for your students.

- L1 Level 1 activities should be appropriate for students with learning difficulties.
- L2 Level 2 activities should be within the ability range of all students.
- L3 Level 3 activities are designed for above-average students.
- ELL ELL activities should be within the ability range of English Language Learners.
- COOP LEARN Cooperative Learning activities are designed for small group work.
- LS Multiple Learning Styles logos, as described on page 6T, are used throughout to indicate strategies that address different learning styles.
- P These strategies represent student products that can be placed into a best-work portfolio.
- PBL Problem-Based Learning activities apply real-world situations to learning.

Hands-on Activities

Student Text Lab Worksheet

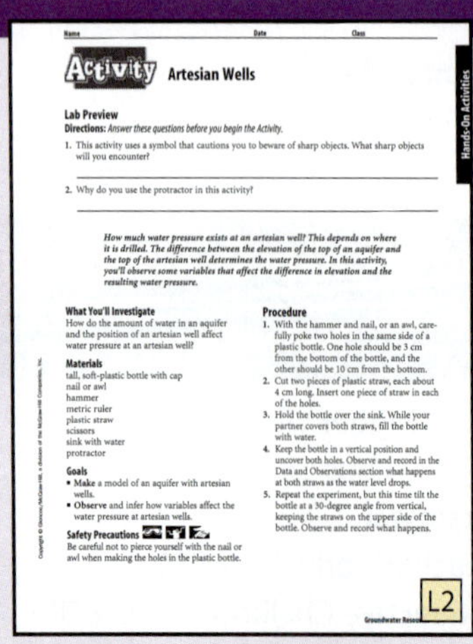
Name Date Class

Activity **Artesian Wells**

Hands-On Activities

Lab Preview

Directions: *Answer these questions before you begin the Activity.*

1. This activity uses a symbol that cautions you to beware of sharp objects. What sharp objects will you encounter?
2. Why do you use the protractor in this activity?

How much water pressure exists at an artesian well? This depends on where it is drilled. The difference between the elevation of the top of an aquifer and the top of the artesian well determines the water pressure. In this activity, you'll observe some variables that affect the difference in elevation and the resulting water pressure.

What You'll Investigate

How do the amount of water in an aquifer and the position of an artesian well affect water pressure at an artesian well?

Materials

tall, soft-plastic bottle with cap
nail or awl
hammer
metric ruler
plastic straw
scissors
sink with water
protractor

Goals

- **Make** a model of an aquifer with artesian wells.
- **Observe** and infer how variables affect the water pressure at artesian wells.

Safety Precautions

Be careful not to pierce yourself with the nail or awl when making the holes in the plastic bottle.

Procedure

1. With the hammer and nail, or an awl, carefully poke two holes in the same side of a plastic bottle. One hole should be 3 cm from the bottom of the bottle, and the other should be 10 cm from the bottom.
2. Cut two pieces of plastic straw, each about 4 cm long. Insert one piece of straw in each of the holes.
3. Hold the bottle over the sink. While your partner covers both straws, fill the bottle with water.
4. Keep the bottle in a vertical position and uncover both holes. Observe and record in the Data and Observations section what happens at both straws as the water level drops.
5. Repeat the experiment, but this time tilt the bottle at a 30-degree angle from vertical, keeping the straws on the upper side of the bottle. Observe and record what happens.

L2

Laboratory Activities

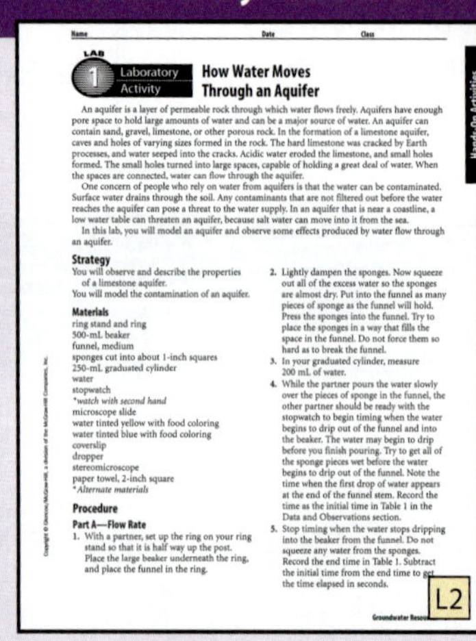
Name Date Class

Lab 1 Laboratory Activity **How Water Moves Through an Aquifer**

Hands-On Activities

An aquifer is a layer of permeable rock through which water flows freely. Aquifers have enough pore space to hold large amounts of water and can be a major source of water. An aquifer can contain sand, gravel, limestone, or other porous rock. In the formation of a limestone aquifer, caves and holes of varying sizes formed in the rock. The hard limestone was cracked by Earth processes, and water seeped into the cracks. Acidic water eroded the limestone, and small holes formed. The small holes turned into large spaces, capable of holding a great deal of water. When the spaces are connected, water can flow through the aquifer.

One concern of people who rely on water from aquifers is that the water can be contaminated. Surface water drains through the soil. Any contaminants that are not filtered out before the water reaches the aquifer can pose a threat to the water supply. In an aquifer that is near a coastline, a low water table can threaten an aquifer, because salt water can move into it from the sea.

In this lab, you will model an aquifer and observe some effects produced by water flow through an aquifer.

Strategy

You will observe and describe the properties of a limestone aquifer.
You will model the contamination of an aquifer.

Materials

ring stand and ring
500-mL beaker
funnel, medium
sponges cut into about 1-inch squares
250-mL graduated cylinder
water
stopwatch
watch with second hand
microscope slide
water tinted yellow with food coloring
water tinted blue with food coloring
coverslip
dropper
stereomicroscope
paper towel, 2-inch square
Alternate materials

Procedure

Part A—Flow Rate

1. With a partner, set up the ring on your ring stand so that it is half way up the post. Place the large beaker underneath the ring, and place the funnel in the ring.
2. Lightly dampen the sponges. Now squeeze out all of the excess water so the sponges are almost dry. Put into the funnel as many pieces of sponge as the funnel will hold. Press the sponges into the funnel. Try to place the sponges in a way that fills the space in the funnel. Do not force them so hard as to break the funnel.
3. In your graduated cylinder, measure 200 mL of water.
4. While the partner pours the water slowly over the pieces of sponge in the funnel, the other partner should be ready with the stopwatch to begin timing when the water begins to drip out of the funnel and into the beaker. The water may begin to drip before you finish pouring. Try to get all of the sponge pieces wet before the water begins to drip out of the funnel. Note the time when the first drop of water appears at the end of the funnel stem. Record the time as the initial time in Table 1 in the Data and Observations section.
5. Stop timing when the water stops dripping into the beaker from the funnel. Do not squeeze any water from the sponges. Record the end time in Table 1. Subtract the initial time from the end time to get the time elapsed in seconds.

L2

Meeting Different Ability Levels

Content Outline

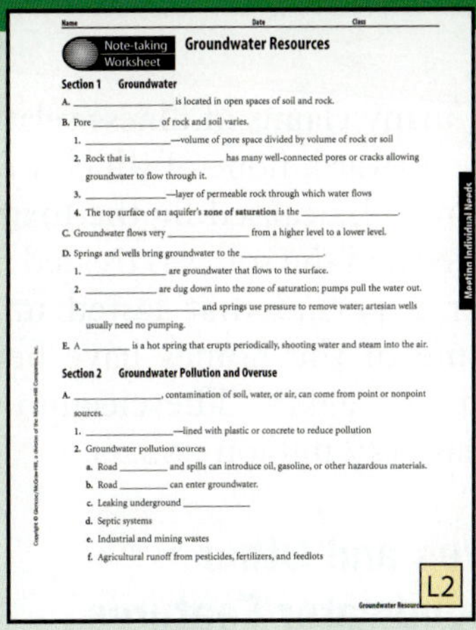

Reinforcement

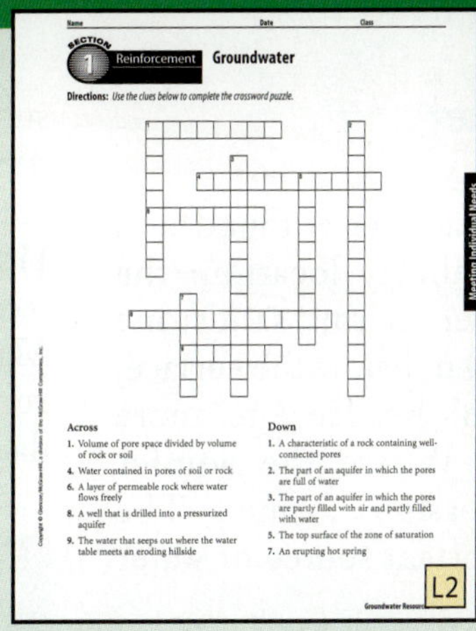

Enrichment

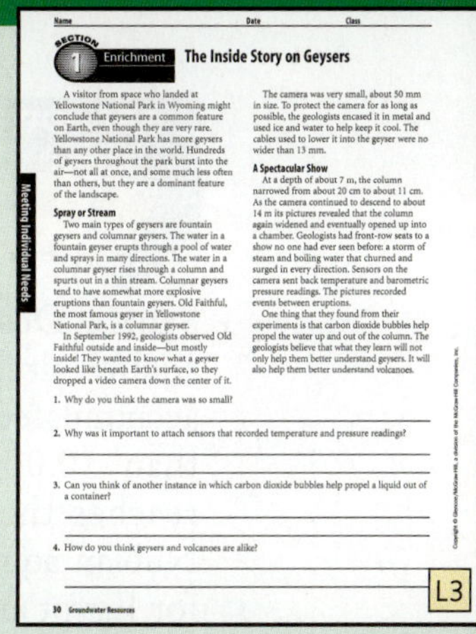

Directed Reading (English/Spanish)

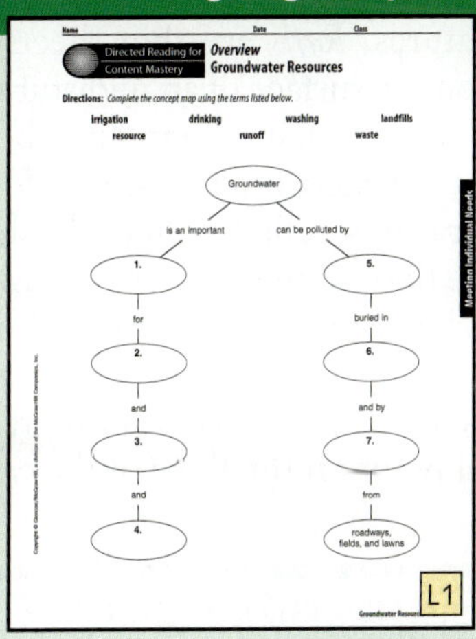

Study Guide

Reading Essentials

Assessment

Test Practice Workbook

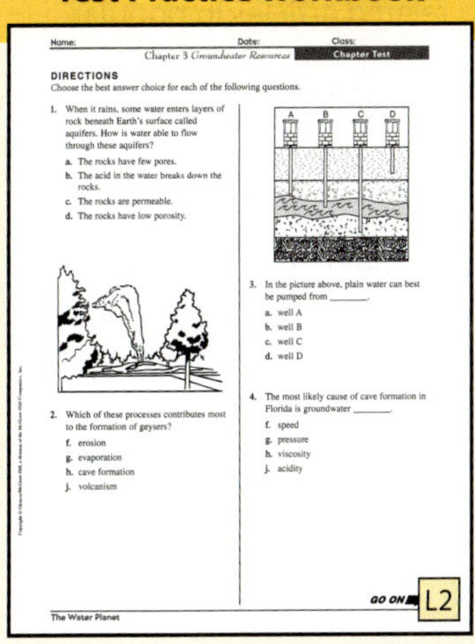

Chapter Review

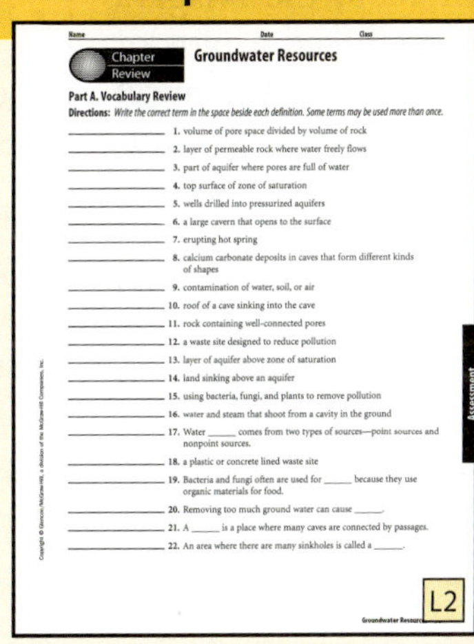

Chapter Tests

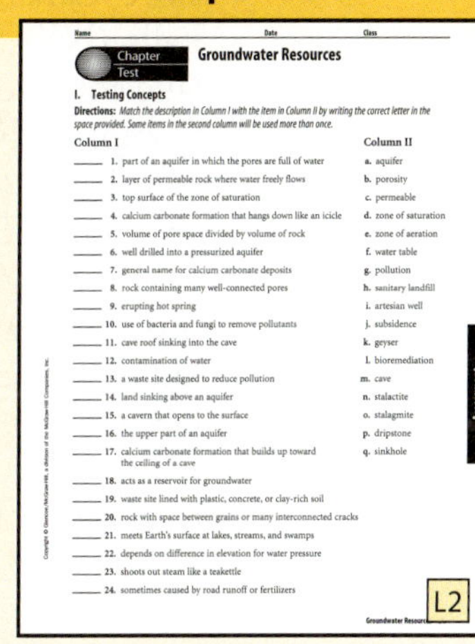

Science Content Background

section 1 Groundwater

Aquifers

The Nubian aquifer is found in a seemingly unlikely location: the Sahara Desert. The water in this sandstone aquifer, situated about 100 m below the surface, accumulated during Earth's last Ice Age, more than 10,000 years ago. Where the aquifer reaches the surface, an oasis is formed. The Nubian aquifer is an important source of water for Egypt and Libya.

Hot Springs, Geysers, and Geothermal Energy

In areas where volcanic activity is occurring, a large reservoir of hot magma heats surrounding rocks and groundwater. This source of heat, combined with a layer of highly porous and permeable rock capped by a layer of impermeable rock, provides the ideal conditions for the production of geothermal energy. In Iceland, geothermal energy provides up to 50 percent of the country's energy needs. It is used to heat water, buildings, and greenhouses.

section 2 Groundwater Pollution and Overuse

Chemical Pollution

One of the more famous examples of groundwater contamination occurred in Love Canal, New York. In 1953, the Niagara Falls (NY) School Board purchased a plot of land for one dollar from the Hooker Chemical Company. For eleven years the chemical company had used the land as a disposal site for waste chemicals. The chemical company had installed a clay cap on the waste disposal site. The school board was warned about the toxicity of the site, but developed the land and built several elementary schools there. In the 1970s, the water table rose at the site. As a result, toxic chemicals (including dioxin) seeped into homes and the schools and contaminated the groundwater. In subsequent years, many claims of illness related to the toxins came forward, none of which were conclusively shown to be caused by the toxins. Homes were vacated and the soil was treated for toxic wastes over a period that lasted until March 2000. Some of the homes have been rehabilitated and sold again. The cleanup to date has cost at least 129 million dollars.

section 3 Caves and Other Groundwater Features

Cave Features

A *speleothem* is the general name given to cave features. *Rafts* are thin sheets of calcite that float on the surface of an undisturbed cave pool. *Helicites*, calcite that has been deposited in thin, twisted branches, can form bushlike configurations called helicite bushes. Growing from the floor of the cave, these bushes can reach 2 m in height. *Flowers*, composed of a combination of calcium and sulfur, are found in the form of curved, petal-like crystals that radiate from a central point on the floor of the cave.

chapter content resources

Internet Resources

For additional content background, visit bookh.msscience.com to:

- access your book online
- find references to related articles in popular science magazines
- access Web links with related content background
- access current events with science journal topics

Print Resources

Physical Geology, by Charles C. Pummer and David McGeary, William C. Brown Publishers, 1993

Living in the Environment, by G. Tyler Miller Jr., Wadsworth Publishing Company, 1994

Groundwater and Wells, by Fletcher G. Driscoll, Smyth Co. Inc. Publishers, 1986

IDENTIFYING Misconceptions

Find Out What Students Think

Students may think that . . .

The supply of groundwater is limitless.

Groundwater is always safe to drink.

When students turn on a water faucet, they rarely stop to think about the source of the water, whether that source will run out, or whether the water is safe to drink. Those students who get their water from a groundwater source are especially likely to ignore these issues, as the source of their water is hidden beneath the surface.

Discussion

- Fill a glass with water from the drinking fountain. Tell students the source of your drinking water (groundwater aquifer, reservoir, or river). Ask students to make a drawing that shows the path they think the water followed as it moved from its source to the fountain.
- Ask a volunteer to present his or her drawing to the class. Ask others to add additional details to the description provided by the volunteer.
- When a fairly complete picture of your water system has emerged, have students discuss their ideas about how the source of the water is replenished. Ask questions such as, "What might happen to our water supplies if there were an extended drought?" "How would a large increase in the number of people utilizing our water source affect supplies?"

Promote Understanding

Activity

Before students begin Section 2, have them discuss how a groundwater source might become polluted. Explain that many groundwater sources are to some extent "self-cleaning."

- Provide each group with 2 plastic jars, 2 coffee filters, a large rubber band, clean sand and pebbles, aquarium filter floss, food coloring, water, and garden soil with bits of leaves, sticks, and other debris in it.

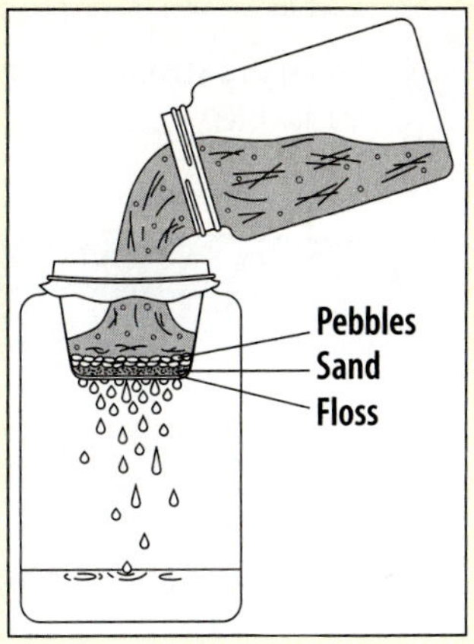

- Have students construct a filter system as shown. Next, have them place the soil and several drops of food coloring in the second jar, add water, cap tightly, and shake. Then have them slowly pour this mixture through the filtration system and observe the results.
- Have students identify what each component of the activity represents in nature. Emphasize that groundwater must be treated before drinking, and that water pumped from a well should be tested regularly for chemical contaminants.

Assess

After completing the chapter, see *Identifying Misconceptions* in the Study Guide at the end of the chapter.

chapter 3

Chapter Vocabulary

groundwater, p. 68
porosity, p. 69
permeable, p. 69
aquifer, p. 70
zone of saturation, p. 70
water table, p. 70
artesian wells, p. 73
geyser, p. 74
pollution, p. 76
sanitary landfill, p. 78
bioremediation, p. 83
subsidence, p. 84
cave, p. 85
dripstone, p. 87
sinkhole, p. 88

Science Journal Student responses will vary, but may include responses such as drinking water, water for bathing, water used for washing clothes, and water used for cooking.

INTERACTIVE CHALKBOARD with Image Bank

PowerPoint® Presentations

This CD-ROM is an editable Microsoft® PowerPoint® presentation that includes:

- a pre-made presentation for every chapter
- interactive graphics
- animations
- audio clips
- image bank
- all new section and chapter questions
- Standardized Test Practice
- transparencies
- pre-lab questions for all labs
- Foldables directions
- links to bookh.msscience.com

chapter 3

Groundwater Resources

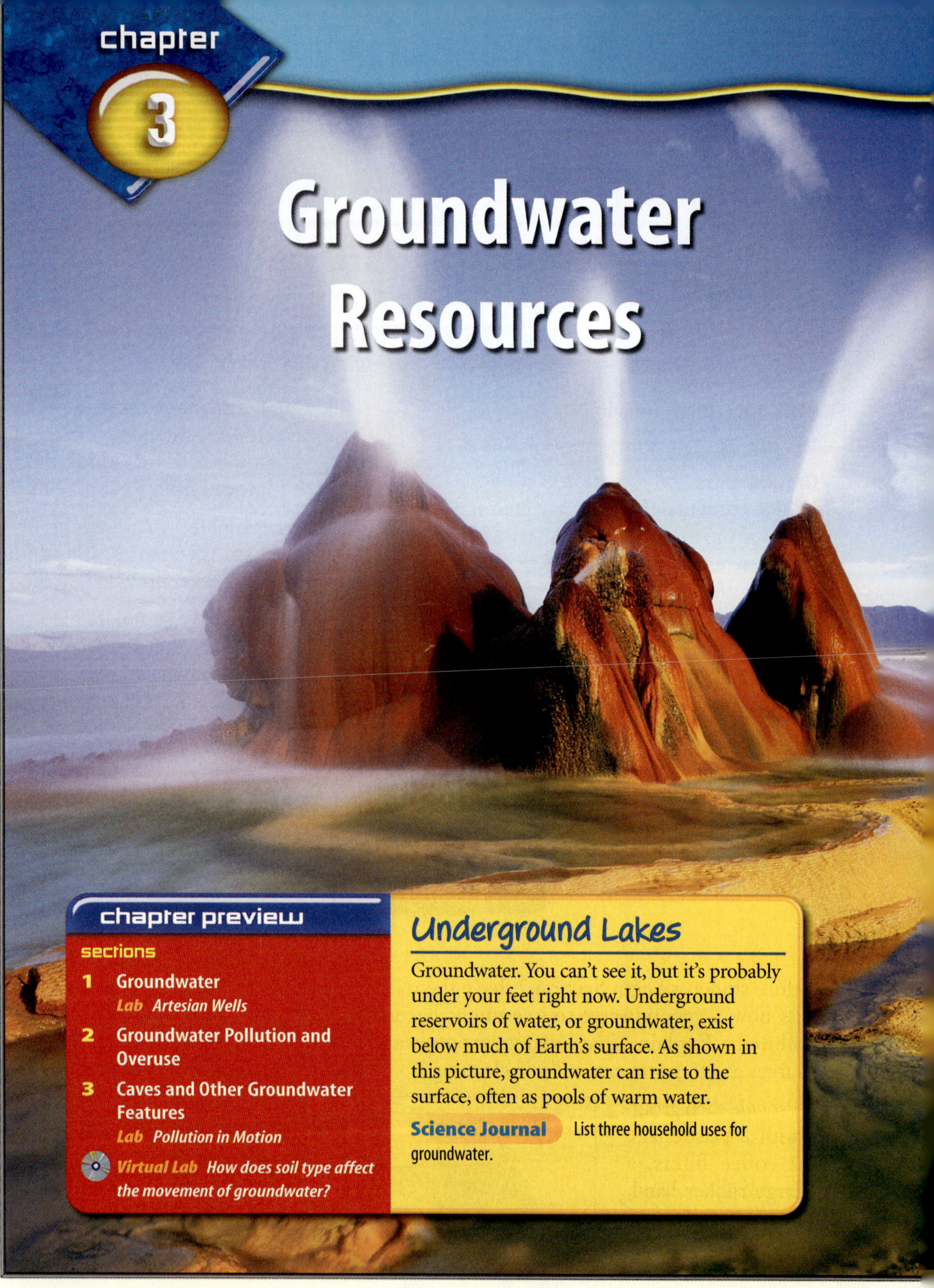

chapter preview

sections

1. Groundwater
   Lab *Artesian Wells*
2. Groundwater Pollution and Overuse
3. Caves and Other Groundwater Features
   Lab *Pollution in Motion*

Virtual Lab *How does soil type affect the movement of groundwater?*

Underground Lakes

Groundwater. You can't see it, but it's probably under your feet right now. Underground reservoirs of water, or groundwater, exist below much of Earth's surface. As shown in this picture, groundwater can rise to the surface, often as pools of warm water.

Science Journal List three household uses for groundwater.

Theme Connection

Systems and Interactions Water flowing through pores in soil and rocks forms the groundwater system. Precipitation and surface water seep into the ground, replenishing these systems. Pollution from the surface also seeps into the ground, affecting groundwater quality.

About the Photo

Hot Springs Pools of hot water like the one shown in the picture above represent multiple geologic process going on at one time. For a phenomenon like this to occur, groundwater must come into contact with heated rock. The water must also be close to the ground surface. Only certain areas of the world, such as the American northwest have these conditions.

Start-Up Activities

Launch LAB

Groundwater Infiltration

When it rains, water sinks into the ground at some locations and runs off the surface at others. What do you think causes these differences? Water soaks into the ground if there are spaces in the soil and rock. Do the following lab to investigate these spaces.

1. Immerse a dry paper towel in a large beaker of water.
2. Remove the towel. Let it drip into the beaker for 10 s.
3. Squeeze the towel into another beaker. Use a graduated cylinder to measure the volume of water held by the towel.
4. Repeat the experiment using the damp towel.
5. Repeat the experiment with a different brand of paper towel.
6. **Think Critically** In your Science Journal, record the amount of water held by each towel. Infer why the damp towels and the dry towels don't hold the same amounts of water. Infer which brand of paper towel has more pore space.

FOLDABLES™ Study Organizer

Groundwater Resources Make the following Foldable to help you identify the main concepts about groundwater resources.

STEP 1 **Collect** 2 sheets of paper and layer them about 3 cm apart vertically.

STEP 2 **Fold** up the bottom edges of the paper to form 4 equal tabs.

STEP 3 **Fold** the papers and staple along the fold. **Label** each tab as shown.

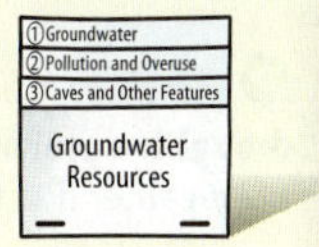

Finding Main Ideas As you read the chapter, summarize the main concepts in sections 1, 2, and 3.

Science Online — Preview this chapter's content and activities at bookh.msscience.com

Launch LAB

Purpose Use the Launch Lab to help students understand the function of pore space. Inform students that this chapter will discuss how pore space affects groundwater movement. L1 ELL LS Visual-Spatial

Preparation Obtain two different brands of paper towels.

Materials paper towels (two brands, 1 towel of each), large beakers (2), graduated cylinder, magnifying lens

Teaching Strategy Have students examine the towels closely with a magnifying lens to identify pore spaces.

Think Critically

The amount of water held by each towel will vary. Damp towels cannot hold as much water because they have less available pore space. The paper towel that holds the most water has the most pore space.

Assessment

Oral Infer how the amount of pore space in soils and rocks affects their ability to absorb water. The more pore space rocks and soils have, the more water they can absorb. Use **Performance Assessment in the Science Classroom**, p. 89.

FOLDABLES™ Study Organizer **Dinah Zike Study Fold**

Student preparation materials for this Foldable are available in the Chapter *Fast File* Resources.

1 Motivate

INTERACTIVE CHALKBOARD
PowerPoint® Presentations

Bellringer

Section Focus Transparencies also are available on the Interactive Chalkboard CD-ROM. L2 ELL

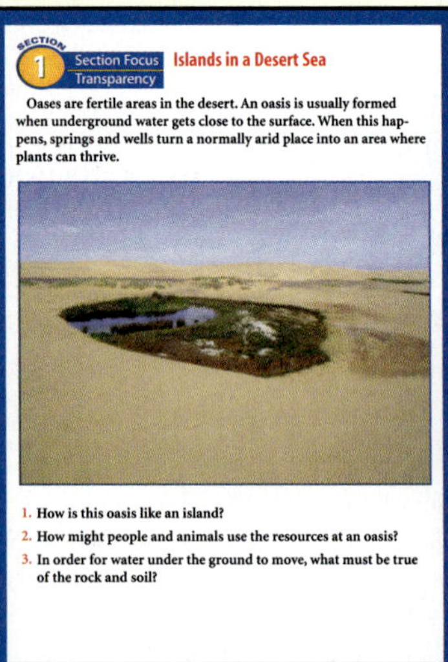

Tie to Prior Knowledge

Rainwater Runoff Have students recall watching rain fall onto a field or lawn. Ask students where the water goes when it doesn't become runoff. It soaks into the soil. Tell students that in this section they will find out what happens to the water underground and how people use it.

section 1

Groundwater

as you read

What You'll Learn

- **Explain** the importance of groundwater.
- **Describe** how groundwater moves.
- **Compare and contrast** the formation of springs, artesian wells, and geysers.

Why It's Important

Groundwater is one of Earth's most important resources for drinking, washing, and irrigation.

Review Vocabulary

drought: a prolonged period of dry weather due to a lack of rain

New Vocabulary

- groundwater
- porosity
- permeable
- aquifer
- zone of saturation
- water table
- artesian well
- geyser

Importance of Groundwater

What do washing clothes, brushing teeth, and taking a bath have in common? These activities wouldn't be possible without freshwater from Earth. Look at the man in **Figure 1.** He is pumping water from deep underground. Every day, people depend on groundwater and other sources of freshwater for drinking, cooking, washing dishes, flushing toilets, and watering lawns. It's used in swimming pools, in industrial plants, and to irrigate crops.

One of the most important sources of freshwater is groundwater. **Groundwater** is water contained in the open spaces, or pores, of soil and rock. There is about 30 times more groundwater than all the surface freshwater on Earth. Water in streams and lakes are examples of surface freshwater. In the United States, 23 percent of the water used by people comes from groundwater. And, more than half of all drinking water in the United States comes from underground sources. Almost everyone who lives in rural areas uses groundwater for drinking and other household uses. If the human population continues to increase, even more groundwater will be used.

Figure 1 Every day each person in the United States directly or indirectly uses more than 5,600 L of water.

Section 1 Resource Manager

Chapter FAST FILE Resources

- Transparency Activity, p. 44
- Directed Reading for Content Mastery, pp. 19, 20
- MiniLAB, p. 3
- Note-taking Worksheets, pp. 33–35
- Enrichment, p. 30
- Lab Activity, pp. 9–12
- Reinforcement, p. 27

Mathematics Skill Activities, p. 1

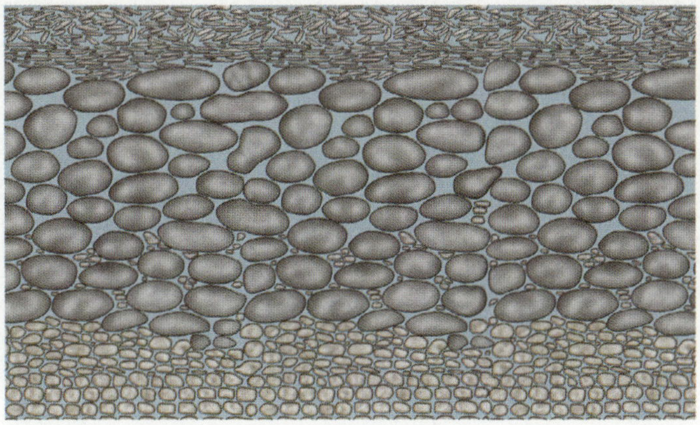

Figure 2 Pore sizes and shapes vary in different kinds of rocks.

Water Beneath Earth's Surface

Unlike Earth's surface water, groundwater normally is not found in pools or channels. Instead, it flows through a series of interconnected pores in soil and some rock. As you learned if you did the Launch Lab, the amount of pore space determines how much water can be held. The pores in soils are the spaces between the fragments and pieces of rock. The pores in rocks are the spaces between the grains and crystals, and also cracks in the rock. Pore size varies, as shown in **Figure 2.** Rocks such as sandstone and loose deposits of sand and gravel can hold large amounts of groundwater.

Porosity Not all soil and rock have the same amount of pore space. The volume of pore space divided by the volume of a rock or soil sample is its **porosity.** Soil or rock with many large pores has high porosity. Even soil or rock with small openings between grains can be porous. But soil or rock with few, small pores has low porosity.

Reading Check *What is porosity?*

Permeability When soil or rock has high porosity and the pores are interconnected, water passes through it easily. The ability of rock and soil to transmit water and other fluids is called permeability. Rock that is **permeable** contains many well-connected pores or cracks and allows groundwater to flow through it. Although limestone might contain very little space between grains, it usually has many interconnected cracks. Therefore, limestone is permeable. Impermeable rock, such as shale and granite, has few pores or pores that are not well connected. Groundwater cannot pass through impermeable rock.

Mini LAB

Measuring Porosity

Procedure

1. Put 100 mL of dry **gravel** into a **250-mL beaker** and 100 mL of dry **sand** into **another 250-mL beaker.**
2. Fill a **graduated cylinder** with 100 mL of **water.**
3. Pour the water slowly into the beaker with the gravel. Stop pouring the water when it just covers the top of the gravel.
4. Record the volume of the water that is used. This amount is equal to the volume of the pore spaces.
5. Repeat steps 2 through 4 with the beaker containing the sand.

Analysis

1. Which substance has more pore space—gravel or sand?
2. Using the formula below, calculate the porosity of each sediment sample.

$$\frac{\text{Volume of pore spaces}}{\text{100 mL of sediment}} \times 100 = \text{Percent porosity}$$

Active Reading

Pair of Pairs This strategy encourages students to brainstorm together, recording on one paper their ideas. After a few minutes, they combine with another pair and share responses. Finally, groups of four share responses. Have students use the Pair of Pairs strategy to answer the Analysis questions in the MiniLAB.

2 Teach

Reading Check

Answer the volume of pore space divided by the volume of a rock or soil sample

Mini LAB

Purpose Students contrast the pore space of different materials. L1 LS **Logical-Mathematical**

Materials 250-mL beakers (2), 100 mL dry sand, 100 mL dry gravel, 100-mL graduated cylinder, water

Teaching Strategy Remind students that they will have to subtract the final volume reading on the graduated cylinder from the initial reading of 100 mL to find the volume of water used.

Troubleshooting Sand and gravel should be dry and of fairly uniform grain size for best results.

Analysis

1. Answers will vary depending on type of sand and gravel used. In general, gravel has relatively few, but large spaces. Sand has many small spaces. For this reason, their porosities may be similar.
2. Answers will vary with student results. Review how to determine percent.

Assessment

Content Have students draw magnified images of the pore spaces and sediments in their two beakers. Use **Performance Assessment in the Science Classroom,** p. 127.

Make a Model

Aquifers Have students make model aquifers by doing the following: Place a piece of aluminum foil on the desktop. Position a sponge on top of the foil. Use a small beaker to pour water slowly onto the sponge. Lift the sponge and examine the foil. Lift the foil, and examine the desktop. Compare the sponge to a porous layer of soil or rock and the foil to an impermeable rock layer.

Caption Answer

Figure 3 There is less precipitation in the west.

IDENTIFYING Misconceptions

Groundwater Supply Students may think the supply of groundwater is limitless and always safe to drink. Refer to page F at the beginning of this chapter for teaching strategies that address this misconception.

Discussion

High Water Table How might the water level in a stream be affected by a heavy rainfall if the water table is very high? The stream is more likely to run full and overflow if the water table is high.

Quick Demo

Make a Geyser

Materials Large test tube, baking soda to represent hot rock, vinegar to represent ground water

Estimated Time 10 minutes

Procedure Pour some baking soda into the test tube. Pour some vinegar over the baking soda. Watch as the mixture bubbles and overflows. Explain that as the underground pressure relieves the process begins again.

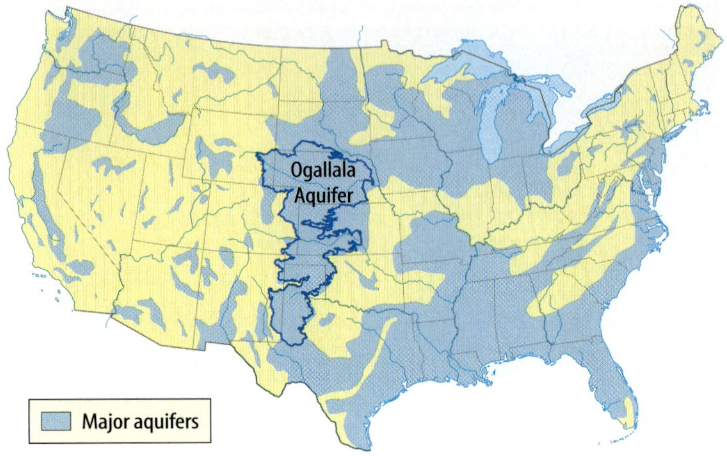

Figure 3 This map shows the major aquifers in the United States. The large Ogallala Aquifer stretches from the Texas Panhandle to South Dakota.
Explain *why large aquifers are found throughout the eastern two-thirds of the country while smaller ones are found in the west.*

Aquifers

When it rains or when snow melts, water seeps into permeable soil and rock. You might wonder how deep it can go. Groundwater will keep moving down until it reaches an impermeable layer of rock or sediment. When this happens, the impermeable layer acts like a dam. The water stops seeping downward and begins filling up the pores in the rock and soil above the impermeable layer.

An **aquifer** is a layer of permeable rock through which water flows freely. Aquifers act as reservoirs, or storage areas, for groundwater. **Figure 3** shows some of the larger aquifers in the United States. Sand, gravel, sandstone, porous limestone, and highly fractured bedrock of any type can make good aquifers. Shale, mudstone, clay, or other impermeable rocks or sediments that don't contain fractures are not good aquifers. Layers made up of these materials are called aquitards. Aquifers and surface water occur in watersheds, which are large regions drained by a particular river system.

Science Online

Topic: Groundwater Interactions

Visit bookh.msscience.com for Web links to information about the relationship between groundwater and surface water in a watershed.

Activity List major rivers in the Ogallala aquifer region.

The Water Table

The upper part of an aquifer is called the zone of aeration. The pores in this area are partially filled with water, mainly concentrated around the surfaces of grains and fractures. The rest of the pore space in the zone of aeration is filled with air. Below this zone is the **zone of saturation,** where the pores of an aquifer are full of water. The top surface of the zone of saturation is the **water table. Figure 4** shows that the water table meets Earth's surface at lakes, streams, and swamps. Much of the water in some streams is groundwater that has flowed into the stream channels. How deep below ground the water table is depends on rainfall. During heavy rainfall, the water table rises toward Earth's surface. During a drought, the water table falls, and some streams and ponds might dry up.

Differentiated Instruction

Learning Disabled Fill a plastic shoe box or small aquarium with sand. Scoop out a deep depression to represent a swamp. Pour colored water so that it fills the box to the bottom of the "swamp." Mark the side of the box to identify the "water table." Have students identify the "zone of saturation." Add more water and have students observe what happens to the "water table" and "swamp."

Challenge Challenge students to do research and compare maps of the United States large aquifers and annual precipitation in major cities. Students should notice that large aquifers typically occur where rain is most prevalent. Students can design a map showing the location of large aquifers and a contour map showing rainfall rates across the country. Overlay the two maps. P

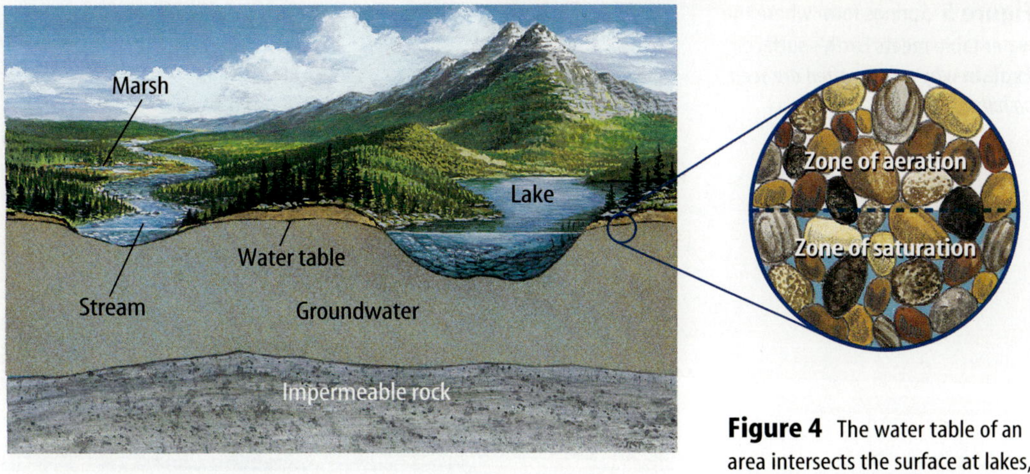

Figure 4 The water table of an area intersects the surface at lakes, streams, and other wetlands. **Describe** *some relationships between groundwater and surface water in a watershed.*

Groundwater Flow

Water below the surface doesn't just sit there—it's on the move. How fast groundwater moves depends on the permeability of the rock and soil that it flows through and the groundwater's pressure. Where permeability is high, groundwater speed is faster. Where permeability is low, its speed is slower. Groundwater pressure is caused by gravity. Groundwater flows from higher-pressure regions toward lower-pressure regions.

You can imagine the shape of the water table as being much like the surface of Earth, with hills and valleys in some areas. In fact, the water table may mimic the shape of the land surface, with higher levels located under hills and lower levels located under valleys. Just as water in a river flows from higher elevations to lower elevations, groundwater flows from where the water table is relatively high to where it is low. The greater the slope of the water table is, the faster groundwater flows. Groundwater can flow toward or away from streams or lakes, depending on the slope of the water table.

Speed of Groundwater Flow Water in rivers or streams flows at speeds ranging from about 3 km to 18 km per hour. Groundwater moves much more slowly. What would happen if you could move only a few centimeters per day? It might take you a year to get from your bedroom to the breakfast table. At that rate, it would take you the rest of your life to walk to school. It's hard to picture something moving so slowly, but that's about the average speed of groundwater. The speed of groundwater moving through an aquifer ranges from about 1.5 m per year to 1.5 m per day. Its speed averages about 2 cm per day.

INTEGRATE Chemistry

Hard Water Water in some aquifers contains dissolved calcium and magnesium. It is called hard water because it is hard to form suds or lather with soap. Find out if you live in an area that has hard water and how water is treated to make it soft.

Caption Answer

Figure 4 The surface of a lake or river is the water table; groundwater often feeds rivers or lakes through springs.

IDENTIFYING Misconceptions

Ground Composition Students may think that the ground under their feet is solid and completely dry. Emphasize that even "dry" ground may have groundwater trapped in it.

Use an Analogy

Groundwater Speed In one day, the average distance groundwater travels is equal to the length of a human thumbnail.

Fun Fact

In the United States, groundwater sources supply approximately 34 percent of the water used for agriculture.

INTEGRATE Chemistry

Hard Water Students can contact the local water utility to find out whether the water is considered hard. One method used to soften water is to treat the water with soda ash and lime.

Differentiated Instruction

English-Language Learners In pairs, have students try to draw a generalized hydrologic cycle. Help students to illustrate that water falls as precipitation and collects in streams, lakes, oceans, or infiltrates as groundwater. Water collects in surface water bodies and evaporates to form clouds. The clouds condense and form precipitation, completing the cycle.

Teacher FYI

U.S. Freshwater The largest source of freshwater in the United States is groundwater. It is estimated that more than 95 percent of all freshwater in the United States is stored in porous rocks and sand.

Caption Answer

Figure 5 On a hillside, Earth's surface might intersect the water table.

Discussion

Spring Location Is a spring more likely to be located high or low on a hillside? Explain. Low; the water table is more likely to meet Earth's surface near the bottom of a hill.

Reading Check

Answer place where the water table meets the surface and water seeps out of the ground

Activity

Bottled Water Have students research different brands of spring water sold at a local grocery. What is the source of the water? What do the labels claim? Are they misleading? How is the water monitored to ensure safety? Have students work in pairs and report what they find to the rest of the class. Invite students to use their newly discovered information to debate the benefits and disadvantages of drinking bottled spring water. COOP LEARN LS Interpersonal

Figure 5 Springs form where the water table meets Earth's surface. **Explain** *why springs often are seen on hillsides.*

Springs and Wells

How do people obtain groundwater to use? In some regions, groundwater flows freely out of the ground onto the surface. Where the water table meets Earth's surface, water seeps out to form a spring, shown in **Figure 5.** Some communities get their water from flowing springs.

Reading Check *What is a spring?*

Digging Wells Where springs do not occur, wells must be dug or drilled to obtain water. Most modern wells are drilled using drill rigs mounted on large trucks like the one in **Figure 6.** A drill head, or bit, is placed at the end of a metal pipe. As the bit drills deeper, more pipe is added at the top. Wells more than 300 m deep can be drilled in this way.

Figure 6 This truck has drilling equipment that can drill deep water wells in solid rock.

The hole, or shaft, for the well must be drilled into the zone of saturation. The lower end of the shaft is lined with screens. Then a well casing—a pipe that has slots on the lower part—is set in place inside the hole. The slots allow water in, but keep sand and gravel out. The well casing extends about 0.3 m above ground. The upper part of the shaft around the casing and the surface area around the shaft are sealed with concrete. This prevents surface water from flowing into the well. After such a well has been dug, water can be pumped to the surface. At one time, windmills or hand-operated pumps were used to pump water. Today, most wells have electric-powered pumps.

Curriculum Connection

Health Hot springs have been used for thousands of years to help remedy health problems. Some people claim that hot springs can cure everything from sore muscles and arthritis to digestive ailments and dry skin. Have students research why President Franklin Delano Roosevelt frequented Warm Springs, Georgia. Roosevelt hoped to rebuild the muscles in his legs that had been affected by polio.

Teacher FYI

World Water Distribution Many people do not have enough water. The daily amount needed per person is about 80 L, but three-fourths of the world's people have access to just 50 L per day. Many Kenyans have access to only 5 L per day. The average American uses over 600 L per day.

Artesian Springs You learned earlier that pressure on groundwater is caused by gravity. Pressure on groundwater can be high when water is trapped between formations of rock or sediment. Where aquifers are trapped between two layers of impermeable material, water in the sandwiched aquifer is under pressure. The impermeable layers act like the wall of a pipe. Water enters the sandwiched aquifer at the highest part of the aquifer, which puts pressure on water at lower elevations. If a crack or fault cuts through the impermeable layers, groundwater from the aquifer rises toward the surface along the crack. This groundwater can form artesian springs. Artesian springs are common in the Great Plains of the United States, as well as in Australia and northwestern Africa.

Artesian Wells Some communities get water from wells without using pumps. These wells, called **artesian wells,** are drilled into pressurized aquifers, as shown in **Figure 7.** The amount of water pressure depends on the difference in elevation between the highest part of the aquifer and the well. The greater this difference in elevation, the greater the pressure. However, some artesian wells do require pumping. Sometimes the top of the well is above the surface to which water under pressure will rise. Then water will not flow freely from the well.

Artesian springs and wells helped shape the history of the United States. The largest aquifer in the United States covers more than 450,000 km^2. Look back at **Figure 3.** Notice that the Ogallala Aquifer stretches from the Texas Panhandle northward to South Dakota. Using artesian springs and wells from this aquifer, pioneers had easy access to water. Early settlements developed around these water supplies.

Science Online

Topic: Springs

Visit bookh.msscience.com for Web links to information about hot springs.

Activity Prepare a travel brochure about one of the sites that would encourage people to visit the springs.

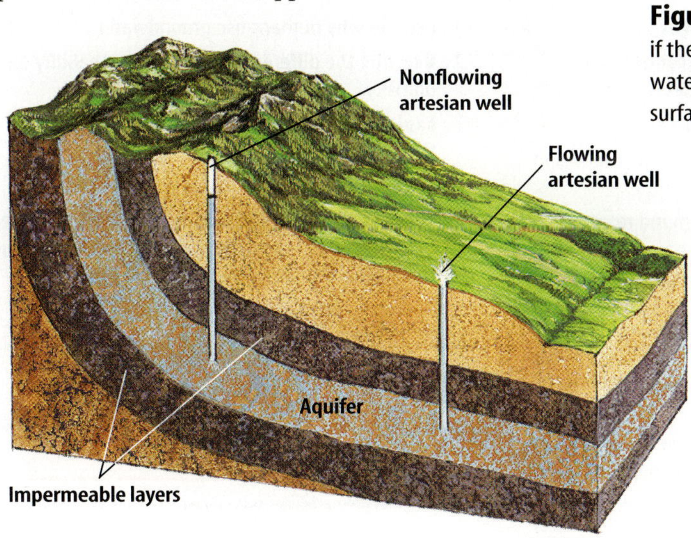

Figure 7 In an artesian well, if the pressure is great enough, water will be forced up to the surface and into the air.

Use Science Words

Word Usage The term *artesian well* originated in the French region of Artois, where the oldest artesian well in Europe was bored in 1126. Have students use *artesian well* in a sentence describing its characteristics. Possible answer: An artesian well is a well in which water under pressure rises toward the surface.

Discussion

Digging a Well What would you want to know about an area before digging a well for your drinking water? Possible answers: the water quality of the aquifer; the existence of nearby water-polluting facilities; the depth of the water table

Virtual Labs

Soil Type *How does soil type affect the movement of groundwater?*

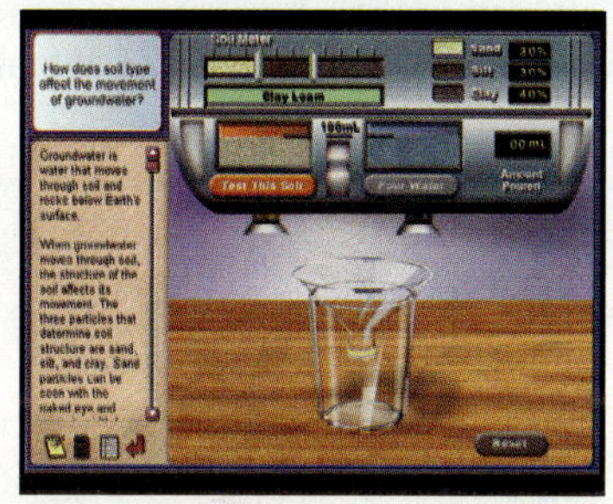

Visual Learning

Figure 7 Ask students why the mountains are an important part to the formation of artesian wells. Mountains provide a water infiltration area higher than the surrounding lowlands. The difference in elevation creates a pressure differential that can create artesian wells and springs.

Science Journal

Ogallala Aquifer Have students read an article about the Ogallala aquifer and then write an article in their Science Journals explaining its importance for drinking water and agriculture in certain parts of the United States. Encourage students to draw a map that shows the aquifer and includes some of the major cities that draw water from it. L2 LS **Visual-Spatial**

SECTION 1 Groundwater H ◆ 73

Reading Check

Answer a hot spring that erupts periodically, shooting water and steam into the air

3 Assess

DAILY INTERVENTION

Check for Understanding

Logical-Mathematical Using a large, two-column chart, brainstorm with the class how to conserve water at home and at school. Discuss community-wide procedures to reduce water consumption.

Reteach

Modeling Have groups of students make models that show how the position of the water table affects the formation of a well, geyser, hillside spring, stream, lake, hot spring, and artesian well. L2 ELL LS **Visual-Spatial**

Assessment

Process Have students compare and contrast wells, geysers, and hot springs. All bring water to Earth's surface. In most wells, water must be pumped to the surface. In geysers and hot springs, the water is hot. A geyser is a hot spring that periodically erupts. Use **Performance Assessment in the Science Classroom,** p. 89.

Figure 8 Old Faithful is the most famous geyser in the United States. The time between its eruptions can vary from 35 min to 120 min. As it erupts, it shoots out about 40,000 L of water and steam high into the air.

Geysers

Molten material below Earth's surface can reach temperatures above 650°C. Rock in contact with molten material can heat groundwater to high temperatures. Water heated naturally in this way can form geysers, such as the one shown in **Figure 8.** A **geyser** is a hot fountain that erupts periodically, shooting water and steam into the air.

Reading Check *What is a geyser?*

Geyser Eruptions Below the surface, a network of fractures and cavities fills with groundwater. Groundwater is heated to temperatures higher than the boiling temperature at the surface. The water expands, forcing its way out at the top of a cavity. As some of the water escapes, the pressure inside the cavity drops. This allows the remaining water to boil quickly because of the drop in pressure. Much of the water turns to steam. The steam shoots out the opening, much like steam out of a teakettle, forcing the remaining water out with it. Some of the geyser water reenters the channels to begin the process again. Although geysers erupt repeatedly, many are not entirely predictable. Old Faithful in Wyoming's Yellowstone National Park is more predictable than most geysers.

Groundwater is an important water resource, but in many locations, it is polluted. In the next section, you'll learn about the sources of pollution and how they can be cleaned up.

section 1 review

Summary

Importance of Groundwater

- As the human population continues to increase, more groundwater is used.

Water Beneath Earth's Surface

- Groundwater is contained in open spaces in rocks called pores.
- Aquifers are large underground reservoirs of water.
- Groundwater moves much more slowly than surface water.

Springs, Wells, and Geysers

- Wells are used to collect groundwater from deep underground.
- Geysers shoot hot water from the ground when water comes in contact with molten material.

Self Check

1. **Explain** why humans use groundwater.
2. **Describe** the difference between permeability and porosity.
3. **Explain** why some rocks are more permeable than others.
4. **Summarize** the eruption process of a geyser.
5. **Think Critically** What happens to the water table if the amount of water pumped from an aquifer is greater than the amount of water seeping in?

Applying Math

6. **Use Percentages** Of the 289.2 billion liters of groundwater used in the United States each day, 185.5 billion liters are used for irrigation. Calculate the percentage of groundwater used for irrigation.

Science Online bookh.msscience.com/self_check_quiz

section 1 review

1. because it is a major source of freshwater for humans
2. Permeability is the ability of water to pass through soil and rock. Porosity is the volume of pore space divided by the total volume of a rock or soil sample.
3. some have many connected pores, while others do not
4. Water at the bottom of a fracture in rock is heated by a nearby magma source. The water expands, forcing some water out at the surface. As the water escapes, pressure inside the fracture drops, causing some of the remaining water to quickly turn to steam. Hot water and steam shoot out of the fracture. The fracture then refills and the process repeats.
5. The water table drops.
6. 185.5 billion L/289.2 billion L × 100% = 64.1%

LAB

Artesian Wells

How much water pressure exists at an artesian well? This depends on where it is drilled. The difference between the elevation of the top of an aquifer and the top of the artesian well determines the water pressure. In this lab, you'll observe some variables that affect the difference in elevation and the resulting water pressure.

Real-World Question

How do the amount of water in an aquifer and the position of an artesian well affect water pressure at an artesian well?

Goals

- **Model** an aquifer with artesian wells.
- **Observe** and infer how variables affect the water pressure at artesian wells.

Materials

tall, soft-plastic bottle with cap
nail or awl
hammer
protractor
sink with water
plastic straw
scissors
metric ruler

Safety Precautions

WARNING: *Be careful not to pierce yourself with the nail or awl when making the holes in the plastic bottle.*

Procedure

1. With the hammer and nail, or an awl, carefully poke two holes in the same side of a plastic bottle. One hole should be 3 cm from the bottom of the bottle, and the other should be 10 cm from the bottom.
2. Cut two pieces of plastic straw, each about 4 cm long. Insert one piece of straw in each of the holes.
3. Hold the bottle over the sink. While your partner covers both straws, fill the bottle with water.
4. Keep the bottle in a vertical position and uncover both holes. Observe and record in your Science Journal what happens at both straws as the water level drops.
5. Repeat the experiment, but this time tilt the bottle at a 30-degree angle from vertical, keeping the straws on the upper side of the bottle. Observe and record what happens.

Conclude and Apply

1. Which part of your bottle apparatus represents the impermeable rock layers? The aquifer? Artesian wells?
2. How does the amount of water in an aquifer affect the water pressure at an artesian well?
3. How does tilting the aquifer affect the water pressure at an artesian well?
4. How does the position of an artesian well in an aquifer affect the pressure of the water at the well?

Communicating Your Data

Make a diagram of your bottle apparatus. Compare your diagram with those of other students.

BENCH TESTED

Real-World Question

Purpose Students make a model of an artesian well and infer how variables affect well pressure. L2 ELL LS **Kinesthetic**

Process Skills make models, identify and manipulate variables, compare and contrast, draw conclusions

Time Required 40 minutes

Alternate Materials If sink space is limited, students can work in dish pans and use beakers of water to fill the plastic bottles.

Procedure

Safety Precautions Caution students to use care when handling the nail or awl.

Teaching Strategy Review with students how artesian wells work before beginning the activity.

Troubleshooting Secure straws with clay to prevent water from leaking from the bottle.

Conclude and Apply

1. rock—walls of the container; aquifer—water; well—straws
2. The greater the amount of water, the greater the water pressure.
3. The greater the angle, the greater the water pressure.
4. The lower the well in relation to the top of the aquifer, the greater the water pressure.

Assessment

Process Have students design an experiment to see if the temperature of water in an aquifer affects water pressure at an artesian well. Use **Performance Assessment in the Science Classroom,** p. 95.

Communicating Your Data

Suggest that students use word processing programs and graphics programs to prepare their diagrams.

section 2 Groundwater Pollution and Overuse

1 Motivate

INTERACTIVE CHALKBOARD PowerPoint® Presentations

Bellringer

Section Focus Transparencies also are available on the Interactive Chalkboard CD-ROM. L2 ELL

Section 2 Section Focus Transparency **Land Slick**

What happens when oil or gasoline is spilled on the ground? These products aren't easily broken down, and when it rains, they can wash off the roads and into the soil. The rings in the puddle below are petroleum products floating in rainwater.

1. What do you think happens to gas and oil after they are washed off the roads?
2. What other dangerous substances might seep through the soil?

Groundwater Resources

Tie to Prior Knowledge

Polluted Groundwater Ask students to think back to how rainwater seeps into the ground to become part of the groundwater system. What would happen to the supply of groundwater if the water that seeped into the ground was polluted? Groundwater also would become polluted. Why might cleaning up polluted groundwater be difficult? Possible answer: Access is difficult because the system is underground.

as you read

***What* You'll Learn**

- **Identify** the sources of groundwater pollution.
- **Explain** the ways groundwater is cleaned.
- **Explain** what happens when too much water is pumped from an aquifer.

***Why* It's Important**

Good health depends on having clean water.

Review Vocabulary
toxic: pertaining to or caused by poison

New Vocabulary
- pollution
- sanitary landfill
- bioremediation
- subsidence

Sources of Pollution

It's easy to spot some kinds of water pollution. Polluted rivers and lakes can smell bad or have oil slicks. However, it's harder to detect the many chemicals that find their way into groundwater from polluted surface water and soil. One reason so many chemicals get into groundwater is that many substances dissolve in water that enters the groundwater system.

Pollution comes in many forms and from many different sources. **Pollution** is the contamination of soil, water, air, or other parts of the environment by something harmful. Water pollution comes from two types of sources—point sources and nonpoint sources. An example of point-source pollution is waste dumped directly into water from a pipe or a channel. Nonpoint-source pollution originates over larger areas, such as roadways, shown in **Figure 9,** agricultural areas, and industrial sites. Most groundwater pollution comes from nonpoint sources. These pollutants can seep into soil and eventually travel down into aquifers. **Figure 10** shows some examples of where groundwater pollution comes from.

Figure 9 Salt spread on this road might eventually seep down into groundwater.

Section 2 Resource Manager

Chapter *FAST FILE* Resources
- Transparency Activity, p. 45
- Directed Reading for Content Mastery, p. 20
- MiniLAB, p. 4
- Enrichment, p. 31
- Reinforcement, p. 28
- Lab Activity, pp. 13–16

Reading and Writing Skill Activities, p. 23
Life Science Critical Thinking/Problem Solving, p. 6
Home and Community Involvement, p. 23

NATIONAL GEOGRAPHIC

VISUALIZING SOURCES OF GROUNDWATER POLLUTION

Figure 10

Pollutants can enter the groundwater of an aquifer in a number of different ways. Pollutants can come from factories, farms, or homes. The three-dimensional cutaway below shows some of the major sources of groundwater pollution.

Salt used to melt ice and snow on roads can seep into groundwater.

Toxic substances from mines can dissolve in rain-water or melted snow and seep into groundwater.

In cities, sewer lines can break or leak. Pipelines that leak oil or other toxic substances also contribute to groundwater pollution.

Fertilizers and other chem-icals used on cropland wash into the soil when it rains and eventually enter groundwater.

Chemicals and bacteria from leaking landfills can pollute groundwater.

Gasoline and other toxic chemicals oozing from underground storage tanks can poison ground-water supplies.

Some septic systems contaminate groundwater with harmful bacteria.

Animal wastes from feed lots and farms can pollute groundwater.

NATIONAL GEOGRAPHIC

Visualizing Sources of Groundwater Pollution

Have students examine the pictures and read the captions. Then ask the following questions.

How is people's use of chemicals related to the pollution of groundwater? Possible answer: Except for septic systems, sewer lines, and animal wastes, most groundwater pollution comes from chemicals that people use to grow crops, power vehicles, manufacture products, and make travel safe (for example, using salt on icy roads).

Could any activities in your own community add to the pollution of groundwater? Explain. Possible answer: If students live in a rural area they might mention the use of fertilizers and other chemicals on crops; from animals wastes; from septic systems. If they live in cities or suburbs they might mention leaking gasoline in underground storage tanks; leaking sewer lines; salt used on highways to melt snow and ice; or leaking landfills.

Activity

Board Game Have small groups of students design and make a board game that deals with sources and prevention of groundwater pollution.

Teacher FYI

Decreasing Water Pollution The Safe Drinking Water Act established federal standards to ensure that drinking water is safe. The Clean Water Act allocates money to states for building sewage and wastewater treatment facilities to clean effluent that flows into surface waters. The United States Environmental Protection Agency oversees the enforcement of such environmental laws. As a result of these regulations, water pollution in the United States has decreased.

Discussion

Effects of a Landfill How might a landfill affect drinking water taken from a private well? If the landfill is not a sanitary landfill, wastes could seep into the ground, polluting the groundwater from which the well water is drawn.

Quick Demo

Oil Pollution

Materials 250-mL beaker, water, vegetable oil

Estimated Time 10 minutes

Procedure Pour equal amounts of water and vegetable oil into the beaker. Observe the separation. Shake the mixture and observe again. Explain that the vegetable oil represents a gasoline spill and the water is a drinking water source. Have students infer what might happen if a well was drilled into an area where a gasoline spill had occurred.

Figure 11 About 55 percent of the solid wastes collected in cities and towns is put in sanitary landfills. The landfills must be checked constantly for leaks so that the soil and the groundwater below it are not polluted.

Landfills Most of the materials that people throw away end up in landfills. These wastes include food wastes, paper, metals, oils, sprays, batteries, baby diapers, cleaners, cloth, and many other items. Some of these substances are poisonous and cause cancer. Because landfills are dug into the ground when they are constructed, there is a possibility that landfill waste can seep into the ground. To help reduce groundwater pollution, communities construct sanitary landfills, shown in **Figure 11.**

In a **sanitary landfill,** there is a lining of plastic or concrete, or the landfill is located on clay-rich soils that trap the liquid waste. Although sanitary landfills greatly reduce the chance that hazardous substances will leak into the surrounding soil and groundwater, some still gets into the environment. The best solution to limiting the hazardous substances getting to groundwater is to reduce the waste disposed in landfills. You can help by contacting your local waste management office about where to safely dispose of your family's hazardous wastes. Examples of household hazardous wastes are insect sprays, weed killers, batteries, drain cleaners, bleaches, medicines, and paints.

Road Runoff and Waste Spills Another source of groundwater pollution is runoff from roadways and parking lots. When cars leak oil or gasoline is spilled, groundwater pollution can result. Another risk to groundwater involves the transportation of toxic materials along roadways and railways. Many communities require roadway carriers of such materials to use outer belt freeways. Outer belts often are located far enough from towns and cities that a spill would not pose as much of a risk to city wells.

LAB DEMONSTRATION

Purpose to model road runoff

Materials plastic shoe box, sand, metric ruler, water, scissors, food coloring, cardboard, aluminum foil

Preparation Cut the cardboard to make a 3-cm × 20-cm long strip. Cover with foil.

Procedure Fill the box with sand and half-fill it with water to model an aquifer. Place the strip on top of the sand to model a road. Place several drops of food coloring on the "road." Slowly pour water onto the "road" to simulate precipitation.

Expected Outcome Food coloring flows with the water into the "aquifer."

Assessment

How do salts, oil, gasoline, and other toxic materials contaminate groundwater? When it rains or snow melts, these materials become part of runoff that seeps into soil.

Road Salt Many states use salt to melt ice and snow on roads and highways. Often, salty water runs off roads, where it seeps into the soil and eventually reaches the groundwater. Snowplowing speeds up the process when plows push salt-containing snow to the sides of roads where it melts and more readily seeps into the ground.

Storage Tanks Gasoline and other substances often are stored in large underground tanks. Storing hazardous materials underground can cause groundwater contamination if the tanks leak. This is being reduced by using double-walled containers for hazardous substances that resist underground weathering. If left to rust or corrode, leaking pipelines and underground storage tanks pose a risk to aquifers. Federal laws now require that all underground storage tanks be monitored continuously for leaks. If a leak is found, action must be taken for cleanup. Sometimes, this involves removing the tank from the ground.

Septic Systems About one-third of all homes in the United States dispose of their wastewater through septic systems. Septic systems consist of a septic tank and an absorption area, shown in **Figure 12.** Solids are removed from the wastewater in the tank. Pipes carry wastewater from the tank to the absorption area. The absorption area then uses the soil and crushed rock to filter and treat the water before it reaches the water table.

If septic systems are not properly designed or maintained, sewage can get into surrounding soil and rock. As it migrates, sewage can carry harmful bacteria. If the bacteria get into drinking water, they can make people ill. Proper installation of septic systems and pumping out tank wastes on a regular basis greatly reduce the chances that groundwater will be polluted.

INTEGRATE Career

Health Department Local health department officials often test residential wells for contamination when complaints are filed. They often test water for bacteria, typhoid, dysentery, and hepatitis. If you have a well, why should you have your well water tested on a regular basis?

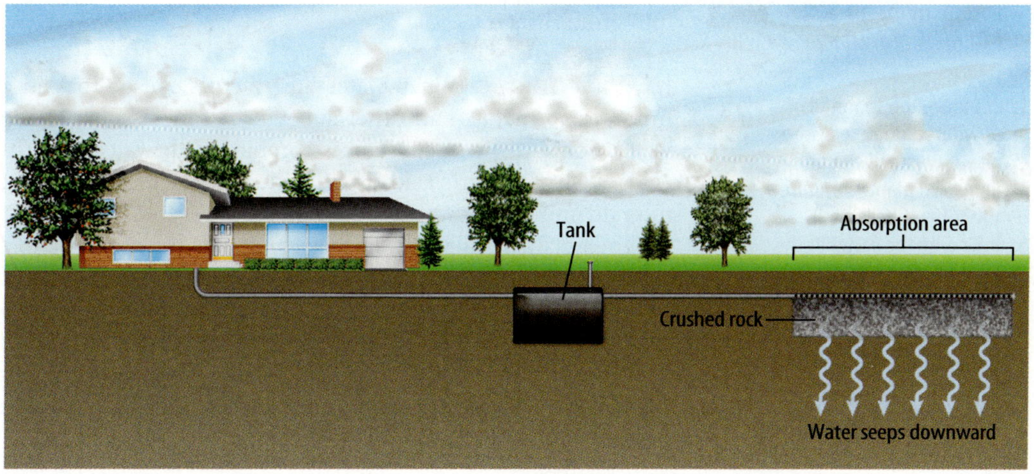

Figure 12 Although newer septic systems work efficiently, many older systems can allow pollutants to enter the groundwater.

INTEGRATE Career

Health Department Testing allows a well owner to monitor water quality and to stop using the water if the well becomes contaminated.

Research Have students research where their local health department is located. Have students mail requests to the health department for information or brochures regarding health issues and private water well supplies.

Discussion

Protecting Groundwater Because of possible leakage into groundwater supplies, should there be restrictions on the use of materials such as salts on roads? Some students will think that this would be a good way to prevent pollution. Others will think the banning of these materials would cause safety problems, and that such substances should be used with care while groundwater supplies are monitored.

Teacher FYI

Human Wastewater On average, each person living in the United States produces 500 L of wastewater each day. This wastewater includes suspended solids, dissolved organic and inorganic substances, and bacteria and viruses.

Visual Learning

Figure 12 Have a student review for the class how a septic system works, using this figure as a visual aid. The septic tank removes solids from the wastewater. Pipes carry the wastewater from the tank to the absorption area. In the absorption area, soil and crushed rock filter and treat the water before it reaches the water table.

Mini LAB

Purpose Students model how groundwater becomes polluted. L2 ELL LS **Visual-Spatial**

Materials rubber band, coffee filter, clear drinking glass, powdered drink (coffee or chocolate), spoon, water

Teaching Strategy To prevent the powder and water from flowing down the outside of the glass, have students push a few centimeters of the filter down into the glass before securing it with a rubber band.

Analysis

1. Both are sources of possible "pollution" as water seeps through them and picks up substances; both are porous and allow water to pass through them.
2. Some powdered drink dissolved in the water and passed through the filter. Materials from the waste pile dissolve in the water and are carried with the water into soil.

Assessment

Process Infer what can be done to prevent groundwater pollution from stockpiles and waste piles. Piles could be placed on top of liners that prevent substances from entering groundwater. Use **Performance Assessment in the Science Classroom,** p. 89.

Try at Home

Reading Check

Answer Toxic substances can leak through soil into groundwater. Lined holding ponds can overflow during heavy rains.

Figure 13 Properly lined holding ponds reduce the chances of groundwater pollution.

Mini LAB

Modeling Groundwater Pollution

Procedure

1. Use a **rubber band** to secure a **coffee filter or a sheet of porous paper** to the top of a **clear drinking glass.**
2. Place one spoonful of powdered **coffee, hot chocolate, or other powdered drink** onto the filter.
3. Slowly pour **water** over the powdered liquid and examine the water at the bottom of the glass. *Do NOT drink the water.*

Analysis

1. Compare the spoonful of powdered drink to a stockpile or waste pile at an industrial plant. Compare the coffee filter to soil.
2. Infer why the water at the bottom of the glass is discolored. Explain how groundwater below a stockpile or waste pile can be polluted after a heavy rain.

Try at Home

Industrial Wastes One of the major sources of groundwater pollution is from holding ponds like the one in **Figure 13.** These are shallow depressions in the ground that are used for holding industrial chemicals. These ponds sometimes contain concentrated toxic solutions that leak through the soil and into groundwater. If holding ponds are lined with plastic, concrete, clay or other fine-grained sediments, the liquids are less likely to seep into the groundwater supply. However, even lined ponds can overflow during heavy rains, spilling liquid wastes directly into the surrounding soil.

Reading Check *How can chemical holding ponds cause groundwater pollution?*

Another source of industrial groundwater pollution is from stockpiles and waste piles. A stockpile is a pile of materials that will be used at a later time. A waste pile is a mound of debris that is to be disposed of in the future. When rain falls and dissolves materials in stockpiles and waste piles, toxic materials can get into groundwater.

Mining Wastes When water flows across soil or rock in a mine, toxic metals and other substances can dissolve in the water. Some of these substances combine with water to form acids, such as sulfuric acid. These toxic substances can, in turn, migrate into groundwater.

Today, laws require that mining companies make every effort to prevent the flow of pollutants into groundwater. Some of these efforts include changing the pathway of water entering a mine after it rains or snows. Channels and pipes can divert water so that it does not flow across mining wastes. If some water does migrate across mining wastes, it can be captured and treated before entering an aquifer.

Curriculum Connection

Social Studies Cleaning up a pollution source is expensive. Have students research this issue. Then form groups to debate who should pay for cleanups when a huge and costly underground leakage problem is found, especially if the polluter refuses to pay or has left the area. Possible answers: the polluter should be located and forced to pay; the current property owner should pay; tax money should be used for cleanup.

Teacher FYI

Leaking Gasoline The EPA estimates that a leak of only 4 L of gasoline can contaminate water for 50,000 people. The greatest cause of contamination is leaky underground storage tanks. The cost to clean up an underground gasoline storage tank can range from $20,000 to $1 million.

Agricultural Runoff When rain falls on fields and lawns that have been treated with pesticides and fertilizers, these substances can move through soil and pollute groundwater. Fertilizers are especially harmful. When ammonia fertilizers decompose, they create nitrates. Water that is high in nitrates creates a serious health problem for infants and elderly people. In some rural areas, people must use bottled water for drinking because of nitrates in their well water.

Feedlots Another source of groundwater pollution is from animal feedlots. A feedlot is where a large number of animals like chickens, cows, sheep, or pigs are raised for food. In these areas, large amounts of animal wastes, or manure, collect. Manure contains large concentrations of nitrates and bacteria that can pollute aquifers if the manure is not stored and handled properly. Monitoring wells can be installed around manure storage ponds or tanks. These wells allow groundwater to be constantly monitored for pollution.

Science Online

Topic: Groundwater Pollution

Visit bookh.msscience.com for Web links to information about what farmers are doing to reduce groundwater pollution.

Activity List some of the major chemicals that farmers apply to the land and their purpose.

Applying Science

Can stormwater be cleaned and reused for irrigation?

The council of Parfitt Square in Bowden has called a meeting of all residents to discuss implementing a new stormwater management system. Plans will be presented that call for the storage of all stormwater runoff in the aquifer. The water will be cleaned to remove pollutants before storage and then reused for irrigation.

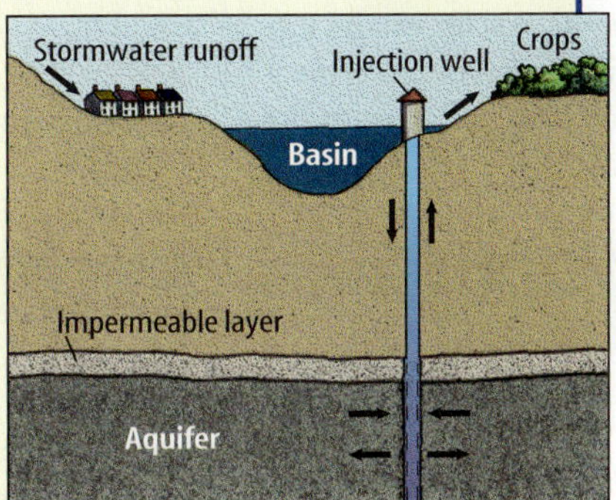

Identifying the Problem

All of the stormwater will be collected in a basin. It will pass through a sediment trap, where most of the sediment will settle out. The water then will be treated and sent to the aquifer, where it will be stored until it is needed for irrigation. The aquifer is large and contains coarse river gravel. Water in the aquifer has a high concentration of salt, which currently makes it unusable for irrigation. Residents can examine the diagram of the stormwater system and ask questions. What questions would you ask the council members?

Solving the Problem

1. If the objective is to collect all of the stormwater runoff, how will this affect the aquifer?
2. What would be the advantages of this stormwater management system to the neighborhood? Are there any disadvantages?

Applying Science

Answers

1. The addition of all storm water runoff will help recharge the aquifer. If all the pollutants are not removed from the runoff, this additional water could pollute the aquifer. Also, the addition of water may dilute the salt so that the water is usable for irrigation.
2. Advantages: Provides a way to capture and use water that is currently "lost"; helps increase available water supplies; may help in flood control. Possible disadvantages: May deprive areas downstream of water.

Teacher FYI

Anaerobic Digesters Anaerobic digesters are devices that are used to reduce environmental problems associated with feedlots and to generate electricity. The digesters use anaerobic bacteria to decompose animal waste. The process helps control odor, kills many harmful microorganisms, and reduces nutrients in the manure. The methane gas produced can be used to generate electricity.

Science Journal

Nitrates Have students research and write in their Science Journals about the health problems caused when infants and elderly people drink water contaminated by nitrates. Students should include that nitrates interfere with the ability of blood to carry oxygen. In infants, the disorder is called methemoglobinemia or "blue baby syndrome." If the condition is untreated, an infant could die.

SECTION 2 Groundwater Pollution and Overuse H ◆ 81

Use Science Words

Word Meaning Have students research and explain why people who study groundwater and its pollution are called *hydrologists.* Hydrology is the science that deals with water on the surface, in soil and rocks, and in the atmosphere. "Hydro" means "water," "-logy" means "science." A hydrologist is a person who practices hydrology. LS **Linguistic**

Activity

Cleaning Up Groundwater Pollution To demonstrate how hard it is to clean pollution from soil and groundwater, provide each student group with a dry, flat sponge. Have them put the sponge into a shallow pan of water so that only the bottom half gets wet. The wet bottom will represent an aquifer. The dry top represents porous soil. Have students put the sponge on a piece of plastic wrap, dry side up. Then have them put one drop of food coloring in the middle of the dry part of the sponge to represent point source pollution in soil. Without squeezing the sponge, have students try to remove the dye. They might use water or soap. If they use soap, they will need to remove that as well, since soap also pollutes soil and groundwater. Other suggestions might include using an absorbent material, such as paper towels or cornstarch. L2 ELL LS **Logical-Mathematical**

Discussion

Increasing Population How could the increasing world population affect available groundwater? As population increases, there is a greater chance that more groundwater will be removed than can be replenished. So groundwater supplies could dry up.

Caption Answer

Figure 14 The control system will not shut off the water pump, thereby allowing the oil to seep to a lower level.

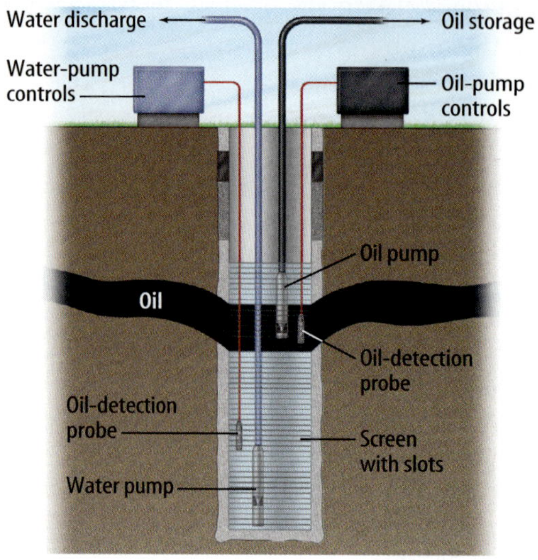

Figure 14 Pumping causes water and oil to move into a well. One pump removes groundwater, while the other pump removes the oil. A control system shuts off the water pump if oil is detected by the probe just above the water pump. **Infer** *What would happen if the lower oil-detection probe was broken?*

Groundwater Cleanup

Imagine trying to clean paint or oil out of a sponge. You would have to soak the sponge in chemicals, squeeze it, rinse it, and repeat this process several times. Even after this, you may never get all the paint or oil out because some will cling to the sides of the pores. Now, imagine trying to clean the polluted pores in rocks. You aren't allowed to use dangerous chemicals. You can't squeeze rocks. In addition, water moves through the pores slowly. Cleaning groundwater isn't easy.

In many ways, it's easier to prevent groundwater pollution than it is to clean it up. The slow speed of groundwater movement often makes groundwater cleanup a slow and difficult process. Surface water and groundwater is constantly being used for human consumption, recreation, and industrial purposes. Therefore, water that is polluted should be cleaned.

Removing Pollution Sources The first step in cleaning groundwater is to remove the source of the pollution. In 1980, Congress established a program called Superfund to eliminate the worst hazardous-waste sites in the United States. The Environmental Protection Agency (EPA) works in cooperation with states to locate, investigate, and clean up these sites. These sites include abandoned warehouses, manufacturing facilities, processing plants, and landfills. The cleanup operation for Superfund sites is expensive and time consuming, but it often is possible to clean these and many other point-source pollution sites. Nonpoint-source pollution is much more difficult to eliminate. If the pollution source can be removed, the next task is to clean the groundwater itself.

Cleanup Methods One way of cleaning groundwater is to pump the water out, remove the pollutants, and then return the water to the aquifer. Another way is to treat water while it remains in an aquifer. The method used for treatment of water left in an aquifer depends on the type of pollutant. In one method, gas is injected under pressure into wells to push the pollutants to the surface, where they can be removed. Another method uses steam, heated water, or electricity to increase the flow of pollutants to a well, where they can be removed. **Figure 14** shows one way that oil polluting groundwater can be removed.

Active Reading

ReQuest Have students listen carefully as you read an interesting piece or story aloud. Have students construct questions for a discussion. Assign students to read the story and participate in the questioning with other students. Have students participate in a ReQuest with the chapter feature or another interesting piece related to chapter content.

Differentiated Instruction

Challenge Have students research and explain the EPA's brownfields program. Encourage students to include information on any nearby brownfields. Brownfields are abandoned or underused industrial and commercial facilities where possible contamination is preventing redevelopment. The EPA brownfields program provides money for brownfields cleanup. L3 LS **Linguistic**

Bioremediation Another way to clean groundwater is by bioremediation. **Bioremediation** is a process that uses living organisms to remove pollutants. Bacteria and fungi often are used in bioremediation because they use organic materials for food. In water that is polluted with sewage and other organic materials, these organisms break down harmful substances into harmless substances.

Some kinds of plants also can be used for bioremediation. The roots of some plants are able to absorb certain dangerous metals. Where groundwater is within 3 m of the ground surface and soil contamination is within 1 m of the ground surface, plants have been used successfully to remove pollutants. Poplar trees like those shown in **Figure 15** have been successfully used to remove petroleum products from groundwater in Ogden, Utah.

Reading Check *What is bioremediation?*

Figure 15 In addition to sunflowers, alfalfa, and clover, poplar trees also have been used for bioremediation. Because tree roots grow more deeply into the ground than the roots of smaller plants, trees can be used to remove deeper pollution.

Groundwater Shortages

Besides pollution, another groundwater problem exists. It's caused when large populations overpump aquifers. Overpumping means that the amount of groundwater removed from an aquifer is greater than the amount of water flowing into the aquifer. In some parts of the southwestern United States, increased demand for water has caused groundwater to be removed faster than rain and snow can replenish it. When this happens, the water table drops and some wells go dry, as shown in **Figure 16.** This creates water shortages in some regions. You learned earlier about the Ogallala Aquifer. Its water has supported life and agriculture on the High Plains for many years. However, over the years more water has been pumped out of this aquifer than has been replaced by rainfall. It's one of many aquifers that are drying up.

Figure 16 When the amount of water pumped out of an aquifer is greater than the amount flowing in, some wells go dry.

Fun Fact

Unlike most small plants, alfalfa roots can grow 4.5 m into the soil, making them useful in removing pollutants from groundwater.

Reading Check

Answer a process that uses living organisms to remove pollutants

Visual Learning

Figure 16 Have a volunteer use **Figure 16** to explain why the well on the left has gone dry. Pumping from the aquifer has lowered the water table beneath the well's intake.

Inquiry Lab

Bioremediation

Purpose to illustrate bioremediation as a groundwater cleanup technique

Possible Materials bowls, food coloring (red or blue), water, celery stalks with leaves

Estimated Time 20 minutes

Teaching Strategies

- Students can make simple bioremediation systems by placing freshly cut celery sticks into a bowl of colored water.
- Students can observe the colored water (pollution) moving up the celery stalks.
- Students can observe the pollution moving into the leaves and infer where the pollution might go after it reaches the leaves.
- Allow students to explore other questions that arise from this lab.

For additional inquiry activities, see *Science Inquiry Labs*.

Differentiated Instruction

Learning Disabled In pairs, have students write down the consequences of a groundwater shortage and how it might impact their lives. Answers could include no drinking water and no water for bathing or cooking. Trees and plants could become stressed or die.

Curriculum Connection

Social Studies Mexico City, Bangkok, and Jakarta are sinking as groundwater is withdrawn to provide water for their growing populations. Have students research the populations of these cities in 1950, 1975, and 2000, and make bar graphs of their data. Encourage students to make a display of their graphs that includes maps showing the location of each city and its population data. P

Caption Answer

Figure 17 Salt water will seep into the well, making the water unfit for drinking.

3 Assess

DAILY INTERVENTION

Check for Understanding

Visual-Spatial Draw a fictitious map of a city. Include businesses, commercial areas, factories, roads, rivers, and streams. Have students identify potential pollution sources and pollution migration routes to surface water bodies.

Reteach

Pollution Source Have students describe ways groundwater becomes polluted. As students share their answers, make a large spider concept map on the board or on an overhead. L2 LS **Visual-Spatial**

Assessment

Process Have students design models that show the causes and results of subsidence in one area. Use **Performance Assessment in the Science Classroom,** p. 123.

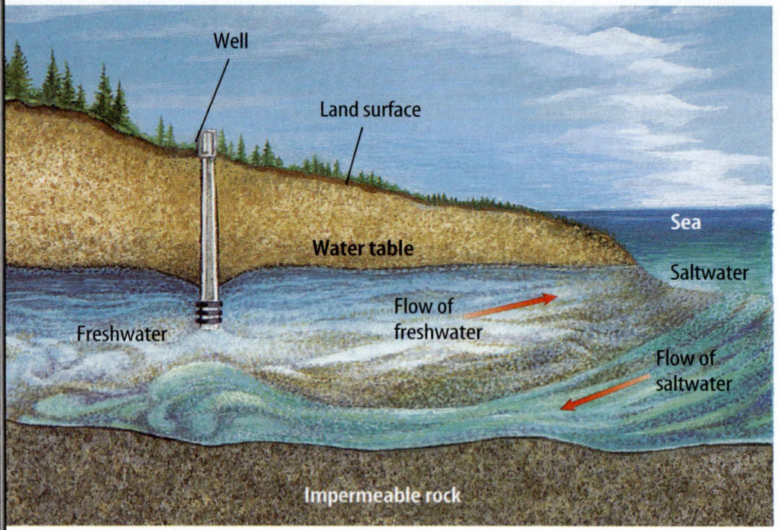

Figure 17 Pumping of groundwater can reduce freshwater flow toward coastal areas and cause salt water to be drawn toward the freshwater zones of the aquifer.
Describe *what will happen when the saltwater layer reaches the bottom of the well.*

Sinking Land

When water no longer fills the pores in an aquifer, the land above the aquifer sinks. This is called **subsidence** (sub SI dents). So much groundwater has been removed in parts of California, Florida, and Texas that land surfaces have dropped 5 m or more. In coastal regions, the result of sinking land is a rise in sea level. As a result, flooding along Galveston Bay in Texas has caused some people to abandon their homes.

Seawater Pollution Another problem can arise when too much water is removed from an aquifer along a coast. Seawater can seep into the aquifer and contaminate wells, as shown in **Figure 17.** Seawater is more dense than fresh groundwater because it contains dissolved salts. As a result, salt water tends to flow slowly into the lower levels of a coastal aquifer. Notice how the layer of salt water forms an upward cone near the pumping well. This process, called saltwater intrusion, has happened in Florida and Texas, as well as in other places.

section 2 review

Summary

Sources of Pollution

- Pollution comes from both point and nonpoint sources.
- Fertilizers that are put on fields and lawns are toxic and can contaminate groundwater.

Groundwater Cleanup

- Groundwater cleanup can be a slow, difficult, and expensive process.
- Preventing groundwater contamination is often easier than cleaning polluted water.

Groundwater Shortages and Sinking Land

- Water shortages and subsidence can be caused when large populations use aquifers.

Self Check

1. **Name** three sources of groundwater pollution. How do these pollutants get into groundwater?
2. **Explain** why it is difficult to clean groundwater.
3. **Name** two different ways to clean groundwater.
4. **Describe** four things that can happen when too much water is pumped from an aquifer.
5. **Think Critically** Benjamin Franklin once said, "When the well's dry, we'll know the worth of water." What do you think he meant?

Applying Math

6. **Calculate** If a 500-gallon gasoline spill has occurred and 300 gallons of gasoline have been removed, what percentage of the spill remains in the ground?

Science Online bookh.msscience.com/self_check_quiz

section 2 review

1. Possible answers: landfills, road runoff, waste spills, holding ponds, septic tanks, industrial wastes, mining wastes, agricultural and feedlot runoff; pollutants seep into soil, and then into groundwater.
2. It is difficult to remove pollution in the pores of underground rock and soil, and water moves through these pores very slowly.
3. Remove water from the aquifer, clean it, and return it to the aquifer; clean the water while it is still in the aquifer.
4. wells go dry, subsidence, coastal flooding, saltwater intrusion
5. People take water for granted until there isn't any.
6. $(500 \text{ gal} - 300 \text{ gal}) \div 500 \text{ gal} = 0.4 = 40\%$

section 3

Caves and Other Groundwater Features

Formation of Caves

Some of the most beautiful natural features on Earth are caves like the one shown in **Figure 18.** A **cave,** or cavern, is an underground chamber that opens to the surface. A cave system is a place where many caves are connected by passages. Carlsbad Caverns and Mammoth Cave are two very famous cave systems in the United States. Carlsbad Caverns, in New Mexico, has at least 48 km of passages. One of the caves in the system is so big it could hold 11 football fields. This cave is 25 stories high. The Mammoth Cave system, in Kentucky, has about 540 km of mapped passages. It is the longest recorded cave system in the world. Scientists estimate that there could be as many as 900 km of passages that have yet to be discovered and explored in this cave system.

About 17,000 caves have been found in the United States. Probably many more remain to be discovered and explored. Did you ever wonder what happens to make caves form?

as you read

What You'll Learn

- **Recognize** how caves change through time.
- **Explain** how cave features form.
- **Explain** what causes sinkholes and disappearing streams.

Why It's Important

As caves evolve, surface features sometimes change.

Review Vocabulary

evaporate: the physical change of a material from the liquid state to the vapor state

New Vocabulary

- cave
- dripstone
- sinkhole

Figure 18 This cave near Austin, Texas, is one of more than 100 caves in the United States that people can visit to experience the work of groundwater.

1 Motivate

INTERACTIVE CHALKBOARD PowerPoint® Presentations

Bellringer

Section Focus Transparencies also are available on the Interactive Chalkboard CD-ROM. L2 ELL

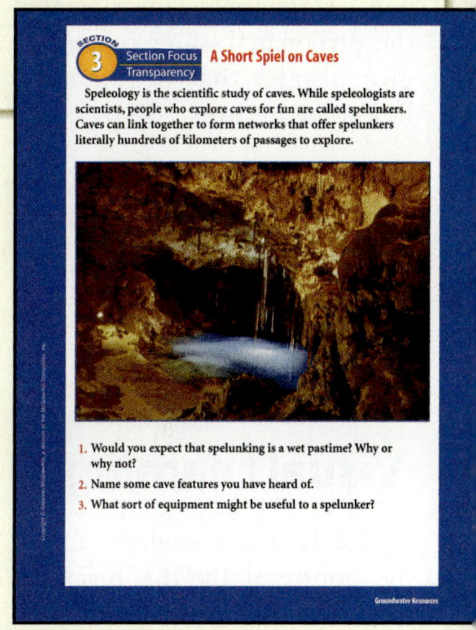

SECTION 3 Section Focus Transparency **A Short Spiel on Caves**

Speleology is the scientific study of caves. While speleologists are scientists, people who explore caves for fun are called spelunkers. Caves can link together to form networks that offer spelunkers literally hundreds of kilometers of passages to explore.

1. Would you expect that spelunking is a wet pastime? Why or why not?
2. Name some cave features you have heard of.
3. What sort of equipment might be useful to a spelunker?

Groundwater Resources

Tie to Prior Knowledge

Caves Have students recall visiting or seeing pictures of caves. Ask them to list the features of the caves as you write them on the board. Tell students that in this section they will learn how these cave features form.

Section 3 Resource Manager

Chapter *FAST FILE* Resources

- Transparency Activity, pp. 46, 47–48
- Directed Reading for Content Mastery, pp. 21, 22
- Enrichment, p. 32
- Reinforcement, p. 29
- Lab Worksheet, pp. 7–8

Performance Assessment in the Science Classroom, p. 44

Lab Management and Safety, p. 38

2 Teach

Quick Demo

Making Caves

Materials 250-mL beaker, carbonated water (clear soda), and limestone chips to represent weathering bedrock

Estimated Time ten minutes

Procedure Half-fill a beaker with limestone chips. Pour carbonated water over the chips, and have students observe what happens. Help them conclude that the carbonated water represents groundwater that weathers the limestone in caves.

Fun Fact

The hobby of exploring caves is called spelunking.

Visual Learning

Figure 19 Some students might be confused by the bricklike pattern used to represent rock in this figure. Ask students to speculate about why a bricklike pattern might be used by geologists to represent limestone bedrock. Because limestone forms in layers with natural fractures, outcrops of limestone to display a bricklike pattern.

Figure 19 Caves form in regions where groundwater dissolves rock.

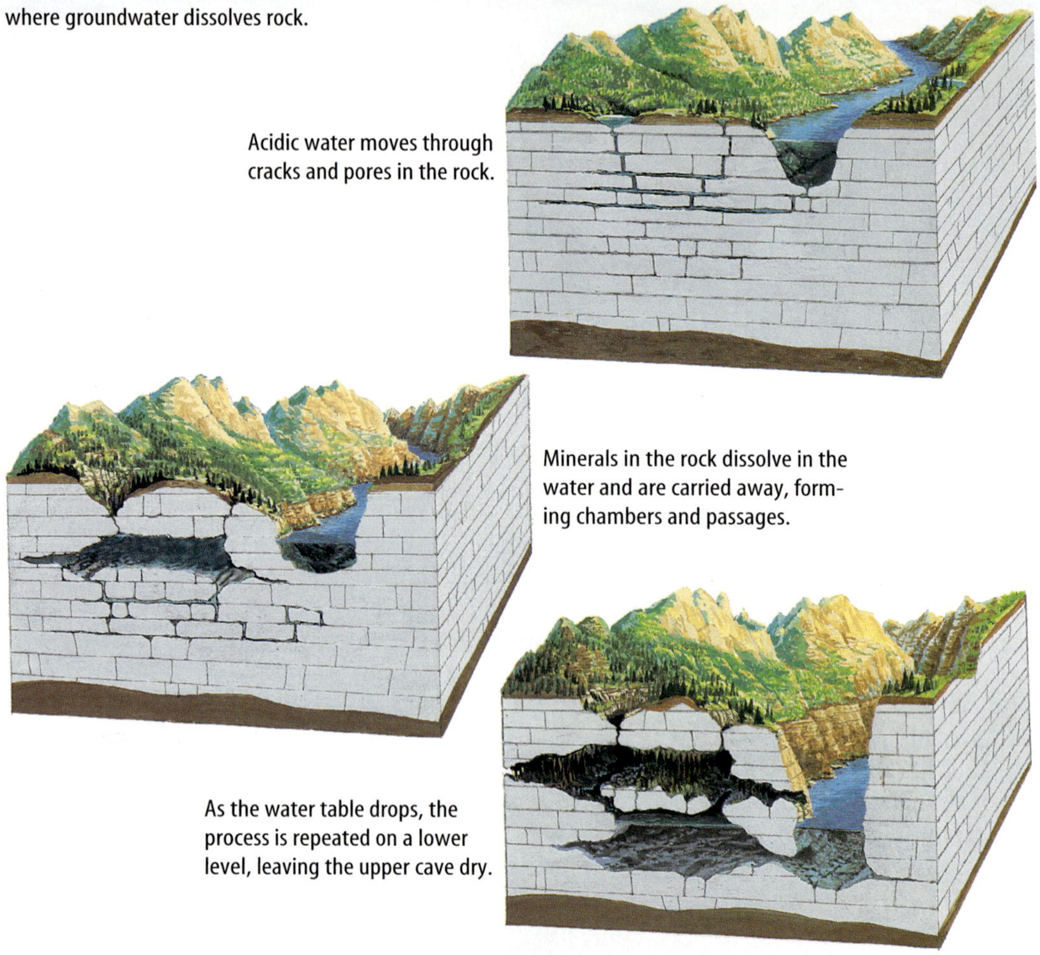

INTEGRATE Chemistry

Effects of Groundwater Caves form because groundwater is slightly acidic. As you learned earlier, groundwater originates as surface water. Surface water forms when rain falls or snow melts. As water from precipitation falls through the atmosphere and then seeps through soil, it chemically combines with carbon dioxide to form carbonic acid.

Carbonic acid makes groundwater naturally acidic. As this weak acid moves slowly through fractures and pores in certain kinds of rock, the acid reacts with the rock. For example, when carbonic acid passes through limestone, marble, and dolostone, the rocks dissolve easily. Some of the compounds forming these rocks are carried away as ions in solution. **Figure 19** shows how caves change over time.

Cultural Diversity

Ancient Cave Over time, people have used Mammoth Cave for many things. Between 3,000 and 4,000 years ago, Native Americans used the cave for shelter. They also used its minerals for medicinal purposes. Have students use the Internet or other sources to learn more about the Native American tribes that lived near the cave and how they viewed it.

Teacher FYI

Speleology Speleology is the scientific discipline concerned with all aspects of caves and cave systems. Speleologists map caves, describe their features, and investigate rates of limestone solution and dripstone formation.

Formation of Cave Features

Groundwater doesn't only dissolve rocks to create caves. It sometimes creates spectacular ceiling, wall, and floor deposits inside the caves. As the water drips and then evaporates, it leaves behind deposits of calcium carbonate called **dripstone. Figure 20** shows different kinds of dripstone formations.

When water drips from the ceiling, it can create long, wavy shapes, much like draperies. It also can form tubes like soda straws that fill with deposits. Some material also is added to the outside of the tubes. Soda-straw formations are the first stage of stalactites. When the soda straws are plugged, water trickling down their outside turns them into larger stalactites. Stalactites are dripstone deposits that hang down from the ceilings of caves, much like icicles. As water drips to the floor, towers of stalagmites can build up toward the ceiling. Where stalactites join stalagmites, columns are created. Round pearl-like objects, called cave pearls, form when water drips into shallow cave ponds.

Reading Check *What is dripstone?*

Science Online

Topic: Caves

Visit bookh.msscience.com for Web links to information about Carlsbad Caverns and Mammoth Cave.

Activity Determine how many miles of explored cave passages have been discovered in your state.

Figure 20 Dripping water creates beautiful formations in some caves.

At times, water contains minerals that result in draperies with dark orange, reddish, or brown bands. These draperies are called cave bacon.

Find the stalactites, stalagmites, and columns.

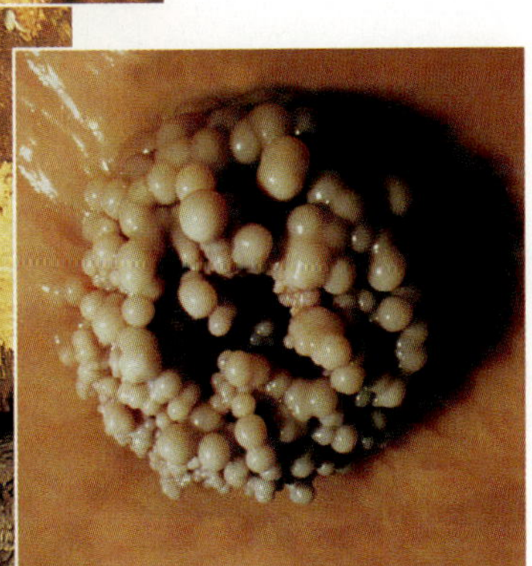

Cave pearls form in a variety of sizes and colors.

Reading Check

Answer the calcium carbonate deposits left behind as water dripping from the roof and down the walls of a cave evaporates

Discussion

Water Table Elevations Suppose you discover a cave with dripstone formations. What does this tell you about the present and past elevations of the water table? The cave was formed when groundwater dissolved the rock. For this to have happened, the water table had to be at least as high as the roof of the cave. The dripstone formations formed when the cave was relatively dry, as it is now. This indicates that the water table is lower than the cave's floor.

Differentiated Instruction

Challenge The term used for cave exploration is *spelunking*. A person who goes spelunking is called a spelunker. Spelunkers across the world have explored miles of new caves. Have students research spelunking and have students list the equipment a typical spelunker uses when he or she explores caves. Typical equipment used are flashlights, hardhats, boots, ropes, carabineers(hooks), backpacks. L2

Visual Learning

Figure 20 Where in this cave is it likely that new columns will form? Students should identify areas where stalagmites and stalactites are building toward one another.

Teacher FYI

Remembering Terms Stalactites or stalagmites, which is which? A good way for students to remember is to use this mnemonic: "A stalactite holds *tight* to the ceiling. A stalagmite *might* reach up to it."

Discussion

Sinkhole Formation In a karst area, why might sinkholes be more likely to form during a drought? During a drought, the water table drops because there is less water to replenish groundwater. When the water table falls, cave roofs might have less support and can collapse.

Visual Learning

Figure 21 Could there be additional danger from the sinkhole you see here? Yes; additional areas around it might collapse. If buildings or roads are nearby, they would have to be closed.

Figure 21 In one day, land collapsed to create this sinkhole in Winter Park, Florida.

Sinkholes In some caves, so much rock dissolves that the cave's roof can no longer support the land above it and the land collapses into the cave. The depression it leaves is called a **sinkhole.** In regions that overlie thin, delicate roofs of caves, dangerous situations can arise. Roadways, housing developments, and agricultural areas are vulnerable when rock and soil collapse rapidly into an underground cavern.

Sinkholes often develop in zones of fractured limestone. When groundwater enters the limestone along the fractures, rock dissolves near the cracks and caves form. **Figure 21** shows a sinkhole that formed in Winter Park, Florida. Sinkholes also occur when so much groundwater is pumped out that the water table drops and flooded caves become dry. When water is no longer in the cave to help support the roof, the roof of the cave can collapse. An area where there are many sinkholes is called a karst area.

Disappearing Streams Walking through a karst area, such as in central Kentucky, you might see sinkholes and lakes. Some of these lakes formed when sinkholes filled with water. You might also see some strange streams. They flow along the surface and then suddenly disappear. Where do they go? These disappearing streams fall into caves and continue flowing as underground streams. **Figure 22** shows one way that scientists track these disappearing streams.

Figure 22 Fluorescent dye is poured into streams where they sink into caves. The dye can be tracked to help determine how and where underground water moves.

Science Journal

That Sinking Feeling Tell students to imagine that a home sits on top of a cave. One day the people inside awake to the feeling that the house is moving. Have students write a short story about what people might experience as their home slips into a sinkhole. Have them share their stories in class.
LS **Intrapersonal**

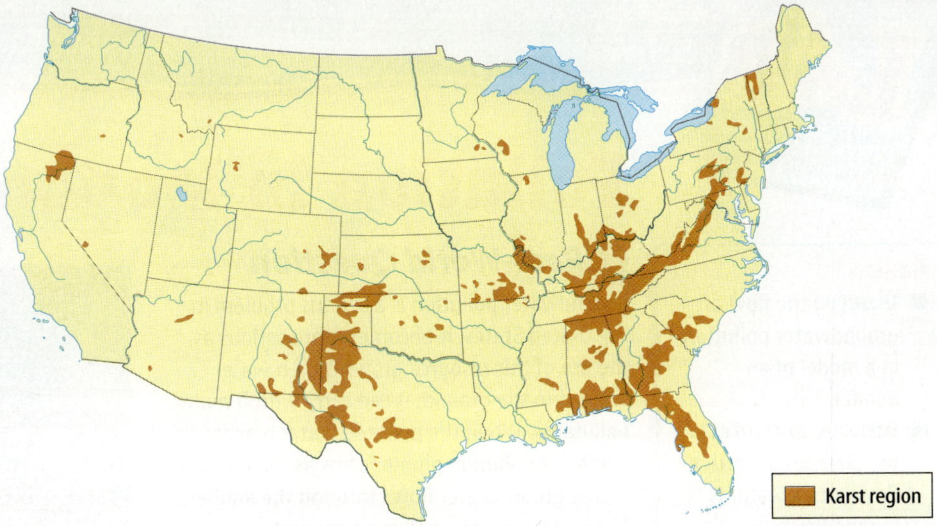

Figure 23 Most karst regions are in central and southern states where limestone is found and rainfall is abundant.
Discuss *Why do you think there are very few karst areas in Arizona and Nevada?*

Location of Karst Areas In the United States, karst areas with caves, sinkholes, and disappearing streams are common in Missouri, Tennessee, Kentucky, Indiana, Alabama, Florida, and Texas, among other places, as shown in **Figure 23.** Worldwide, karst areas are found in southern Europe, southern Australia, southeastern Asia, Puerto Rico, and Cuba.

As you've learned in this chapter, groundwater is one of Earth's most important resources. It provides many communities with water, and creates caves. Unfortunately, groundwater also can transport pollutants. Understanding how groundwater moves through soil and rock is a first step in learning how to clean, conserve, and protect groundwater resources that are vital for life.

section 3 review

Summary

Formation of Caves

- A cave system is a series of many caves that are interconnected by passages.
- Slightly acidic groundwater can dissolve certain types of bedrock.

Cave Features

- Stalactites and stalagmites are formed over time by the dripping of groundwater that leaves deposits of calcium carbonate.
- Cave pearls form in shallow cave ponds.
- An area where there are many sinkholes is called a karst area.

Self Check

1. **Describe** how carbonic acid forms caves.
2. **Explain** how dripstone forms cave features.
3. **Explain** what causes a sinkhole.
4. **Compare** disappearing and surface streams.
5. **Think Critically** Why don't some areas with limestone in the western United States contain karst features?

Applying Math

6. **Calculate** If dripping water from a cave ceiling produces a stalactite at the rate of 1 cm per month, how long will it take to produce a stalactite of 3 m?

Science Online bookh.msscience.com/self_check_quiz

Visual Learning

Figure 23 Do we live in a karst region? Answers will vary; depending on your location.

Caption Answer

Figure 23 Rainfall is not abundant in Arizona and Nevada.

3 Assess

DAILY INTERVENTION

Check for Understanding

Visual-Kinesthetic Have students use their index fingers to demonstrate the orientations of cave features such as stalactites, stalagmites, and columns.

Reteach

Sinkholes Have students draw a series of sketches that show the steps in sinkhole formation. L2
LS **Visual-Spatial**

Assessment

Content Assess students' ability to summarize by asking them to write a report about how groundwater can cause the formation of caves and other underground formations. Use **Performance Assessment in the Science Classroom,** p. 159. P

section 3 review

1. The acid seeps through the rock, reacts with some of the minerals in the rock, and creates new compounds that dissolve in water. These cavities are caves.
2. Deposits of calcium carbonate are left behind as water dripping into the cave evaporates.
3. As a cave continues to enlarge, the surface above the cave might become unsupported. If this happens, the roof of the cave collapses, creating a sinkhole.
4. A surface stream flows along Earth's surface. A disappearing stream flows along the surface and suddenly plunges into a cave.
5. The climate is too dry.
6. $300 \text{ cm} \times \frac{1 \text{ month}}{1 \text{ cm}} \times \frac{1 \text{ year}}{12 \text{ months}} = 25 \text{ years}$

Real-World Question

Purpose Students observe how pollution moves through an aquifer. L2 ELL LS **Kinesthetic**

Process Skills make models, observe and infer, draw conclusions, relate cause and effect, predict, communicate

Time Required 40 minutes

Procedure

Teaching Strategy Have students pour the salt water very slowly onto the gravel.

Ask students if the gravel used in this experiment can be used again, and if so, how? Encourage them to think of wise choices for conservation of materials; the gravel can be dried out and reused year after year.

Troubleshooting Be sure students add enough dye to the salt water so that it is brightly colored.

Analyze Your Data

Expected Outcome At first, the salt water will sink into the "groundwater." After pumping begins, the dye will show that the salt water begins to travel toward the pump.

Error Analysis Errors may result if the pump is not working properly or if students do not put enough salt water into the box.

LAB: Pollution in Motion

Goals

- **Observe** the flow of groundwater pollution in a model of an aquifer.
- **Observe and infer** the direction pollution flows when water is pumped from an aquifer.

Materials

clear-plastic box
aquarium gravel
water
pump (from a bottle of non-aerosol hair spray or liquid soap)
30 mL of salt water
food coloring
5-cm × 5-cm piece of cloth
1,000-mL beaker
small rubber band
50-mL graduated cylinder

Safety Precautions

Real-World Question

Groundwater pollution is a serious problem in the United States. It becomes more serious as the use of this resource grows. When water seeps into the ground, it can carry with it many pollutants. When the polluted water reaches an aquifer, a pollution plume forms as the contaminated groundwater flows through the aquifer. The movement of a pollutant through an aquifer depends on the permeability of the aquifer, the water pressure, whether the pollutant dissolves in water, and other factors. In this lab, you'll investigate and observe the movement of pollution through a model aquifer. How does a pollutant move through an aquifer?

Procedure

1. Add clean gravel to a clear-plastic box until it is three-fourths full.
2. Pour water into the box until it's just below the top of the sediment.
3. Cover the bottom end of the pump with cloth and secure with a rubber band. Push the pump into the sediment near one end of the box.

Differentiated Instruction

English-Language Learners Ask students to explain orally how the pumps in this activity model wells drilled into aquifers. Possible answer: Both are long tubes that force liquids to the surface.

Alternative Inquiry Lab

Your Water Supply Have students investigate where their community obtains their drinking water. Have them contact local water treatment facilities to determine if water comes from surface or ground water sources, how it is treated, and what chemicals are used to make it drinkable. See if there are any local pollution prevention plans in place to help protect the water source from potential spills.

Using Scientific Methods

4. Use a few drops of food coloring to dye the salt water.
5. Pour salt water on top of the sediment at the end of the box away from the pump.
6. Observe the salt water's movement.
7. Sketch the box, the sediment, and the direction that the salt water moves inside the box.
8. **Predict** what will happen to the movement of the salt water when you pump water out of the box.
9. Begin pumping water into the empty beaker and observe the movement of the salt water.
10. Complete your drawing of the saltwater movement.

Conclude and Apply

1. **Infer** why the salt water moved in the direction it did before you began pumping water out of the box.
2. **Infer** what caused the movement of the salt water after you began pumping the water.
3. **Predict** what would happen to the direction of the saltwater flow if you used two pumps located next to each other.
4. **Predict** what would happen to the direction of the saltwater flow if you used two pumps that were located at different ends of the box.
5. **Infer** how pumping water from an aquifer affects the movement of pollutants.
6. **Predict** how an increasing population in an area could affect the movement of pollutants in an aquifer and the water quality in people's wells.

Communicating Your Data

Compare your drawings with drawings made by other students. Compare your conclusions with those of other students.

Conclude and Apply

1. The salt water flows downward, because salt water is denser than fresh water.
2. Water moves from high pressure to low pressure areas. When the pump removes water, a low-pressure area is produced.
3. The flow would continue toward the pumps, but it would be faster.
4. The direction of flow would depend on which of the pumps was pumping more water. The salt water would flow toward the area of lowest pressure in the box.
5. Pollutants are pulled with the water toward the well as water is pumped.
6. Increased population could result in more toxic runoff that could seep into groundwater. The added population could also mean increased pumping from wells, with pollution moving into wells with drinking water.

Assessment

Process Have students design an experiment to determine whether the depth of a well affects the flow of water in an aquifer. Use **Performance Assessment in the Science Classroom,** p. 95.

Communicating Your Data

Post students' drawings in the classroom. Students' conclusions should be similar.

SCIENCE Stats

Caves

Did you know...

...Wind Cave, South Dakota, is one of the world's oldest caves. It was formed more than 320 million years ago. Wind Cave doesn't have many stalactites or stalagmites. It's known for its boxwork, an unusual cave formation comprised of thin calcite fins resembling honeycombs.

...Mammoth Cave is the longest cave in the world. Beneath central Kentucky, 540 km of twisting passages snake through Mammoth Cave. Straightened out, these passageways would stretch the distance from Chicago, Illinois, to Cleveland, Ohio.

Applying Math According to the above graph, Mammoth Cave is about how many times longer than Jewel Cave?

...The deepest cave in the U.S. is in New Mexico. The Lechuguilla (lay chew GEE uh) Cave, in Carlsbad Caverns National Park, is more than 475 m deep. Cave explorers, called spelunkers, are still discovering deeper parts of this cave.

Write About It

Go to bookh.msscience.com/science_stats or a library to find out about the size and history of a cave not discussed in this feature. Write about characteristics of this new cave you've found.

SCIENCE Stats

Content Background

Cave formation is a very long process because groundwater moves so slowly. It may take millions of years for a large cave to form. As long as the water table remains constant, cave passages continue to expand along joints or faults in the rock. When the water table drops, a cave drains, and cave deposits from precipitated minerals begin to form. Once formed, caves may last only a few million years because walls and ceilings weaken and are not able to support the overlying rock.

Discussion

Cave Animals Explain to students that in the absence of light, many cave animals have evolved into creatures without eyes. Nearly 1,000 species, mostly crustaceans, insects and arachnids live exclusively in caves. Ask students to infer what these animals eat and how they find food. Some live on the vegetation in bat droppings, others are carnivores and eat each other. Since they are blind, they rely on their other senses to find food. They feel vibrations and smell their prey. Spiders trap food in their webs.

Activity

Calcium Carbonate Have students place a piece of chalk in a small amount of seltzer water. Explain to them that seltzer water contains carbonic acid and that chalk is a form of calcium carbonate, as is limestone. Have students observe what happens to the chalk. Ask them to relate their observations to how carbonic acid reacts with some minerals to create compounds that dissolve in water.

Applying Math

Answer About 3 times longer; 540 km/180 km = 3

Differentiated Instruction

Challenge Ask students to find out approximately how many kilometers longer Mammoth Cave (563 km) is than Holloch Cave (Switzerland). 563 − 183 = 380 km and 540 − 150 = 390 km, respectively. Ask them if they think Mammoth Cave might be found to be longer than 540 km. Possibly, because new passages might still be found and explored in the cave. L3

Write About It

Many of the measurements that students will find regarding the sizes of caves may not be given in metric measurements. Have students convert all measurements to metric for additional math practice.

Reviewing Main Ideas

Section 1 Groundwater

1. Groundwater moves slowly through pores in soil and rock. An aquifer is a formation of rock or sediment in which water flows freely through pore spaces.
2. In the zone of saturation, pores in an aquifer are full of water. The water table is the top surface of this zone.
3. Springs form where the water table is exposed on hillsides. When groundwater is heated, geysers can form.

Section 2 Groundwater Pollution and Overuse

1. Sources of groundwater pollution include landfills, spills, holding ponds and septic systems, industrial wastes, mining wastes, and runoff.
2. Sometimes, polluted groundwater can be cleaned, after the source of pollution is removed. Bioremediation uses organisms to clean water.
3. Too much pumping of water from an aquifer can result in water shortages, subsidence, and saltwater intrusion.

Section 3 Caves and Other Groundwater Features

1. Water can combine with carbon dioxide in air or soil to form a weak acid that dissolves some types of rock to form caves.
2. As water drips within a cave, it leaves behind calcium carbonate. This forms dripstone structures.
3. When the ceiling in a cave can no longer support the ground above it, it collapses. The hole that forms is a sinkhole.

Visualizing Main Ideas

Copy and complete the following concept map on groundwater shortages.

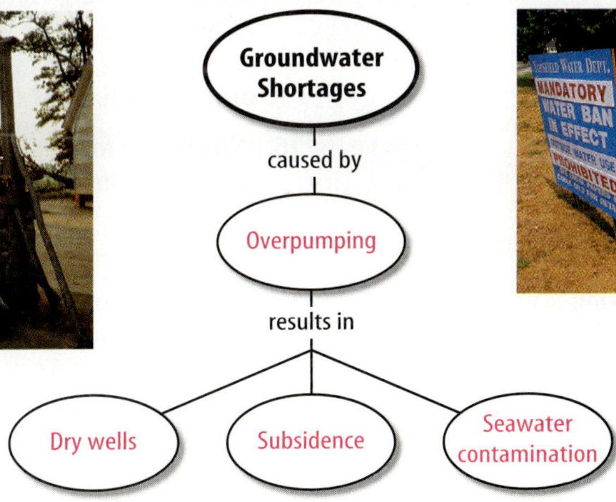

Science Online bookh.msscience.com/interactive_tutor

Identifying Misconceptions

Assess

Materials 8-oz clear plastic cups, rice, food coloring, water, small turkey baster, small square of gauze, rubber band

Procedure Cover the end of the turkey baster with gauze and secure with a rubber band (this is the water well pump). Half-fill the plastic cup with rice. Add water to just below the surface of the rice. Add two drops of food coloring and observe. Every 5 minutes for 30 minutes, pump out some water and observe.

Expected Outcome Students observe the long time for pollution to move through the groundwater.

Reviewing Main Ideas

Summary statements can be used by students to review the major concepts of the chapter.

Visualizing Main Ideas

See student page.

Science Online

Visit bookh.msscience.com
/self_check_quiz
/interactive_tutor
/vocabulary_puzzlemaker
/chapter_review
/standardized_test

Assessment Transparency

For additional assessment questions, use the *Assessment Transparency* located in the transparency book.

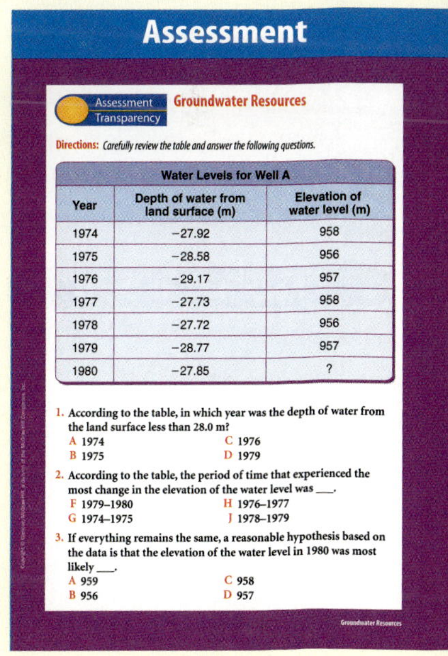
Assessment

Assessment Transparency **Groundwater Resources**

Directions: *Carefully review the table and answer the following questions.*

Water Levels for Well A		
Year	Depth of water from land surface (m)	Elevation of water level (m)
1974	−27.92	958
1975	−28.58	956
1976	−29.17	957
1977	−27.73	958
1978	−27.72	956
1979	−28.77	957
1980	−27.85	?

1. According to the table, in which year was the depth of water from the land surface less than 28.0 m?
   A 1974 C 1976
   B 1975 D 1979
2. According to the table, the period of time that experienced the most change in the elevation of the water level was ____.
   F 1979–1980 H 1976–1977
   G 1974–1975 J 1978–1979
3. If everything remains the same, a reasonable hypothesis based on the data is that the elevation of the water level in 1980 was most likely ____.
   A 959 C 958
   B 956 D 957

Groundwater Resources

Chapter 3 Review

Using Vocabulary

1. water table: top of zone of saturation; zone of saturation: where pores of an aquifer are full of water
2. artesian wells: forms where pressure pushes water to surface of sloping aquifer; geyser: forms where groundwater is heated and water and steam are ejected from an opening
3. subsidence: sunken land that forms when aquifer dries up and pores in aquifer collapse; sinkhole: sunken land that forms when a cave roof no longer supports land above it
4. pollution: substances that produce harmful changes in the environment; groundwater: water in underground pores in soil and rock
5. aquifer: layer of permeable rock through which groundwater flows; groundwater: water in underground pores in soil and rock
6. permeable: rock that permits groundwater to flow through it; zone of saturation: area of permeable rock in an aquifer that's full of water
7. bioremediation: method using living organisms to remove pollutants; subsidence: sunken land that forms when aquifer dries up and its pores collapse
8. dripstone: calcium carbonate deposits in caves; sinkhole: sunken area on surface caused by collapsed cave underneath

Checking Concepts

9. A	14. C
10. B	15. B
11. A	16. D
12. B	17. D
13. D	

chapter 3 Review

Using Vocabulary

aquifer p. 70	pollution p. 76
artesian well p. 73	porosity p. 69
bioremediation p. 83	sanitary landfill p. 78
cave p. 85	sinkhole p. 88
dripstone p. 87	subsidence p. 84
geyser p. 74	water table p. 70
groundwater p. 68	zone of saturation p. 70
permeable p. 69	

Explain the differences between the vocabulary words in each of the following pairs.

1. water table—zone of saturation
2. artesian well—geyser
3. subsidence—sinkhole
4. pollution—groundwater
5. aquifer—groundwater
6. permeable—zone of saturation
7. bioremediation—subsidence
8. dripstone—sinkhole

Checking Concepts

Choose the word or phrase that best answers the question.

9. Which is rock with unconnected pores?
   A) impermeable **C)** aquifers
   B) permeable **D)** saturated
10. Where do most solid wastes end up?
   A) holding ponds **C)** septic systems
   B) landfills **D)** feedlots
11. Which of the following is an example of nonpoint-source pollution?
   A) runoff from a farmer's field
   B) a leaking underground storage tank
   C) a broken pipeline
   D) a leaking landfill

Use the illustration below to answer questions 12–13.

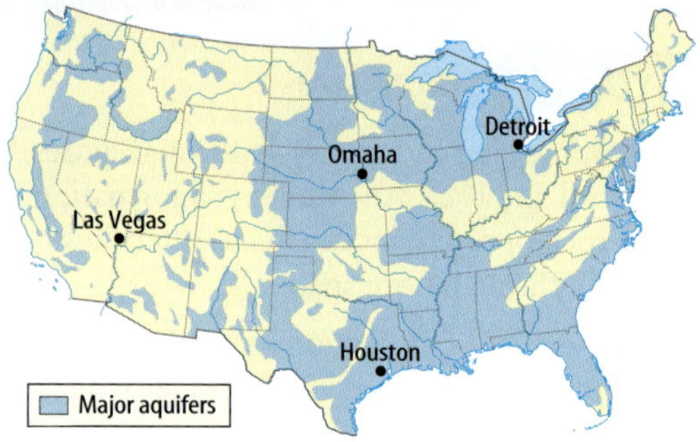

12. Which city is least likely to use groundwater as its primary water source?
   A) Detroit, MI **C)** Houston, TX
   B) Las Vegas, NV **D)** Omaha, NE
13. To which of the following states does the Ogallala Aquifer provide water?
   A) Florida **C)** California
   B) Ohio **D)** Texas
14. Which of the following is an example of an impermeable material?
   A) limestone **C)** clay
   B) sandstone **D)** gravel
15. Which of the following was set up by Congress to clean up hazardous waste sites?
   A) Safe Drinking Water Act
   B) Superfund
   C) The Clean Water Act
   D) bioremediation
16. Which cave formation forms when ceiling structures join with floor structures?
   A) stalactite **C)** cave pearl
   B) stalagmite **D)** column
17. Which layer of an aquifer has its pores filled with both air and water?
   A) water table
   B) impermeable rock
   C) zone of saturation
   D) zone of aeration

Science Online bookh.msscience.com/vocabulary_puzzlemaker

Thinking Critically

18. permeability of the rocks and the slope of the aquifer
19. higher after a heavy rain; lower during a period of drought or if a great deal of water had been pumped from it

Use the ExamView® Pro Testmaker CD-ROM to:

- customize tests
- create multiple versions of tests
- generate tests in Spanish
- build tests aligned with state standards

Thinking Critically

18. Discuss What affects the speed of groundwater flowing through rock?

19. Predict Under what conditions would the water table in an aquifer be higher than normal? What could cause the water table to be lower than normal?

20. Interpret Scientific Illustrations This picture shows stalagmites and stalactites in an underwater cave. Was the water there when these cave features were forming? Explain your answer.

21. Predict A homeowner decides not to have his septic system pumped out regularly. How could this create health problems?

22. Infer why a town might choose to spread sand, rather than salt, on icy roadways.

Use the illustration below to answer question 23.

23. Draw Conclusions A property owner's well runs dry when a nearby housing development drills several new wells. Explain what might have caused this to occur.

24. Classify Which sources of groundwater pollution involve sewage?

25. Compare and contrast porosity and permeability.

26. Test Hypotheses Water tests at a city well indicate high levels of an organic pollutant. Some people have suggested that the pollution is coming from an industrial site east of town. Describe how the city could determine if this is true.

Performance Activities

27. Make a Diorama Use clay or a mixture of flour, salt, and water to make a diorama of a cave. Include at least three different kinds of cave formations in your diorama.

Applying Math

Use the table below to answer questions 28–29.

Geotechnical Sample Results	
Site Location:	Big City, USA
Soil Type:	Dark brown silty/sand
Sample Depth:	3–4 meters below surface
Porosity:	35%
Bulk Density:	1.3 g/mL

28. Sample Density The sample data presented in the table shows a term called bulk density. Bulk density represents the mass (or weight) of the sample divided by the volume of the sample. As shown in the table the bulk density is 1.30 g/mL. If the volume of the sample was 100 mL, what was the mass (weight)?

29. Pore Space The porosity of the sample is 35 percent. If the sample volume is 100 mL, what volume of water will the sample hold?

A) 65 mL **C)** 50 mL
B) 35 mL **D)** 100 mL

20. No; dripstone forms only when water drips, evaporates, and leaves behind calcium carbonate deposits. This could not have happened underwater.

21. The sewage in the tank is more likely to leak out, carrying with it bacteria that can pollute soils and groundwater and create health problems.

22. Sand would provide traction for cars without polluting groundwater.

23. The water table probably dropped below the bottom of the homeowner's well. This could have happened because the new wells are extracting more water than normal from the aquifer.

24. Possible answers: landfills (diapers), septic tanks, sewer lines

25. Both terms relate to characteristics of soil and rock. Porosity is the volume of pore space divided by the total volume of a rock or soil sample. Permeability refers to the ability of water to pass through soil and rock. When soil or rock has high porosity and the pores are interconnected, the soil or rock is permeable.

26. Possible answer: Put a harmless tracking dye into the groundwater at the industrial site and observe to see If it shows up in the well water.

Performance Activities

27. Suggest that students label the dripstone formations in their dioramas. Use **PASC,** p. 123.

Applying Math

National Math Standards

1, 2, 3, 4, 9

28. 130 grams

29. 35 mL

Assessment Resources

Reproducible Masters

Chapter *Fast File* Resources
- Chapter Review, pp. 37–38
- Chapter Tests, pp. 39–42
- Assessment Transparency Activity, p. 49

Glencoe Science Web site
- Chapter Review Test
- Standardized Test Practice

Glencoe Technology
- Assessment Transparency
- Exam*View*® Pro Testmaker
- MindJogger Videoquiz
- Interactive Chalkboard

Part 1 Multiple Choice

Record your answers on the answer sheet provided by your teacher or on a sheet of paper.

Use the illustration below to answer questions 1–2.

National Groundwater Use

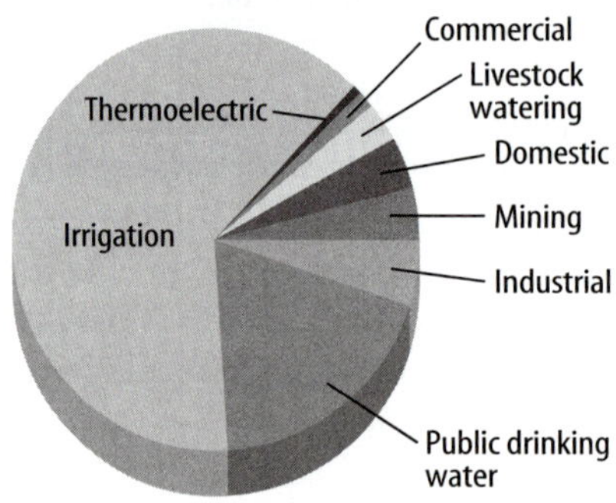

1. Based on the circle graph, which uses the least amount of groundwater?
A) industrial **C)** mining
B) irrigation **D)** thermoelectric

2. According to the circle graph, approximately what percentage of total groundwater is used for irrigation?
A) 30% **C)** 65%
B) 50% **D)** 75%

3. Which determines the speed at which groundwater flows?
A) depth **C)** permeability
B) gravity **D)** porosity

4. Why was the 1980 Superfund established?
A) elimination of hazardous waste sites
B) protection of rare cave systems
C) replenishment of aquifers
D) discovery of new aquifers

Test-Taking Tip

Breakfast The morning of the test, eat a healthy breakfast with a balanced amount of protein and carbohydrates.

5. Which rock is impermeable to water?
A) granite **C)** sandstone
B) limestone **D)** siltstone

6. Soda-straw formations develop into
A) cave pearls. **C)** stalactites.
B) dripstone. **D)** stalagmites.

7. Permeable rock in which water flows is
A) an aquifer. **C)** groundwater.
B) a geyser **D)** the water table.

Use the illustration below to answer question 8 and 9.

8. Which is shown in this illustration?
A) bioremediation
B) nonpoint source pollution
C) point source pollution
D) sanitary landfill

9. Which step is first in groundwater cleanup?
A) Pump the water out, remove pollutants, and return it to the aquifer.
B) Remove the source of pollution.
C) Treat the water in the aquifer.
D) Use living organisms to remove pollutants.

10. Which is bioremediation?
A) Pump the water out, remove pollutants, and return it to the aquifer.
B) Remove the source of pollution.
C) Treat the water in the aquifer.
D) Use living organisms to remove pollutants.

chapter 3 Standardized Test Practice

FAST FILE

Answer Sheet A practice answer sheet can be found at bookh.msscience.com/answer_sheet.

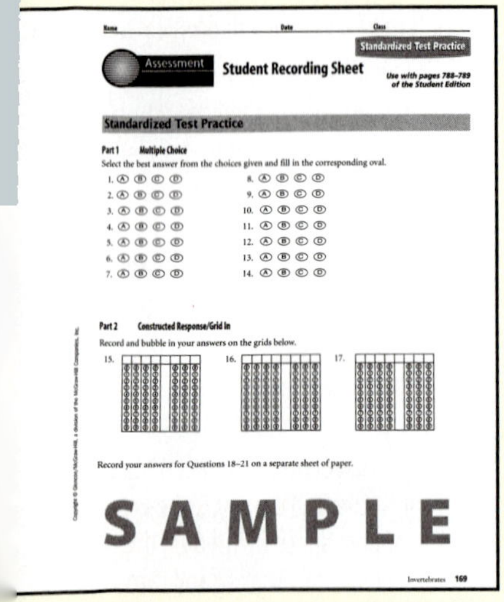
Name Date Class
Standardized Test Practice
Assessment Student Recording Sheet
Use with pages 788–789 of the Student Edition
Standardized Test Practice
Part 1 Multiple Choice
Select the best answer from the choices given and fill in the corresponding oval.
Part 2 Constructed Response/Grid In
Record and bubble in your answers on the grids below.
Record your answers for Questions 18–21 on a separate sheet of paper.
SAMPLE
Invertebrates 169

Part 1 Multiple Choice

1. D	6. C
2. C	7. A
3. C	8. C
4. A	9. B
5. A	10. D

Part 2 Short Response

11. $\frac{90 \text{ mL}}{350 \text{ mL}} = 0.257 \times 100 = 25.7\%$

12. Both deal with land sinking. Subsidence occurs when an aquifer has been drained of all its water and the land above the aquifer sinks. Sinkholes result when a cave's roof can no longer support the land above it and it collapses.

13. During bioremediation, bacteria feed on organic materials and have the ability to break down harmful substances in polluted waterways. The roots of some plants can absorb dangerous metals and other harmful substances in the ground that pollute groundwater.

14. A karst area develops when the dissolution of underlying rock, such as limestone, forms many sinkholes, caves, and underground streams.

15. cave formation

16. As water from precipitation falls through the atmosphere and seeps through the ground, it chemically combines with carbon dioxide to form carbonic acid. As the slightly acidic groundwater moves through the permeable layers of rock, the acid reacts with the rock and it slowly dissolves. Over time holes, chambers, and eventually caves form.

17. As the water level drops, the process repeats itself on a lower level. The upper cave dries out and the water erodes away a lower level.

18. stalactites—deposits of calcium carbonate hanging from the cave ceiling; stalagmites—deposits of calcium carbonate forming on cave floor; columns—deposits of calcium carbonate formed when stalactites and stalagmites merge; cave pearls—round pearl-like objects formed when water drips into shallow cave ponds

Part 2 Short Response/Grid In

Record your answers on the answer sheet provided by your teacher or on a sheet of paper.

11. Calculate the porosity of a 350-mL soil sample having a pore space volume of 90 mL.

12. Compare and contrast subsidence and a sinkhole.

13. How do bacteria and plants play a role in cleaning groundwater?

14. How does a karst area develop?

Use the illustration below to answer questions 15–18.

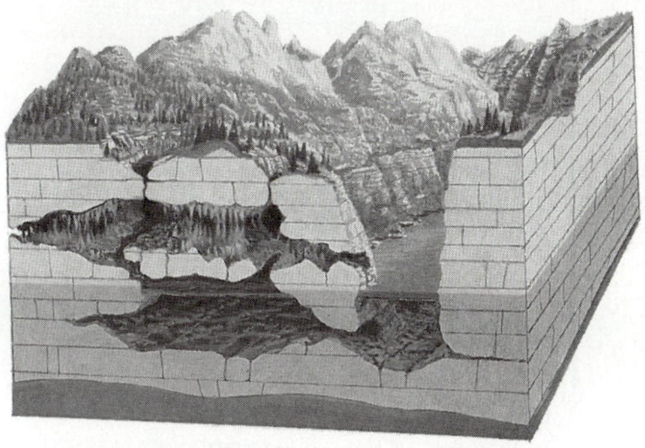

15. What process is occurring above?

16. What role does groundwater play in the formation of caves?

17. What will happen as the water table continues to drop?

18. List and describe at least two formations one might find in a cave.

19. What is the zone of saturation?

20. What traps liquid waste in a sanitary landfill?

21. Describe the zone of aeration.

22. How are the water table and the zone of saturation related?

23. What happens to the flow of groundwater if the slope of the water table increases?

Part 3 Open Ended

Record your answers on a sheet of paper.

24. What are the differences between groundwater and surface water? Explain how groundwater could become surface water.

25. List ways our everyday activities affect groundwater.

Use the illustration below to answer questions 26–28.

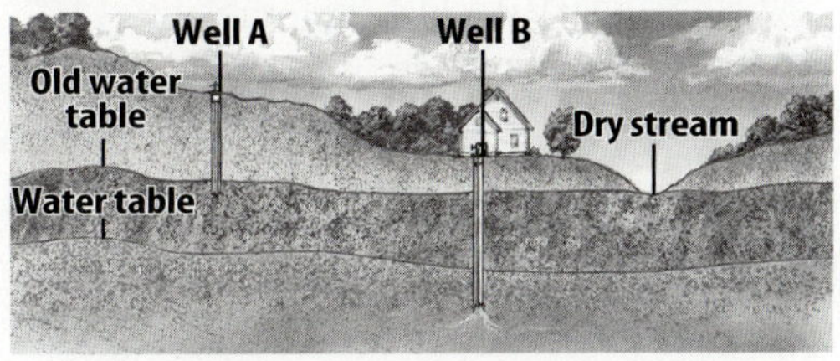

26. Why did well A dry up?

27. How did the homeowners decide where to drill well B?

28. What does the dry stream indicate about the water table?

29. Compare and contrast point-source and nonpoint-source pollution.

30. How does pouring oil on the ground to kill weeds cause groundwater pollution?

31. How is the water table related to the surfaces of streams, lakes, and other wetlands?

32. Traveling at 2 cm per day (the average speed at which groundwater moves through an aquifer), how long would it take you to move to the door of your classroom?

33. How do pesticides and fertilizers pollute groundwater?

34. Explain how columns form in caves.

19. It is the part of an aquifer where the pores are full of water.

20. Liquid wastes from a sanitary landfill are trapped by liners made of plastic, concrete, or clay-rich soils.

21. It is the upper part of an aquifer where the pores are partially filled with water, with the remaining pore space filled with air.

22. The water table is the upper surface of the zone of saturation.

23. The groundwater flow increases.

Part 3 Open Ended

24. Groundwater is rain or snow that leaches down into the ground and settles in the permeable layers of rock and sediment. Surface water is water located on the surface of the Earth in the form of streams, lakes, and oceans. Sometimes the pressure on groundwater is so great it rises to the surface and enters streams, lakes, and oceans. If an area below the water table is exposed, springs can develop.

25. Trash deposited in landfills can seep into the ground and contaminate groundwater. Toxic substances used in mining can be transported into our groundwater with rainfall. Pesticides and fertilizers applied to lawns and fields can seep through soil and pollute groundwater. Nitrates and bacteria contained in animal waste can pollute groundwater. The demand for water for human consumption is sometimes greater than the amount of available groundwater and aquifers can dry up if water is removed faster than it can be replaced.

26. The water table lowered due to demand for water exceeding water replacement.

27. The location of well B had to be lower than the original well A so the homeowners looked for an area where they could drill the least amount but reach the water table.

28. It could be used as an indicator of a dropping water table.

29. Both are sources of pollution. Point source pollution is traceable to a specific source. Nonpoint source pollution is harder to track because it comes from a large area and spreads rapidly through ground and surface water.

30. The oil is pulled by gravity and, if unimpeded, will eventually reach the groundwater.

31. Since water flows to lower elevation, the surface of a stream, lake, or wetland is the water table at that location.

32. If the door is fully open at 1.5 m, it would take about 2.5 months to open it.

33. The pesticides and fertilizers will eventually be dissolved and washed into the ground by rainwater and will be carried to the groundwater.

34. When calcium-rich water drips and evaporates in caves, it leaves behind deposits of calcium carbonate, forming stalactites and stalagmites. Where stalactites join stalagmites, columns are created.

Rubrics

For more help evaluating open-ended assessment questions, see the rubric on p. 10T.

chapter 4 Organizer

Section/Objectives	Standards		Labs/Features
	National	**State/Local**	
Chapter Opener	See pp. 9T–10T for a Key to Standards.		**Launch Lab:** What causes ocean currents?, p. 99 **Foldables,** p. 99 A data-collection lab using Probeware technology can be found in the **Probeware Lab Manual,** pp. 25–28
Section 1 Ocean Water 2 sessions 1 block 1. **Identify** the origin of the water in Earth's oceans. 2. **Explain** how dissolved salts and other substances get into seawater. 3. **Describe** the composition of seawater.	National Content Standards: UCP.1–3, UCP.5, A.1, A.2, B.1, D.1		
Section 2 Ocean Currents 2 sessions 1 block 4. **Explain** how winds and the Coriolis effect influence surface currents. 5. **Discuss** the temperatures of coastal waters. 6. **Describe** density currents.	National Content Standards: UCP.1–3, UCP.5, A.1, A.2, D.1		**Science Online,** p. 105 **MiniLAB:** Modeling a Density Current, p. 107 **Applying Math:** Calculate Density, p. 108 **Integrate Career,** p. 108
Section 3 Ocean Waves and Tides 3 sessions 1.5 blocks 7. **Describe** wave formation. 8. **Distinguish** between the movement of water particles in a wave and the movement of the wave. 9. **Explain** how ocean tides form.	National Content Standards: UCP.1–3, UCP.5, A.1, A.2, D.1, D.3, E.1, E.2		**MiniLAB:** Modeling Water Particle Movement, p. 111 **Visualizing Wave Movement,** p. 112 **Science Online,** p. 113 **Integrate Life Science,** p. 115 **Lab:** Wave Properties, p. 117 **Lab:** Sink or Float, p. 118 **Science and Language Arts:** The Jungle of Ceylon, p. 120

Glencoe Exclusive! TeacherWorks™ All-In-One Planner and Resource Center

Lab Materials	Reproducible Resources	Section Assessment	Technology
Launch Lab: beaker, bowl, ice, warm water, dropper, food coloring	**Chapter *FAST FILE* Resources** Foldables Worksheet, p. 13 Directed Reading Overview, p. 15 Note-taking Worksheets, pp. 29–31	Glencoe's Assessment Advantage	TeacherWorks includes: • Interactive Teacher Edition • Lesson Planner with calendar • Access to all program blacklines • Correlations to standards • Web links
Need materials? Contact Science Kit at 1-800-828-7777 or www.sciencekit.com on the Internet.	**Chapter *FAST FILE* Resources** Transparency Activity, p. 40 Lab Activity, pp. 9–10 Enrichment, p. 26 Reinforcement, p. 23 Directed Reading, p. 16 **Cultural Diversity,** p. 65	**Portfolio** Assessment, p. 103 **Performance** Applying Math, p. 103 **Content** Section Review, p. 103	Section Focus Transparency Virtual Labs CD-ROM Guided Reading Audio Program Interactive Chalkboard CD-ROM
MiniLAB: clear-plastic storage box, water, table salt, glass, food coloring, spoon	**Chapter *FAST FILE* Resources** Transparency Activity, p. 41 MiniLAB, p. 3 Lab Activity, pp. 11–12 Enrichment, p. 27 Reinforcement, p. 24 Directed Reading, p. 16 **Mathematics Skill Activities,** p. 9	**Portfolio** Curriculum Connection, p. 106 **Performance** MiniLAB, p. 107 Applying Math, p. 108 Applying Skills, p. 109 **Content** Section Review, p. 109	Section Focus Transparency Virtual Labs CD-ROM Guided Reading Audio Program Interactive Chalkboard CD-ROM
MiniLAB: clear-plastic storage box, tape, water, cork, spoon **Lab:** 11-in × 14-in white paper, 3-speed electric fan, gooseneck lamp, clock or watch, clear-plastic storage box, water, metric ruler **Lab:** small, uncooked potato; teaspoon; salt; large glass bowl; water; balance; large graduated cylinder; metric ruler	**Chapter *FAST FILE* Resources** Transparency Activity, p. 42 MiniLAB, p. 4 Enrichment, p. 28 Reinforcement, p. 25 Directed Reading, pp. 17, 18 Transparency Activity, pp. 43–44 Lab Worksheets, pp. 5–6, 7–8 **Lab Management and Safety,** p. 65	**Portfolio** Differentiated Instruction, p. 112 **Performance** MiniLAB, p. 111 Applying Skills, p. 116 **Content** Section Review, p. 116	Section Focus Transparency Teaching Transparency Virtual Labs CD-ROM Guided Reading Audio Program Interactive Chalkboard CD-ROM Video Lab Probeware Lab

End of Chapter Assessment

Glencoe's Assessment Advantage

Blackline Masters	Technology	Professional Series
Chapter *FAST FILE* Resources Chapter Review, pp. 33–34 Chapter Tests, pp. 35–38 **Standardized Test Practice,** pp. 19–22	MindJogger Videoquiz Virtual Labs CD-ROM Exam*View*® Pro Testmaker TeacherWorks CD-ROM Interactive Chalkboard CD-ROM	**Performance Assessment in the Science Classroom (PASC)**

chapter 4 Ocean Motion

Transparencies

Section Focus

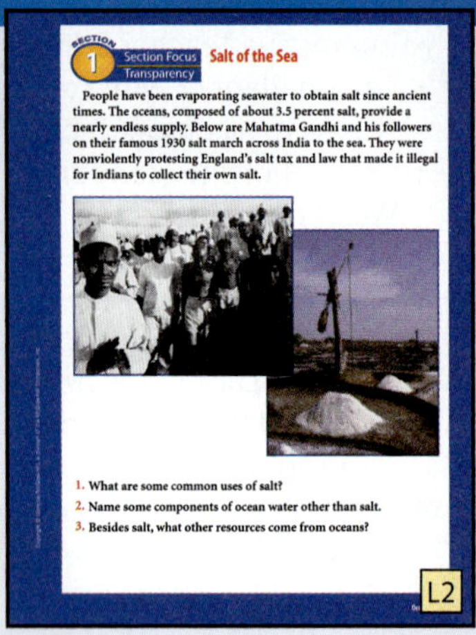
SECTION 1 Section Focus Transparency **Salt of the Sea**

People have been evaporating seawater to obtain salt since ancient times. The oceans, composed of about 3.5 percent salt, provide a nearly endless supply. Below are Mahatma Gandhi and his followers on their famous 1930 salt march across India to the sea. They were nonviolently protesting England's salt tax and law that made it illegal for Indians to collect their own salt.

1. What are some common uses of salt?
2. Name some components of ocean water other than salt.
3. Besides salt, what other resources come from oceans?

L2

SECTION 2 Section Focus Transparency **That Could Cool a Lot of Lemonade**

Once in the water, roughly six-sevenths of an iceberg is below the surface. Since there is so much of the iceberg below the waterline, deeper currents move the mass more than winds or waves. One of the currents that moves icebergs in the North Atlantic is the Labrador Current.

1. Why are deeper currents, more than waves or wind, the moving force behind icebergs?
2. What factors might affect whether or not an iceberg moves far enough south to enter the shipping lanes?

L2

SECTION 3 Section Focus Transparency **Run, Grunion, Run**

Grunion are small fish that know how to take advantage of California tides. When tides are the highest, they swim with each wave as far up the beach as possible and lay their eggs, as shown below. About two weeks later, when tides are again peaking, the hatched baby grunion are carried into the ocean as waves wash over the spawning area.

1. When would be a good time to catch grunion?
2. Why do you think grunion lay their eggs in an area that is temporarily out of reach of the tides?
3. Why do sand castles built on the beach get washed away?

L2

This is a representation of key blackline masters available in the Teacher Classroom Resources. See Resource Manager boxes within the chapter for additional information.

Assessment

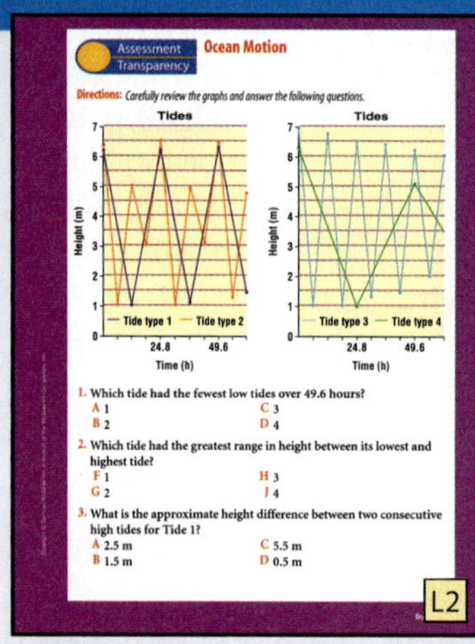
Assessment Transparency **Ocean Motion**

Directions: *Carefully review the graphs and answer the following questions.*

1. Which tide had the fewest low tides over 49.6 hours?
   A 1 C 3
   B 2 D 4
2. Which tide had the greatest range in height between its lowest and highest tide?
   F 1 H 3
   G 2 J 4
3. What is the approximate height difference between two consecutive high tides for Tide 1?
   A 2.5 m C 5.5 m
   B 1.5 m D 0.5 m

L2

Teaching

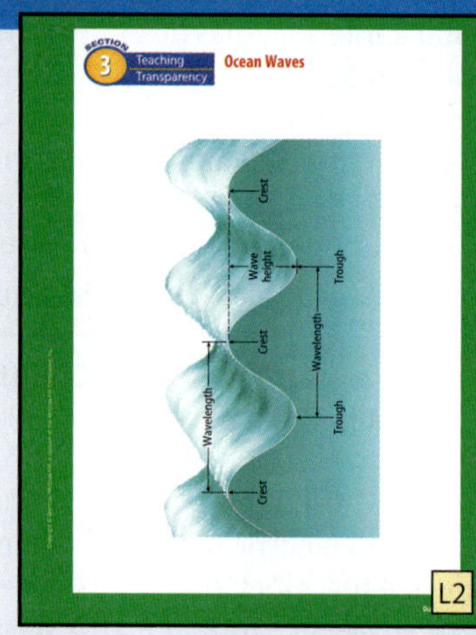
SECTION 3 Teaching Transparency **Ocean Waves**

L2

Key to Teaching Strategies

The following designations will help you decide which activities are appropriate for your students.

- L1 Level 1 activities should be appropriate for students with learning difficulties.
- L2 Level 2 activities should be within the ability range of all students.
- L3 Level 3 activities are designed for above-average students.
- ELL ELL activities should be within the ability range of English Language Learners.
- COOP LEARN Cooperative Learning activities are designed for small group work.
- LS Multiple Learning Styles logos, as described on page 6T, are used throughout to indicate strategies that address different learning styles.
- P These strategies represent student products that can be placed into a best-work portfolio.
- PBL Problem-Based Learning activities apply real-world situations to learning.

Hands-on Activities

Student Text Lab Worksheet

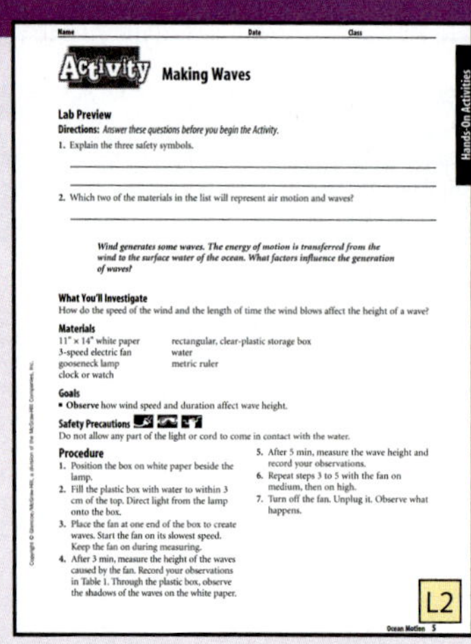
Activity **Making Waves**

Lab Preview

Directions: *Answer these questions before you begin the Activity.*

1. Explain the three safety symbols.
2. Which two of the materials in the list will represent air motion and waves?

Wind generates some waves. The energy of motion is transferred from the wind to the surface water of the ocean. What factors influence the generation of waves?

What You'll Investigate

How do the speed of the wind and the length of time the wind blows affect the height of a wave?

L2

Laboratory Activities

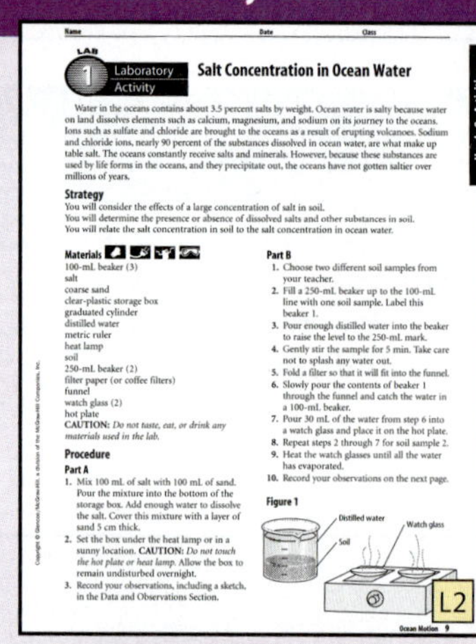
LAB 1 Laboratory Activity **Salt Concentration in Ocean Water**

L2

Meeting Different Ability Levels

Content Outline

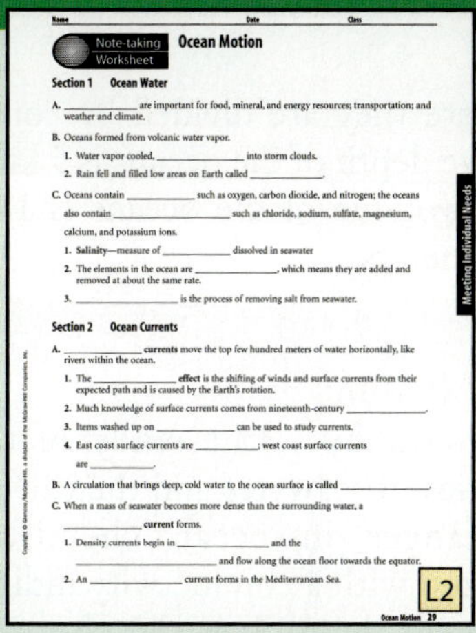

Reinforcement

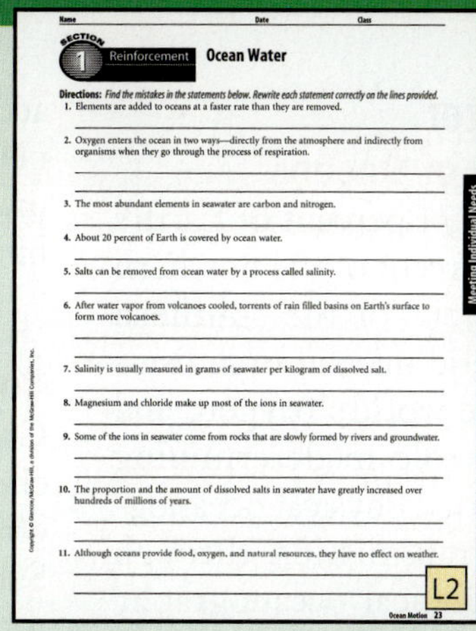

Enrichment

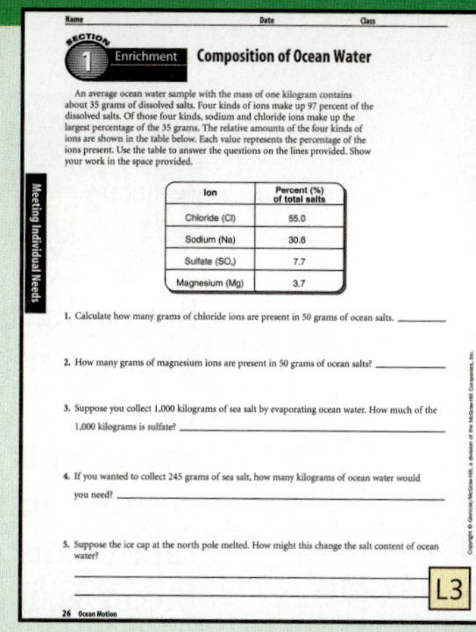

Directed Reading (English/Spanish)

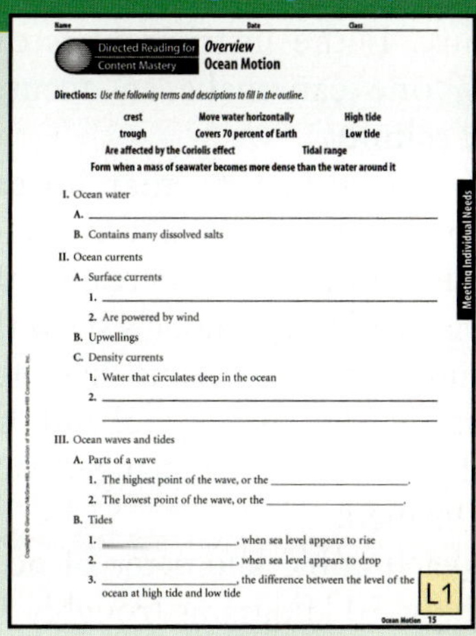

Study Guide

Reading Essentials

Assessment

Test Practice Workbook

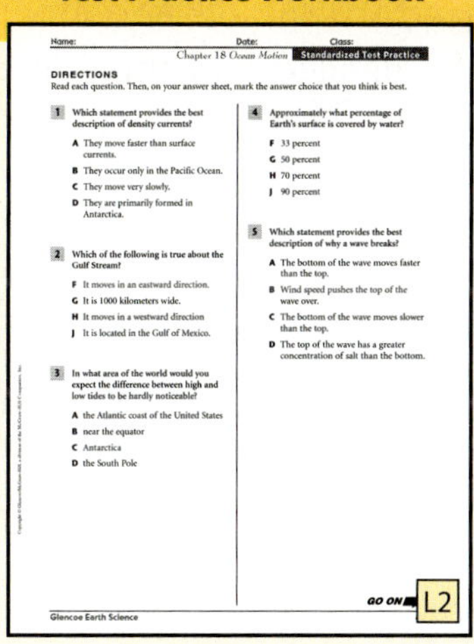

Chapter Review

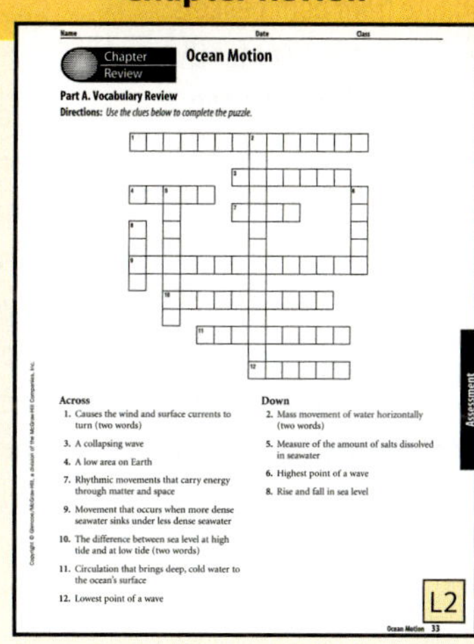

Chapter Tests

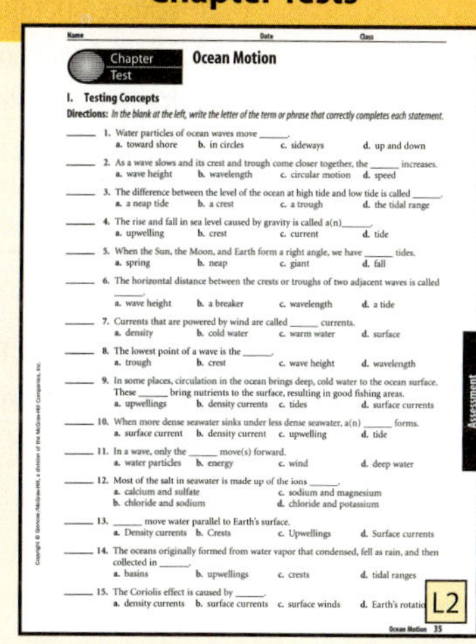

Science Content Background

section 1

Ocean Water

Importance of Oceans

Oceans cover 71 percent of Earth's surface, 61 percent of the northern hemisphere, and 81 percent of the southern hemisphere. Photosynthetic marine organisms produce 70 percent of the world's oxygen, and the oceans play a major role in determining weather and climate patterns. Physical oceanography concerns energy transmission through ocean water. Topics in physical oceanography include wave formation and movement, currents, and tides.

Although oceans are important to life on Earth, they make up only a thin skin on its surface. Isaac Asimov made the following analogy: If Earth were the size of a billiard ball, the oceans would be an unnoticeable film of dampness.

Teacher to Teacher
Angelica Gonzalez
Eastwood Academy
Houston, TX

"To show convection as ice melts, fill two beakers with 250 mL of tepid water. Stir in and dissolve 15 g of NaCl in one beaker. Begin by having students record the temperature of the water in each beaker. Gently place an ice cube dyed with food coloring in the water of each beaker. Start a timer and observe how the dye travels as the ice melts. Have students take temperature readings each minute from both beakers. The dye will float in the salt water. The time it takes for the ice to melt will vary. Ask students why they think the dye floats and the ice cube takes longer to melt in the salt water."

Angelica Gonzalez

Origin of Oceans

The locations of the ocean basins have changed throughout geologic time. Presently, all oceans connect. Geographers have given them names according to where they are divided by continents. The average depth of the oceans is 5 kilometers; the total volume of the oceans is 1.35 billion cubic kilometers.

Composition of Oceans

Chemical oceanography is the study of the chemical properties of seawater and the causes, effects, and changes in ocean chemistry. Salinity is measured with a conductivity indicator. This instrument determines the electrical conductivity of the water when a voltage is applied. The higher the conductivity, the higher the salinity. If the instrument is calibrated properly, one can make an accurate determination of salinity.

Scientists once thought that oceans increased in salinity over time. As our understanding of oceanic processes has increased, that idea has been revised. Current models show that the overall salinity of the oceans has been roughly constant for the last several hundred million years and probably hasn't changed much since the oceans formed.

On average, each cubic kilometer of ocean water contains about 50 kilograms (roughly 100 pounds) of gold. At $300 an ounce, that much gold would be worth approximately half a million dollars. Unfortunately, at present technology levels, extracting that much gold would cost much more than it's worth because of equipment and energy expenses.

Oceans can be divided into three layers based on temperature. Thermal energy in the surface layer is evenly distributed because waves mix the water and because of the turbulence caused by currents. The depth of this layer averages 200 m to 300 m. Below the surface layer is the thermocline, where the temperature drops rapidly to about 5°C. Below the thermocline, temperature decreases slightly to between 0°–3°C.

Section 2 Ocean Currents

Upwelling

Most upwellings occur along the eastern shores of oceans where winds blow surface water away from the shore. From space, satellites can detect upwellings. Because the water in an upwelling is cold compared to the surrounding water, infrared cameras can spot the difference. This information is useful to fishers.

Deep Water

North Atlantic Deep Water forms in the Norwegian and Greenland Seas. Warm water carried poleward by the North Atlantic Drift cools and sinks. After sinking, the water mixes with salty water from the Arctic Sea to become a cold, salty bottom water. This water accumulates behind a submarine plateau until it spills over into the Atlantic Ocean and spreads southward.

Antarctic Bottom Water forms around Antarctica. Regions of exposed seawater within winter ice cover, called polynyas, are thought to be important in the formation of Antarctic Bottom Water. Here sea ice forms rapidly, but is quickly blown away from the open region. The formation of relatively fresh sea ice causes the remaining water to become salty and cold. It then can sink to form Antarctic Bottom Water.

Section 3 Ocean Waves and Tides

Waves

The longer wind blows, the higher the waves mount until they reach a maximum height for a given wind velocity. For a given wind velocity, waves tend to be higher on large, deep bodies of water than on shallower bodies of water. Waves on the larger, deeper bodies develop without the interference of drag on the ocean bottom.

Storm waves have longer wavelengths than normal waves. Their erosional effects on the basin bottom extend relatively far from shore.

Tides

The equilibrium theory of tide formation (discussed in this chapter) states that one high tide should form on the side of Earth toward the Moon and one on the opposite side. In reality, the tidal waves break up into cells in the oceans and high tides rotate around the center of each cell.

chapter content resources

Internet Resources

For additional content background, visit bookh.msscience.com to:

- access your book online
- find references to related articles in popular science magazines
- access Web links with related content background
- access current events with science journal topics

Print Resources

Introductory Oceanography, by Harold Thurman and Elizabeth Burton, Prentice Hall, 2001

Descriptive Physical Oceanography, by George Pickard and William Emery, Pergamon Press, 1995

Sea Change, A Message of the Oceans, by Sylvia Earle, G.P. Putnam's Sons, 1995

Dale E. Boyer/Photo Researchers, Inc.

chapter 4

Chapter Vocabulary

basin, p. 101
salinity, p. 102
surface current, p. 104
Coriolis effect, p. 105
upwelling, p. 107
density current, p. 107
wave, p. 110
crest, p. 110
trough, p. 110
breaker, p. 111
tide, p. 113
tidal range, p. 114

Science Journal Student answers will vary but may include facts such as ocean currents affect weather and climate, surface waves are caused by wind, and tides are caused by gravitational forces.

INTERACTIVE CHALKBOARD with Image Bank

PowerPoint® Presentations

This CD-ROM is an editable Microsoft® PowerPoint® presentation that includes:

- a pre-made presentation for every chapter
- interactive graphics
- animations
- audio clips
- image bank
- all new section and chapter questions
- Standardized Test Practice
- transparencies
- pre-lab questions for all labs
- Foldables directions
- links to bookh.msscience.com

chapter 4

Ocean Motion

chapter preview

sections

1. Ocean Water
2. Ocean Currents
3. Ocean Waves and Tides
   - *Lab* *Wave Properties*
   - *Lab* *Sink or Float?*
   - *Virtual Lab* *How are tides affected by phases of the Moon?*

The Power of Waves

Surfers from around the world experience firsthand the enormous power of moving water. Wind blowing across the ocean surface can create small ripples, the wave shown above in Fiji, and even the giant waves of hurricanes.

Science Journal Record in your Science Journal some facts you know about ocean currents, waves, or tides. Include some pictures to show your ideas.

Theme Connection

Energy Students will learn that wind provides the energy for surface currents and most ocean waves, that temperature and salinity differences power deep-ocean circulation, and that the gravitational attraction of the Sun and the Moon supply the force that sets tides in motion.

About the Photo

The Power of Waves The area along continental margins in which breaking waves release their energy is called the surf zone. As a wave reaches the shore, the shallow water depth causes the wave to lose speed and the wavelength shortens. Wave height increases and the wave breaks.

Start-Up Activities

Launch LAB

Explore How Currents Work

Surface currents are caused by wind. Deep-water currents are created by differences in the density of ocean water. Several factors affect water density. One is temperature. Do the lab below to see how temperature differences create deep-water currents.

1. In a bowl, mix ice and cold water to make ice water.
2. Fill a beaker with warm tap water.
3. Add a few drops of food coloring to the ice water and stir the mixture.
4. Use a dropper to place some of this ice water on top of the warm water.
5. **Think Critically** In your Science Journal, describe what happened. Did adding cold water on top produce a current? Look up the word *convection* in a dictionary. Infer why the current you created is called a convection current.

Science Online | Preview this chapter's content and activities at bookh.msscience.com

FOLDABLES Study Organizer

Ocean Motion Make the following Foldable to help you understand the cause-and-effect relationship of ocean motion.

STEP 1 **Fold** a vertical sheet of paper in half from top to bottom.

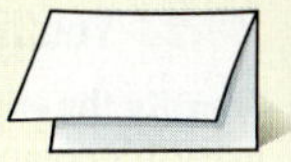

STEP 2 **Fold** in half from side to side with the previous fold at the top.

STEP 3 **Unfold** the paper once. **Cut** only the fold of the top flap to make two tabs.

STEP 4 **Turn** the paper vertically and **label** the front tabs as shown.

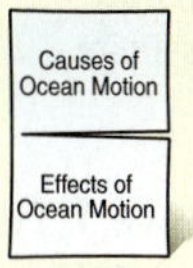

Read and Write As you read the chapter, write what you learn about why the ocean moves and the effects of ocean motion under the appropriate tabs.

Launch LAB

Purpose Use the Launch Lab to introduce students to the way deep-ocean currents work. L2 ELL COOP LEARN LS **Visual-Spatial**

Preparation Prepare ice ahead of time.

Materials ice cubes, warm water, beaker, food coloring, dropper, bowl, cold water

Teaching Strategy Be sure that students understand the term *density*.

Think Critically

When ice water was placed on top of warm water, a current formed as the denser cold water sank through the warmer water, displacing and forcing the warmer water upward. Convection is defined as the movement of fluid between areas that have unequal densities caused by unequal temperatures. It's called a convection current because the difference in density and temperature drives the movement of water.

Assessment

Oral Ask students why a current would not form if warm water was placed on top of ice water. Warm water is less dense than cold water, so it stays on top and there is no movement. Use **Performance Assessment in the Science Classroom**, p. 89. L2

FOLDABLES Study Organizer — Dinah Zike Study Fold

Student preparation materials for this Foldable are available in the Chapter *Fast File* Resources.

Probeware Labs

This chapter's data-collection lab using Probeware technology is included on the *Video Labs CD-ROM*. See the *Probeware Lab Manual* pages 25–28 for student worksheets.

section 1

Ocean Water

as you read

What You'll Learn

- **Identify** the origin of the water in Earth's oceans.
- **Explain** how dissolved salts and other substances get into seawater.
- **Describe** the composition of seawater.

Why It's Important

Oceans are a reservoir of valuable food, energy, and mineral resources.

Review Vocabulary
resource: a reserve source of supply, such as a material or mineral

New Vocabulary
- basin
- salinity

Importance of Oceans

Imagine yourself lying on a beach and listening to the waves gently roll onto shore. A warm breeze blows off the water, making it seem as if you're in a tropical paradise. It's easy to appreciate the oceans under these circumstances, but the oceans affect your life in other ways, too.

Varied Resources Oceans are important sources of food, energy, and minerals. **Figure 1** shows two examples of food resources collected from oceans. Energy sources such as oil and natural gas are found beneath the ocean floor. Oil wells often are drilled in shallow water. Mineral resources including copper and gold are mined in shallow waters as well. Approximately one-third of the world's table salt is extracted from seawater through the process of evaporation. Oceans also allow for the efficient transportation of goods. For example, millions of tons of oil, coal, and grains are shipped over the oceans each year.

Reading Check *What resources come from oceans?*

Figure 1 People depend on the oceans for many resources.

Krill are tiny, shrimplike animals that live in the Antarctic Ocean. Some cultures use krill in noodles and rice cakes.

Kelp is a fast-growing seaweed that is a source of algin, used in making ice cream, salad dressing, medicines, and cosmetics.

1 Motivate

INTERACTIVE CHALKBOARD
PowerPoint® Presentations

Bellringer

Section Focus Transparencies also are available on the Interactive Chalkboard CD-ROM. L2 ELL

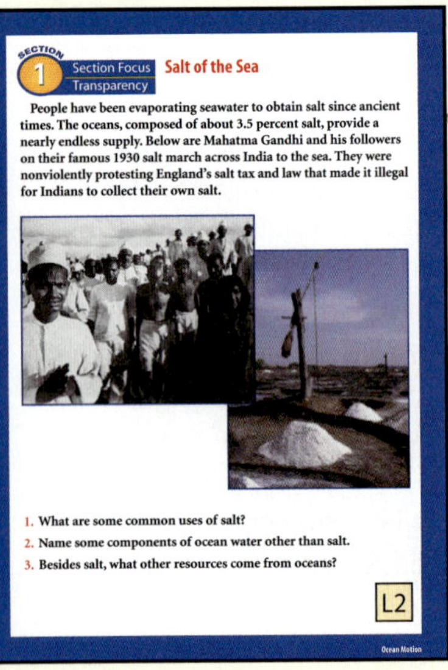

Section 1 Section Focus Transparency — Salt of the Sea

People have been evaporating seawater to obtain salt since ancient times. The oceans, composed of about 3.5 percent salt, provide a nearly endless supply. Below are Mahatma Gandhi and his followers on their famous 1930 salt march across India to the sea. They were nonviolently protesting England's salt tax and law that made it illegal for Indians to collect their own salt.

1. What are some common uses of salt?
2. Name some components of ocean water other than salt.
3. Besides salt, what other resources come from oceans?

L2

Ocean Motion

Tie to Prior Knowledge

Ocean Properties Have students quickly list three things they know about the ocean. How many listed that the water is salty? Where did the salt come from? Most enters the water from surrounding land. Tell students they will discover many things about the ocean in this section, including the reason its water is salty. L2

Reading Check

Answer food; fuels such as oil and natural gas; minerals such as copper, gold, and salt

Section 1 Resource Manager

Chapter *FAST FILE* Resources

Transparency Activity, p. 40
Directed Reading for Content Mastery, pp. 15, 16
Note-taking Worksheets, pp. 29–31
Lab Activity, pp. 9–10
Enrichment, p. 26
Reinforcement, p. 23

Origin of Oceans

During Earth's first billion years, its surface, shown in the top portion of **Figure 2,** was much more volcanically active than it is today. When volcanoes erupt, they spew lava and ash, and they give off water vapor, carbon dioxide, and other gases. Scientists hypothesize that about 4 billion years ago, this water vapor began to be stored in Earth's early atmosphere. Over millions of years, it cooled enough to condense into storm clouds. Torrential rains began to fall. Shown in the bottom portion of **Figure 2,** oceans were formed as this water filled low areas on Earth called **basins.** Today, approximately 70 percent of Earth's surface is covered by ocean water.

Figure 2 Earth's oceans formed from water vapor.

Water vapor was released into the atmosphere by volcanoes that also gave off other gases, such as carbon dioxide and nitrogen.

Condensed water vapor formed storm clouds. Oceans formed when basins filled with water from torrential rains.

Composition of Oceans

Ocean water contains dissolved gases such as oxygen, carbon dioxide, and nitrogen. Oxygen is the gas that almost all organisms need for respiration. It enters the oceans in two ways—directly from the atmosphere and from organisms that photosynthesize. Carbon dioxide enters the ocean from the atmosphere and from organisms when they respire. The atmosphere is the only important source of nitrogen gas. Bacteria combine nitrogen and oxygen to create nitrates, which are important nutrients for plants.

If you've ever tasted ocean water, you know that it is salty. Ocean water contains many dissolved salts. Chloride, sodium, sulfate, magnesium, calcium, and potassium are some of the ions in seawater. An ion is a charged atom or group of atoms. Some of these ions come from rocks that are dissolved slowly by rivers and groundwater. These include calcium, magnesium, and sodium. Rivers carry these chemicals to the oceans. Erupting volcanoes add other ions, such as bromide and chloride.

Reading Check *How do sodium and chloride ions get into seawater?*

2 Teach

Reading Check

Answer Rivers and groundwater slowly dissolve rocks and carry sodium and chloride ions to oceans. Volcanoes also release chloride.

Inquiry Lab

Build a Desalination Plant

Question What is the best design for a desalination plant?

Possible Materials Bunsen burner or hotplate, glass beaker or square metal pan, metal or glass sheets

Estimated Time 1–3 class sessions

Teaching Strategies

- Make it clear to students that they should use a minimum of input energy, or to use free energy. A solar plant would probably be best, but wind or hydro power could also be used.
- Have students research and design a desalination plant. They may then build a scale model to test with sun, wind, or water, or draw up a poster, report, web page, or computer presentation to explain their design.
- Another alternative, is to use Bunsen burners or a hotplate to do desalination in the lab. Have students design roofs of tin, glass, or other plating, and test out different roof shapes.
- Students may enjoy researching initiatives that have been taken locally and globally with this technology, and related political and financial issues.

For additional inquiry activities, see *Science Inquiry Labs.*

Teacher FYI

Manganese Nodules In addition to organic processes, inorganic processes also remove dissolved substances from seawater. One example is the formation of manganese nodules, which are valuable mineral deposits on the ocean floor. The nodules form when manganese precipitates out of seawater.

Differentiated Instruction

Learning Disabled Have groups of students make puzzles of Earth's oceans and landmasses. They can transfer a map of the world onto cardboard and cut it into pieces. Ocean names should be included on the puzzle pieces. Groups should then exchange puzzles and compete to see which group puts the puzzle together fastest. L2 ELL LS **Kinesthetic**

Caption Answer

Figure 3 35g

Discussion

Water Shortages Why are there drinking water shortages in some places if Earth's surface is 70 percent ocean? Although oceans cover most of the planet, ocean water is too salty to drink.

Use Science Words

Word Meaning Have students find out why an industrial plant that removes salt from seawater is called a *desalination* plant. *De-* is a prefix meaning "away"; *saline* relates to salt. L2

Fun Fact

The typical residence time in the ocean for a sodium ion is 260 million years.

Make a Model

Ocean Salinity Have students model how some ions in salt get into oceans. Make a mixture of half salt and half soil. Have pairs of students put a couple of tablespoons of the mixture into a coffee filter and place the filter inside a paper cup with five small holes punched into the bottom. Have students pour three tablespoons of water into the cup and hold it 2 cm above a sheet of black construction paper as the water drips onto the paper. When all the water has dripped out, put the paper in the Sun to dry. Students will find a dried salt residue on the paper. Have them explain their results. As water was poured through the soil, salt dissolved in it and was deposited on the paper. In the same way, runoff picks up and deposits dissolved salts from the land in the ocean. L2

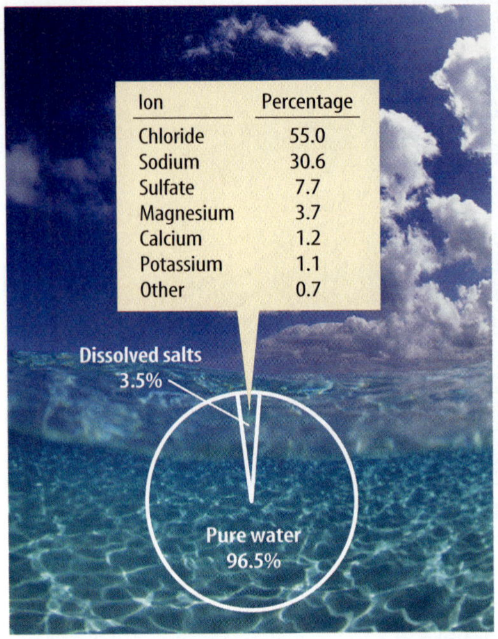

Ion	Percentage
Chloride	55.0
Sodium	30.6
Sulfate	7.7
Magnesium	3.7
Calcium	1.2
Potassium	1.1
Other	0.7

Figure 3 Ocean water contains about 3.5 percent dissolved salts.
Calculate *If you evaporated 1,000 g of seawater, how many grams of salt would be left?*

Salts The most abundant elements in sea water are the hydrogen and oxygen that make up water. Many other ions are found dissolved in seawater. When seawater is evaporated, these ions combine to form materials called salts. Sodium and chloride make up most of the ions in seawater. If seawater evaporates, the sodium and chloride ions combine to form a salt called halite. Halite is the common table salt you use to season food. It is this dissolved salt and similar ones that give ocean water its salty taste.

Salinity (say LIH nuh tee) is a measure of the amount of salts dissolved in seawater. It usually is measured in grams of dissolved salt per kilogram of seawater. One kilogram of ocean water contains about 35 g of dissolved salts, or 3.5 percent. The chart in **Figure 3** shows the most abundant ions in ocean water. The proportion and amount of dissolved salts in seawater remain nearly constant and have stayed about the same for hundreds of millions of years. This tells you that the composition of the oceans is in balance. Evidence that scientists have gathered indicates that Earth's oceans are not growing saltier.

INTEGRATE Life Science

Removal of Elements Although rivers, volcanoes, and the atmosphere constantly add material to the oceans, the oceans are considered to be in a steady state. This means that elements are added to the oceans at about the same rate that they are removed. Dissolved salts are removed when they precipitate out of ocean water and become part of the sediment. Some marine organisms use dissolved salts to make body parts. Some remove calcium ions from the water to form bones. Other animals, such as oysters, use the dissolved calcium to form shells. Some algae, called diatoms, have silica shells. Because many organisms use calcium and silicon, these elements are removed more quickly from seawater than elements such as chlorine or sodium.

Desalination Salt can be removed from ocean water by a process called desalination (dee sa luh NAY shun). If you have ever swum in the ocean, you know what happens when your skin dries. The white, flaky substance on your skin is salt. As seawater evaporates, salt is left behind. As demand for freshwater increases throughout the world, scientists are working on technology to remove salt to make seawater drinkable.

Differentiated Instruction

Challenge Have student teams design and build solar desalination devices. Have each team begin with the same amount and concentration of salt water. Then have teams compete to see which one can collect the most freshwater. L3

Cultural Diversity

Seashell Money Hundreds of years ago, the native people of eastern North America used carved and polished strings of shells, called wampum, as money. They made wampum beads from the white shells of the whelk and the purple shells of a type of clam. When Europeans came to North America, standard-length strings of wampum beads became a unit of currency in their trade with coastal Native Americans.

102 ◆ H CHAPTER 4 Ocean Motion

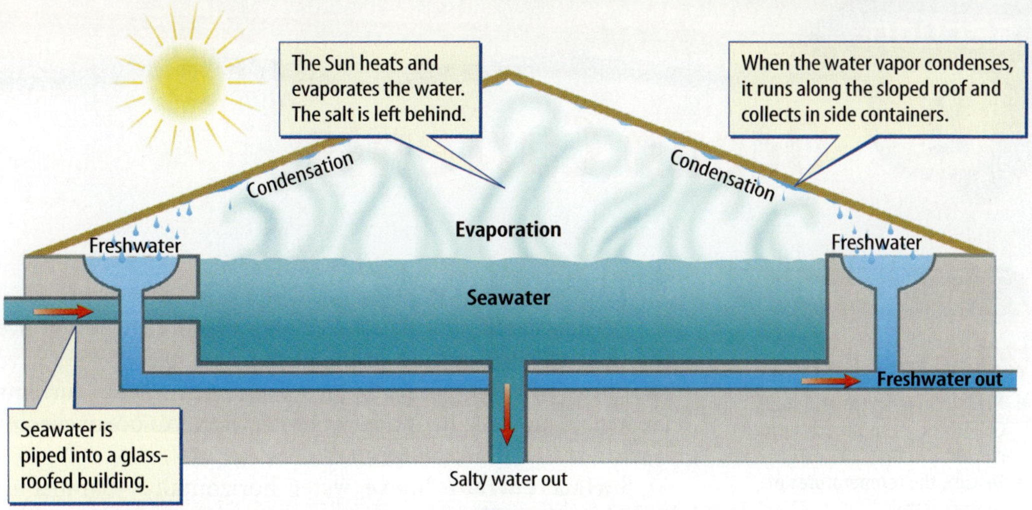

Desalination Plants Some methods of desalination include evaporating seawater and collecting the freshwater as it condenses on a glass roof. **Figure 4** shows how a desalination plant that uses solar energy works. Other plants desalinate water by passing it through a membrane that removes the dissolved salts. Freshwater also can be obtained by melting frozen seawater. As seawater freezes, the ice crystals that form contain much less salt than the remaining water. The salty, unfrozen water then can be separated from the ice. The ice can be washed and melted to produce freshwater.

Figure 4 This desalination plant uses solar energy to produce freshwater.

section 1 review

Summary

Importance of Oceans

- Oceans are a source of food, energy, and minerals.
- Oceans allow for the efficient transportation of goods such as oil, coal, and grains.

Origin of Oceans

- Scientists hypothesize that about 4 billion years ago, water vapor from volcanic eruptions cooled and condensed into storm clouds. Oceans formed as water from torrential rains filled Earth's basins.

Composition of Oceans

- Ocean water contains dissolved gases and salts.
- Oceans are considered to be in a steady state.

Self Check

1. **Describe** five ways Earth's oceans affect your life.
2. **Explain** the relationship between volcanic activity and the origin of Earth's oceans.
3. **Identify** the components of seawater. How do dissolved salts enter oceans? How does oxygen enter oceans?
4. **Think Critically** Organisms in the oceans are important sources of food and medicine. What steps can humans take to ensure that these resources are available for future generations?

Applying Math

5. **Use Proportions** If the average salinity of seawater is 35 parts per thousand, how many grams of dissolved salts will 500 g of seawater contain?

Science Online bookh.msscience.com/self_check_quiz

Quick Demo

Salinity

Materials beaker, water, table salt, balance, glass stirrer, dropper

Estimated Time 5–10 minutes

Procedure To show students that ocean water is too salty to drink dissolve 17.5 g of table salt into 500 mL of water. Have students place a drop of the water on their tongues to taste the saltiness of ocean water.

Visual Learning

Figure 4 Ask students to locate the seawater and the freshwater in the diagram.

3 Assess

DAILY INTERVENTION

Check for Understanding

Visual-Spatial Have students create an events chain map that explains the hypothesis of the origin of the oceans. L2

Reteach

Summaries Direct students to write summaries containing the main points covered in the section. L2 ELL LS **Linguistic**

Assessment

Performance Have each student make a concept map that shows how elements are naturally added to and removed from the oceans. Use **Performance Assessment in the Science Classroom,** p. 161. L2 P

section 1 review

1. Possible answers: source of food, oxygen, minerals, medicines, and energy; means of transportation; effects on weather and climate; recreation
2. Volcanoes released water vapor that accumulated, cooled, and condensed to fall as rain that filled the basins about 4 billion years ago.
3. Ocean water contains dissolved gases and dissolved salts. River water and groundwater carry dissolved salts from soil and rock on land into the ocean. Volcanoes also provide chlorine. Some oxygen comes directly from the atmosphere, and other oxygen is produced when organisms photosynthesize.
4. People can conserve these resources by not overharvesting the organisms and by not polluting the oceans.
5. 17.5 g (1/2 × 35)

1 Motivate

INTERACTIVE CHALKBOARD PowerPoint® Presentations

Bellringer

Section Focus Transparencies also are available on the Interactive Chalkboard CD-ROM. L2 ELL

Section 2 Section Focus Transparency **That Could Cool a Lot of Lemonade**

Once in the water, roughly six-sevenths of an iceberg is below the surface. Since there is so much of the iceberg below the waterline, deeper currents move the mass more than winds or waves. One of the currents that moves icebergs in the North Atlantic is the Labrador Current.

1. Why are deeper currents, more than waves or wind, the moving force behind icebergs?
2. What factors might affect whether or not an iceberg moves far enough south to enter the shipping lanes?

L2

Tie to Prior Knowledge

Currents Ask students whether they have ever gone swimming or wading in a river, a large lake, or the ocean. Have they ever felt tugged or pulled by water moving under the surface? Tell students these were currents. Also tell them that in this section they will learn how currents in the ocean form and how they move. L2

section 2

Ocean Currents

as you read

What You'll Learn

- **Explain** how winds and the Coriolis effect influence surface currents.
- **Discuss** the temperatures of coastal waters.
- **Describe** density currents.

Why It's Important

Ocean currents and the atmosphere transfer heat that affects the climate you live in.

Review Vocabulary
circulation: a water current flow occurring in an area in a closed, circular pattern

New Vocabulary
- surface current
- Coriolis effect
- upwelling
- density current

Surface Currents

When you stir chocolate into a glass of milk, do you notice the milk swirling around in the glass in a circle? If so, you've observed something similar to an ocean current. Ocean currents are a mass movement, or flow, of ocean water. An ocean current is like a river within the ocean.

Surface currents move water horizontally—parallel to Earth's surface. These currents are powered by wind. The wind forces the ocean to move in huge, circular patterns. **Figure 5** shows these major surface currents. Notice that some currents are shown with red arrows and some are shown with blue arrows. Red arrows indicate warm currents. Blue arrows indicate cold currents. The currents on the ocean's surface are related to the general circulation of winds on Earth.

Surface currents move only the upper few hundred meters of seawater. Some seeds and plants are carried between continents by surface currents. Sailors take advantage of these currents along with winds to sail more efficiently from place to place.

Figure 5 These are the major surface currents of Earth's oceans.

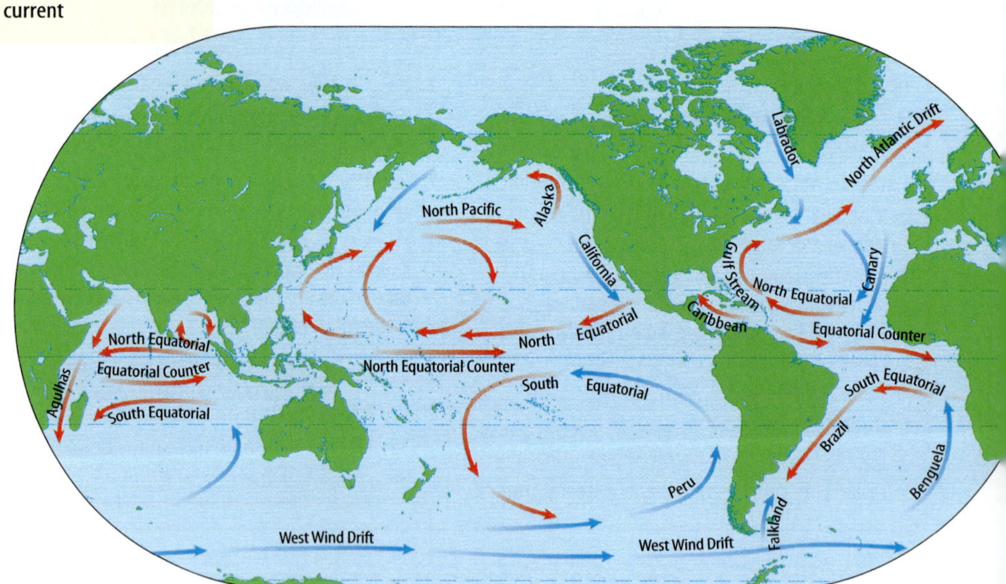

Section 2 Resource Manager

Chapter *FAST FILE* Resources

Transparency Activity, p. 41
MiniLAB, p. 3
Lab Activity, pp. 11–12
Enrichment, p. 27
Reinforcement, p. 24

How Surface Currents Form Surface ocean currents and surface winds are affected by the Coriolis (kor ee OH lus) effect. The **Coriolis effect** is the shifting of winds and surface currents from their expected paths that is caused by Earth's rotation. Imagine that you try to draw a line straight out from the center of a disk to the edge of the disk. You probably could do that with no problem. But what would happen if the disk were slowly spinning like the one in **Figure 6?** As the student tried to draw a straight line, the disk rotated and, as shown in **Figure 6,** the line curved.

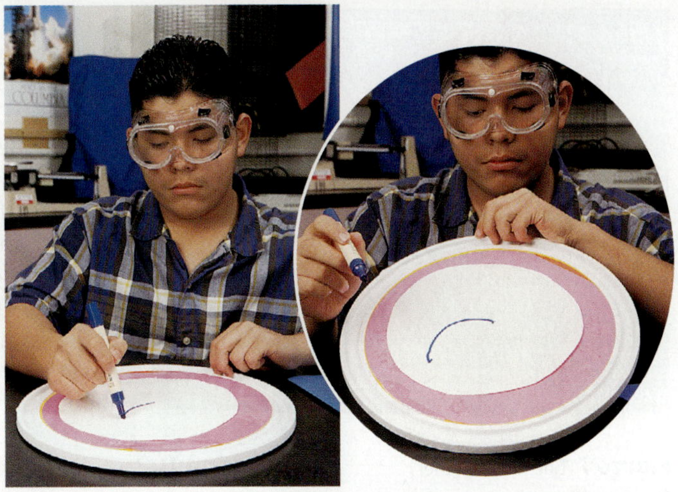

Figure 6 The student draws a line straight out from the center while spinning the disk. Because the disk was spinning, the line is curved.
Describe *how the Coriolis effect influences surface currents in the southern hemisphere.*

A similar thing happens to wind and surface currents. Because Earth rotates toward the east, winds appear to curve to the right in the northern hemisphere and to the left in the southern hemisphere. These surface winds can cause water to pile up in certain parts of the ocean. When gravity pulls water off the pile, the Coriolis effect turns the water. This causes surface water in the oceans to spiral around the piles of water. The Coriolis effect causes currents north of the equator to turn to the right. Currents south of the equator are turned to the left. Look again at the map of surface currents in **Figure 5** to see the results of the Coriolis effect.

The Gulf Stream Although satellites provide new information about ocean movements, much of what is known about surface currents comes from records that were kept by sailors of the nineteenth century. Sailors always have used surface currents to help them travel quickly. Sailing ships depend on some surface currents to carry them to the west and others to carry them east. During the American colonial era, ships floated on the 100-km-wide Gulf Stream current to go quickly from North America to England. Find the Gulf Stream current in the Atlantic Ocean on the map in **Figure 5.**

In the late 1700s, Deputy Postmaster General Benjamin Franklin received complaints about why it took longer to receive a letter from England than it did to send one there. Upon investigation, Franklin found that a Nantucket whaling captain's map furnished the answer. Going against the Gulf Stream delayed ships sailing west from England by up to 110 km per day.

Science Online
Topic: Ocean Currents
Visit bookh.msscience.com for Web links to information about ocean currents.

Activity Choose an area of the world's oceans and make a map of the currents in that area. In which direction do they travel? What are the currents named?

2 Teach

Visual Learning

Figure 5 Tell students that a gyre refers to the circular pattern of currents that occurs in each of the major ocean basins. Which currents make up the gyre in the South Pacific Ocean? South Equatorial current and Peru current

Discussion

Gulf Stream Current Why is the Gulf Stream current not as important to sailors today as it was 200 years ago? Sailors 200 years ago used the strength of the wind in their sails and the force of currents pushing their ships to power them across the oceans. Today, ships have motors that power them, making dependence on currents like the Gulf Stream less necessary.

Caption Answer

Figure 6 Currents south of the equator are turned to the left.

Activity

Ocean Currents Have students find accounts of explorers and sailors that discuss moving through surface currents in different areas of the ocean.

Differentiated Instruction

Learning Disabled Have students who are experiencing difficulty conceptualizing the Coriolis effect carry out the action depicted in **Figure 6.** Have them cut a large circle from a piece of paper. They should begin by placing the point of their pencil or marker in the center of the circle. As one student draws a straight line out from the center, have another student slowly turn the paper clockwise Repeat the steps, this time turning the paper counterclockwise.

Active Reading

Write-Draw-Discuss This strategy encourages students to actively participate in reading and lectures, assimilating content creatively. Have students write about an idea, then make an illustration or drawing. Ask students to share responses with the class and display several examples. Have students Write-Draw-Discuss about surface currents. L2

IDENTIFYING Misconceptions

Gulf Stream Students may think that the Gulf Stream is located in the Gulf of Mexico. The Gulf Stream begins off the east coast of Florida, fed by other currents that flow through the Gulf of Mexico. It sweeps northeastward along the east coast of the United States and eventually turns east toward Europe.

Discussion

Cold Currents Although the climate in the area is warm in summer, the waters of San Francisco Bay never get very warm. Ask students to use what they know about currents to give a possible explanation. Cold currents along the coast of California could influence the temperature of the bay.

Teacher FYI

Tracking Ocean Currents Oceanographers track ocean currents with neutrally buoyant floats and satellites. Neutrally buoyant floats sink to a desired depth and move with the current. Satellites are equipped with devices that detect surface-current velocities using infrared, color, and microwave emissions from the sea's surface.

Caption Answer

Figure 8 near the equator

Figure 7 Bottles and other floating objects that enter the ocean are used to gain information about surface currents.

Tracking Surface Currents Items that wash up on beaches, such as the bottle shown in **Figure 7,** provide information about ocean currents. Drift bottles containing messages and numbered cards are released from a variety of coastal locations. The bottles are carried by surface currents and might end up on a beach. The person who finds a bottle writes down the date and the location where the bottle was found. Then the card is sent back to the institution that launched the bottle. By doing this, valuable information is provided about the current that carried the bottle.

Warm and Cold Surface Currents Notice in **Figure 5** that currents on the west coasts of continents begin near the poles where the water is colder. The California Current that flows along the west coast of the United States is a cold surface current. East-coast currents originate near the equator where the water is warmer. Warm surface currents, such as the Gulf Stream, distribute heat from equatorial regions to other areas of Earth. **Figure 8** shows the warm water of the Gulf Stream in red and orange. Cooler water appears in blue and green.

As warm water flows away from the equator, heat is released to the atmosphere. The atmosphere is warmed. This transfer of heat influences climate.

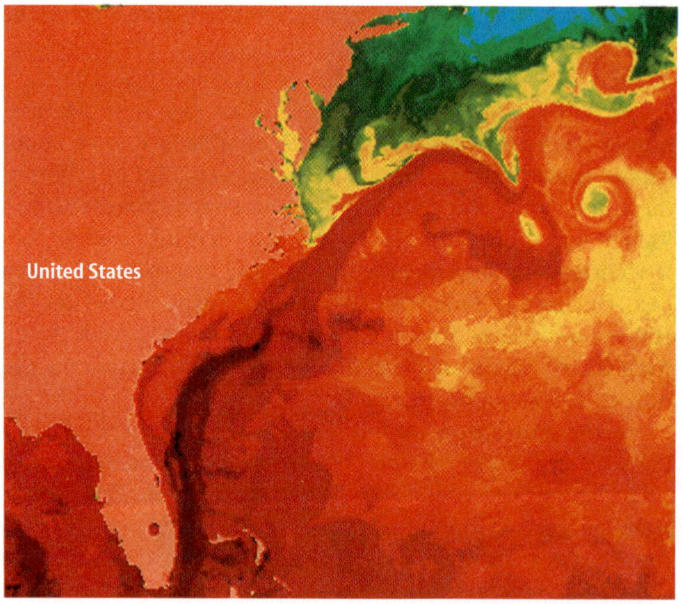

Figure 8 Data about ocean temperature collected by a satellite were used to make this surface-temperature image of the Atlantic Ocean.
Infer *Where does the Gulf Stream originate?*

Curriculum Connection

History Have students research either the way the positions of currents affected ancient trade routes, or Benjamin Franklin's study of the Gulf Stream. Have them write brief reports on what they find. L2 LS **Linguistic** P

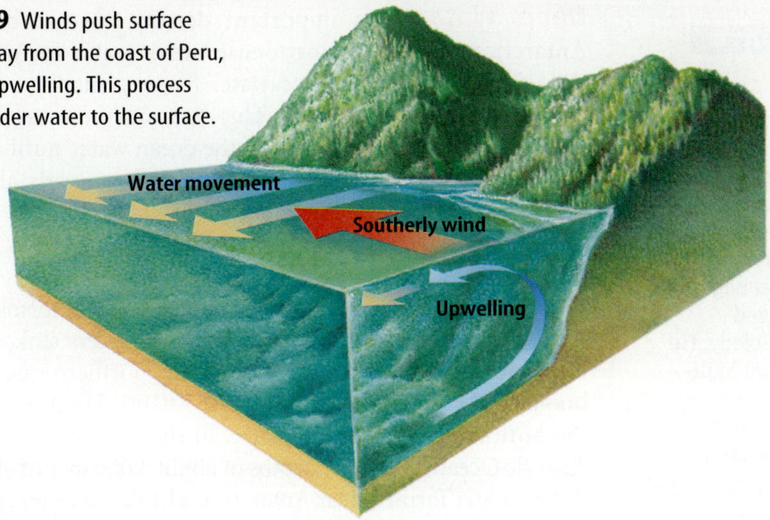

Figure 9 Winds push surface water away from the coast of Peru, causing upwelling. This process brings colder water to the surface.

Upwelling

Upwelling is a vertical circulation in the ocean that brings deep, cold water to the ocean surface. Along some coasts of continents, wind blowing parallel to the coast carries water away from the land because of the Coriolis effect, as shown in **Figure 9.** Cold, deep ocean water rises to the surface and replaces water that has moved away from shore. This water contains high concentrations of nutrients from organisms that died, sank to the bottom, and decayed. Nutrients promote primary production and plankton growth, which attracts fish. Areas of upwelling occur along the coasts of Oregon, Washington, and Peru and create important fishing grounds.

Density Currents

Deep in the ocean, waters circulate not because of wind but because of density differences. A **density current** forms when a mass of seawater becomes more dense than the surrounding water. Gravity causes more dense seawater to sink beneath less dense seawater. This deep, dense water then slowly spreads to the rest of the ocean.

The density of seawater increases if salinity increases, as you can see if you perform the MiniLAB on this page. It also increases when temperature decreases. In the Launch Lab at the beginning of the chapter, the cold water was more dense than the warm water in the beaker. The cold water sank to the bottom. This created a density current that moved the food coloring.

Changes in temperature and salinity work together to create density currents. Density currents circulate ocean water slowly.

Mini LAB

Modeling a Density Current

Procedure

1. Fill a **clear-plastic storage box** (shoe-box size) with room-temperature **water.**
2. Mix several spoonfuls of **table salt** into a **glass of water** at room temperature.
3. Add a few drops of **food coloring** to the saltwater solution. Pour the solution slowly into the freshwater in the large container.

Analysis

1. Describe what happened when you added salt water to freshwater.
2. How does this lab model density currents?

Try at Home

Mini LAB

Purpose Students model a density current. L1 ELL

COOP LEARN **Kinesthetic**

Materials clear-plastic storage box, water, 8-ounce glass or beaker, table salt, food coloring, teaspoon

Teaching Strategy Review density with students before assigning the MiniLAB.

Safety Precautions Caution students not to taste, eat, or drink any lab materials.

Troubleshooting To see the current, have students look into the box from the side, not the top.

Analysis

1. The salt water sank to the bottom and spread out toward the sides of the container.
2. Higher salinity increases the density of ocean water. This can contribute to the formation of density currents.

Assessment

Process Have students draw and label a diagram based on this MiniLAB that shows how density currents in the ocean work. Use **Performance Assessment in the Science Classroom,** p. 89. L2

Try at Home

Differentiated Instruction

Challenge Have students research how the Gulf Stream affects the climate of land areas it passes close to on its trip to northwestern Europe. In general, because of prevailing wind patterns, the Gulf Stream makes the climate in northwestern Europe warmer than the climate of eastern North America. When the warm Gulf Stream water comes in contact with the cooler air near England, fog forms. L3

Cultural Diversity

Fishing Upwelling of cold water off the coast of Peru brings large numbers of anchovies to the coast. Peruvians eat these fish and make them into fish meal, which is used to make tortillas and bread and to feed chickens. Have students research what happens during an El Niño. Wind patterns reverse, blocking the upwelling cold water, and stopping the movement of anchovies to the Peruvian coast. L2

INTEGRATE Career

Oceanographer Students interested in pursuing oceanography as a career should take science and math classes in high school. A bachelor's degree in biology, chemistry, geology, or physics provides a good base to enter a masters or doctoral program in oceanography. Career opportunities include conducting field research and teaching.

Applying Math

National Math Standards

Correlation to Mathematics Objectives

1

Answers to Practice Problems

1. You know the volume and the mass of the sample and the equation $d = m/v$. Substituting these values into this equation gives $d = 79{,}000$ g/78,000 cm^3. This equals a density of 1.01 g/cm^3.
2. In this problem you know the density and the mass, so you must rearrange the equation $d = m/v$ to solve for v. This is $v = m/d$. Then substituting the given values gives $v = 50{,}000$ g/1.03 g/cm^3. Solving this gives a volume of 48,544 cm^3.

INTEGRATE Career

Oceanographer An oceanographer studies the composition of ocean water; ocean currents, waves, and tides; and marine life. Dr. Robert D. Ballard is an oceanographer who revolutionized deep-sea archaeology. He developed several high-tech vessels that can explore ocean bottoms previously out of reach.

Deep Waters An important density current begins in Antarctica where the most dense ocean water forms during the winter. As ice forms, seawater freezes, but the salt is left behind in the unfrozen water. This extra salt increases the salinity and, therefore, the density of the ocean water until it is very dense. This dense water sinks and slowly spreads along the ocean bottom toward the equator, forming a density current. In the Pacific Ocean, this water could take 1,000 years to reach the equator.

In the North Atlantic Ocean, cold, dense water forms around Norway, Greenland, and Labrador. These waters sink, forming North Atlantic Deep Water. In about the northern one-third to one-half of the Atlantic Ocean, North Atlantic Deep Water forms the bottom layer of ocean water. In the southern part of the Atlantic Ocean, it flows at depths of about 3,000 m, just above the denser water formed near Antarctica. The dense waters circulate more quickly in the Atlantic Ocean than in the Pacific Ocean. In the Atlantic, a density current could circulate in 275 years.

Applying Math — Calculate Density

DENSITY OF SALT WATER You have an aquarium full of freshwater in which you have dissolved salt. If the mass of the salt water is 123,000 g and its volume is 120,000 cm^3, what is the density of the salt water?

Solution

1. *This is what you know:*
   - volume: $v = 120{,}000$ cm^3
   - mass of salt water: $m = 123{,}000$ g
2. *This is what you need to find:* density of water: d
3. *This is the equation you need to use:* $d = m/v$
4. *Substitute the known values:* $d = 123{,}000\text{g}/120{,}000\text{cm}^3 = 1.025$ g/cm^3
5. *Check your answer:* Multiply your answer by the volume. Do you calculate the same mass of salt water that was given?

Practice Problems

1. Calculate the density of 78,000 cm^3 of salt water with a mass of 79,000 g.
2. If a sample of ocean water has a density of 1.03 g/cm^3 and a mass of 50,000 g, what is the volume of the water?

Science Online For more practice, visit bookh.msscience.com/math_practice

LAB DEMONSTRATION

Purpose to model a density current

Materials four Erlenmeyer flasks, two 2-hole stoppers, four glass tubes, water, salt, food coloring

Preparation Insert the glass tubes into the stoppers.

Procedure Fill the four flasks with water. Add food coloring to two flasks. Add 35 g of salt to the water in a third flask. Insert stoppers into the two flasks without colored water. Invert each stoppered flask over a flask of colored water. Observe.

Expected Outcome Colored water moves up into the flask with salt water, but does not move into the flask with freshwater.

Assessment

What caused the density current in one set of flasks? Salty, dense water sank, forcing less dense water up. What was the purpose of the other set of flasks? That set was the control.

Intermediate Waters A density current also occurs in the Mediterranean Sea, a nearly enclosed body of water. The warm temperatures and dry air in the region cause large amounts of water to evaporate from the surface of the sea. This evaporation increases the salinity and density of the water. This dense water from the Mediterranean flows through the narrow Straits of Gibraltar into the Atlantic Ocean at a depth of about 320 m. When it reaches the Atlantic, it flows to depths of 1,000 m to 2,000 m because it is more dense than the water in the upper parts of the North Atlantic Ocean. However, the water from the Mediterranean is less dense than the very cold, salty water flowing from the North Atlantic Ocean around Greenland, Norway, and Labrador. Therefore, as shown in **Figure 10,** the Mediterranean water forms a middle layer of water—the Mediterranean Intermediate Water.

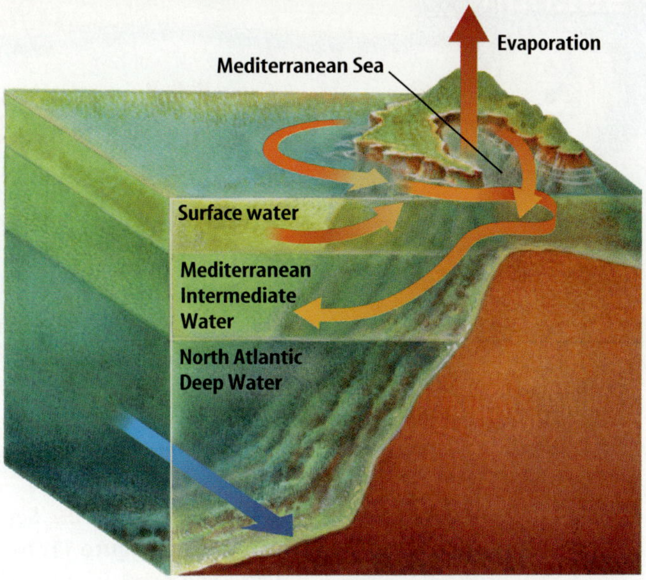

Figure 10 Dense layers of North Atlantic Deep Water form in the Greenland, Labrador, and Norwegian Seas. This water flows southward along the North Atlantic seafloor. Less dense water from the Mediterranean Sea forms Mediterranean Intermediate Water.

Reading Check *What causes the Mediterranean Intermediate Water to form?*

section 2 review

Summary

Surface Currents

- Surface currents are wind-powered ocean currents that move water horizontally.

Upwelling

- Upwelling is vertical circulation that brings deep, cold water to the ocean surface.

Density Currents

- Gravity acts on masses of seawater that are denser than surrounding water, causing the denser water to sink.
- Density currents slowly circulate deep ocean water.

Self Check

1. **Explain** how winds and the Coriolis effect influence surface currents.
2. **Summarize** why upwelling is important.
3. **Describe** how density currents circulate water.
4. **Think Critically** The latitudes of San Diego, California, and Charleston, South Carolina, are exactly the same. However, the average yearly water temperature in the ocean off Charleston is much higher than the water temperature off San Diego. Explain why.

Applying Skills

5. **Predict** what will happen to a layer of freshwater as it flows into the ocean. Explain your prediction.

Science Online bookh.msscience.com/self_check_quiz

Reading Check

Answer Evaporation from the Mediterranean Sea produces a salty water mass that moves out into the Atlantic Ocean, where it sinks beneath the less dense surface water.

3 Assess

DAILY INTERVENTION

Check for Understanding

Visual-Spatial Provide students with an unlabeled diagram of **Figure 9** or **Figure 10.** Ask them to fill in arrows and labels to explain the process of upwelling or deep and intermediate water mass formation respectively. L3

Reteach

Wind Circulation Obtain a map that shows wind circulation patterns. Have students discuss how major wind patterns influence major surface currents as shown in **Figure 5.** L2

Assessment

Process Assess students' abilities to interpret scientific illustrations by having them examine **Figure 5** and speculate about how ocean circulation would be different if the southern tip of South America and the Antarctic Peninsula were connected. The Drake Passage, which separates these two continents, opened about 25 to 30 million years ago and allowed ocean circulation around Antarctica, changing its climate and possibly causing glaciers to form. Use **Performance Assessment in the Science Classroom,** p. 89. L2

section 2 review

1. Winds cause surface currents to begin. As surface currents flow they are acted on by Coriolis which causes them to turn to the right in the northern hemisphere and to the left in the southern hemisphere.
2. Upwelling brings cold, deep water that contains high concentrations of nutrients to the surface. Nutrients promote primary production, which attracts fish. Areas of upwelling along coasts are important fishing grounds.
3. Cold or very salty water is denser than warm or less salty water and sinks in the ocean. This displaces warmer or less salty water, which moves upward. This movement drives density currents.
4. Ocean water off Charleston is influenced by the warm water of the Gulf Stream current, while ocean water off San Diego is influenced by the cold water of the California current.
5. It will flow along the surface because it is less dense than salty seawater.

section 3 Ocean Waves and Tides

1 Motivate

Bellringer

INTERACTIVE CHALKBOARD PowerPoint® Presentations

Section Focus Transparencies also are available on the Interactive Chalkboard CD-ROM. L2 ELL

SECTION 3 Section Focus Transparency **Run, Grunion, Run**

Grunion are small fish that know how to take advantage of California tides. When tides are the highest, they swim with each wave as far up the beach as possible and lay their eggs, as shown below. About two weeks later, when tides are again peaking, the hatched baby grunion are carried into the ocean as waves wash over the spawning area.

1. When would be a good time to catch grunion?
2. Why do you think grunion lay their eggs in an area that is temporarily out of reach of the tides?
3. Why do sand castles built on the beach get washed away?

L1

Ocean Motion

Tie to Prior Knowledge

Wave Formation Ask students whether they have ever seen waves on a lake, a river, or the ocean. Then ask whether they know how waves form. Write students' responses on the board and allow the class to discuss them. Then tell students they will find out in this section how waves form. L2

Caption Answer

Figure 11 Students should identify the highest points of the waves as the crest and the lowest points as the troughs.

---

as you read

***What* You'll Learn**

- **Describe** wave formation.
- **Distinguish** between the movement of water particles in a wave and the movement of the wave.
- **Explain** how ocean tides form.

***Why* It's Important**

Waves and tides affect life and property in coastal areas.

Review Vocabulary

energy: the ability to cause change

New Vocabulary

- wave
- crest
- trough
- breaker
- tide
- tidal range

Waves

If you've been to the seashore or seen a beach on TV, you've watched waves roll in. There is something hypnotic about ocean waves. They keep coming and coming, one after another. But what is an ocean wave? A **wave** is a rhythmic movement that carries energy through matter or space. In the ocean, waves like those in **Figure 11** move through seawater.

Describing Waves Several terms are used to describe waves, as shown in **Figure 11.** Notice that waves look like hills and valleys. The **crest** is the highest point of the wave. The **trough** (TRAWF) is the lowest point of the wave. Wavelength is the horizontal distance between the crests or between the troughs of two adjacent waves. Wave height is the vertical distance between crest and trough.

Half the distance of the wave height is called the amplitude (AM pluh tewd) of the wave. The amplitude squared is proportional to the amount of energy the wave carries. For example, a wave with twice the amplitude of the wave in **Figure 11** carries four times ($2 \times 2 = 4$) the energy. On a calm day, the amplitude of ocean waves is small. But during a storm, wave amplitude increases and the waves carry a lot more energy. Large waves can damage ships and coastal property.

Figure 11 Ocean waves carry energy through seawater.

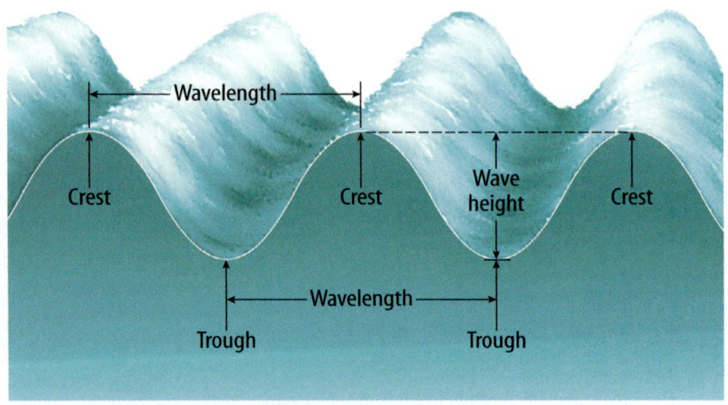

The crest, trough, wavelength, and wave height describe a wave.
Identify *the crests and troughs in the photo on the left.*

Section 3 Resource Manager

Chapter *FAST FILE* Resources

Transparency Activity, pp. 42, 43–44
Directed Reading for Content Mastery, pp. 17, 18
Reinforcement, p. 25
Enrichment, p. 28
MiniLAB, p. 4
Lab Worksheets, pp. 5–6, 7–8

Lab Management and Safety, p. 65

Figure 12 As a wave passes, only energy moves forward. The water particles and the bobber nearly return to their original positions after the wave has passed. **Describe** *what happens to water movement below a depth equal to about half the wavelength.*

Wave Movement You might have noticed that if you throw a pebble into a pond, a circular wave moves outward from where the pebble entered the water, as shown in **Figure 12.** A bobber on a fishing line floating in the water will bob up and down as the wave passes, but it will not move outward with the wave. Notice that the bobber returns to near its original position.

When you watch an ocean wave, it looks as though the water is moving forward. But unless the wave is breaking onto shore, the water does not move forward. Each molecule of water returns to near its original position after the wave passes. **Figure 13** shows this. Water molecules in a wave move around in circles. Only the energy moves forward while the water molecules remain in about the same place. Below a depth equal to about half the wavelength, water movement stops. Below that depth, water is not affected by waves. Submarines that travel below this level usually are not affected by surface storms.

Breakers A wave changes shape in the shallow area near shore. Near the shoreline, friction with the ocean bottom slows water at the bottom of the wave. As the wave slows, its crest and trough come closer together. The wave height increases. The top of a wave, not slowed by friction, moves faster than the bottom. Eventually, the top of the wave outruns the bottom and it collapses. The wave crest falls as water tumbles over on itself. The wave breaks onto the shore. **Figure 13** also shows this process. This collapsing wave is a **breaker.** It is the collapse of this wave that propels a surfer and surfboard onto shore. After a wave breaks onto shore, gravity pulls the water back into the sea.

Reading Check *What causes an ocean wave to slow down?*

Mini LAB

Modeling Water Particle Movement

Procedure

1. Put a piece of **tape** on the outside bottom of a **rectangular, clear-plastic storage box.** Fill the box with **water.**
2. Float a **cork** in the container above the piece of tape.
3. Use a **spoon** to make gentle waves in the container.
4. Observe the movement of the waves and the cork.

Analysis

1. Describe the movement of the waves and the motion of the cork.
2. Compare the movement of the cork in the water with the movement of water molecules in a wave.

2 Teach

Mini LAB

Purpose Students make a wave model that shows water particle movement. L2 COOP LEARN LS **Kinesthetic**

Materials masking tape, rectangular, clear-plastic storage box, water, cork, spoon

Teaching Strategy Students need to make waves, not currents. Thus they must gently tap the bowl of the spoon in one spot on the water.

Analysis

1. The waves move across the box from the point where the spoon is generating them. The cork bobs up and down as waves pass under it.
2. The cork and the water molecules move in small circles as a wave passes. They stay in about the same place, rather than moving across the body of water.

Assessment

Performance Have students devise other ways of creating waves in the box and have them observe the motion of the cork. Use **Performance Assessment in the Science Classroom,** p. 123. L2

Reading Check

Answer friction with the ocean bottom

Caption Answer

Figure 12 water movement stops

Curriculum Connection

Language Arts Explain to students that onomatopoeia is a poetry technique that uses words with sounds that make you think about their meanings. Examples include words like *buzz, hiss, plop, flop, gunk, gushy, swish, splash, zigzag, zing,* and *zip.* Have students write poems using onomatopoeia to describe waves and wave motion. L2 LS **Linguistic**

NATIONAL GEOGRAPHIC

Visualizing Wave Movement

Have students examine the pictures and read the captions. Then ask the following questions.

How does the movement of the dominoes differ from the movement of water molecules in waves? The dominoes simply fall down, while the water particles move around in circles. Both the dominoes and the water particles transfer energy in their movements. (The source of a wave's energy is usually wind.)

How does the degree of slope of the ocean floor near the beach affect the height of waves breaking on the beach? The steeper the slope, the taller the breaking waves; more gradually sloping ocean floor produces shorter waves.

Activity

Moving Energy Have students make diagrams of wave movement and include in their diagrams evidence that energy, not water molecules, is moving forward. Offer the example of a floating seagull bobbing up and down in the waves but not moving closer to the beach. Encourage them to brainstorm as many other examples as they can. L2

NATIONAL GEOGRAPHIC **VISUALIZING WAVE MOVEMENT**

Figure 13

As ocean waves move toward the shore, they seem to be traveling in from a great distance, hurrying toward land. Actually, the water in waves moves relatively little, as shown here. It's the energy in the waves that moves across the ocean surface. Eventually that energy is transferred—in a crash of foam and spray—to the land.

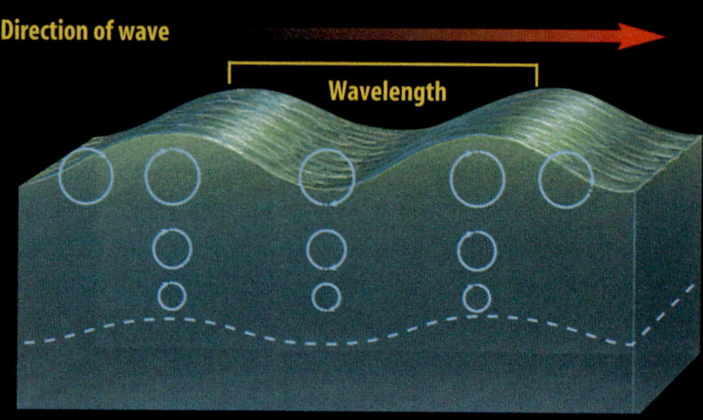

A Particles of water move around in circles rather than forward. Near the water's surface, the circles are relatively large. Below the surface, the circles become progressively smaller. Little water movement occurs below a depth equal to about one-half of a wave's length.

B The energy in waves, however, does move forward. One way to visualize this energy movement is to imagine a line of dominoes. Knock over the first domino, and the others fall in sequence. As they fall, individual dominoes—like water particles in waves—remain close to where they started. But each transfers its energy to the next one down the line.

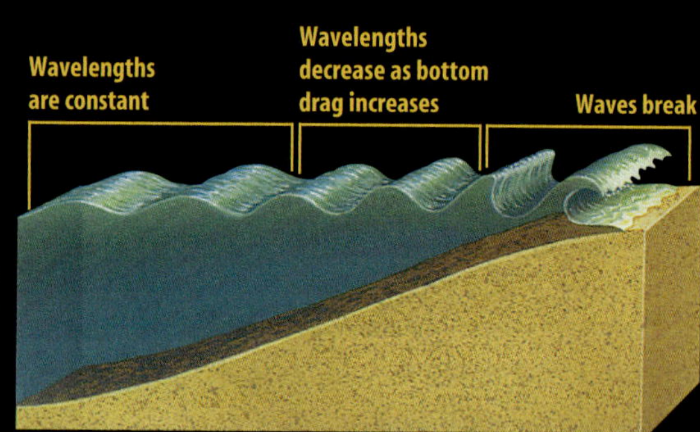

C As waves approach shore, wavelength decreases and wave height increases. This causes breakers to form. Where ocean floor rises steeply to beach, incoming waves break quickly at a great height, forming huge arching waves.

Differentiated Instruction

Challenge Have students find out the location of the world's most popular surfing sites and the reasons for the big waves at those sites. Have them write a report on their findings. L2 LS **Linguistic** P

112 ◆ H CHAPTER 4 Ocean Motion

Figure 14 Waves formed by storm winds can reach heights of 20 m to 30 m.
List *the three factors that affect wave height.*

How Water Waves Form On a windy day, waves form on a lake or ocean. When wind blows across a body of water, wind energy is transferred to the water. If the wind speed is great enough, the water begins to pile up, forming a wave. As the wind continues to blow, the wave increases in height. Some waves reach tremendous heights, as shown in **Figure 14.** Storm winds have been known to produce waves more than 30 m high—taller than a six-story building.

The height of waves depends on the speed of the wind, the distance over which the wind blows, and the length of time the wind blows. When the wind stops blowing, waves stop forming. But once set in motion, waves continue moving for long distances, even if the wind stops. The waves you see lapping at a beach could have formed halfway around the world.

Reading Check *What factors affect the height of waves?*

Tides

When you go to a beach, you probably notice the level of the sea rise and fall during the day. This rise and fall in sea level is called a **tide.** A tide is caused by a giant wave produced by the gravitational pull of the Sun and the Moon. This wave has a wave height of only 1 m or 2 m, but it has a wavelength that is thousands of kilometers long. As the crest of this wave approaches the shore, sea level appears to rise. This rise in sea level is called high tide. Later, as the trough of the wave approaches, sea level appears to drop. This drop in sea level is referred to as low tide.

Science Online
Topic: Tides
Visit bookh.msscience.com for Web links to information about tides.

Activity There are three types of tides: diurnal, semidiurnal, and mixed. Define each type and give an example of where each type occurs.

Use an Analogy

Human Waves Arrange chairs in a long row. Have a group of students sit in the chairs and quickly stand up and sit down in succession to make a human wave while other students watch. Have students compare the movement of each person in the wave to the movement of water molecules in an ocean wave. L2 ELL COOP LEARN LS **Interpersonal**

Reading Check

Answer speed of wind, distance over which wind blows, length of time wind blows

Quick Demo

Tides

Materials a simple model of the Earth-Moon-Sun system in which the pieces can be rearranged easily.

Estimate Time 10 minutes

Procedure Show how all three bodies interact to produce tides in Earth's oceans. Show the arrangement of all three bodies during a new moon. Rearrange the bodies to show the arrangement during a full moon, when the gravity of the Moon and the Sun pull together to generate the greatest tidal range.

Caption Answer

Figure 14 the speed of the wind, the distance over which the wind blows, and the length of time the wind blows

Fun Fact

The height of an average wave is 3.7 m. The largest waves occur in the West Wind Drift beneath the continuous strong winds surrounding Antarctica.

Visual Learning

Figure 15 Tell students that there is a causeway built from the mainland to Mont-Saint-Michel. Ask students if there would be any problem with walking from the mainland to Mont-Saint-Michel at low tide, without using the causeway. Explain. Yes; although it would depend on the time of the walk, the fact that the tide returns quickly would mean someone walking on the tidal flats could become trapped or be drowned by the waters as they return at high tide.

Teacher FYI

Tidal Range Mont-Saint-Michel is located between Normandy and Brittany in France. The island on which the buildings stand is a granite outcrop 78 m high. Great walls surround a Gothic-style Benedictine abbey that was originally built in 966. The walls and towers of the abbey were added much later, as was a causeway that connects the island to land.

Activity

Tides and the Moon If your school is located near an ocean, have students determine the times of today's low and high tides and the phase of the Moon. Have them relate the Moon's phase to the tides. If you do not live near a seacoast, tidal information can be obtained from coastal weather stations. L2

Caption Answer

Figure 16 low tide Have students note the difference between the height of the wharf and sea level. Point out the ladders that allow sailors to access the wharf during low tide. L2

Figure 15 A large difference between high tide and low tide can be seen at Mont-Saint-Michel off the northwestern coast of France.

Mont-Saint-Michel lies about 1.6 km offshore and is connected to the mainland at low tide.

Incoming tides move very quickly, making Mont-Saint-Michel an island at high tide.

Tidal Range As Earth rotates, different locations on Earth's surface pass through the high and low positions. Many coastal locations, such as the Atlantic and Pacific coasts of the United States, experience two high tides and two low tides each day. One low-tide/high-tide cycle takes 12 h, 25 min. A daily cycle of two high tides and two low tides takes 24 h, 50 min—slightly more than a day. But because ocean basins vary in size and shape, some coastal locations, such as many along the Gulf of Mexico, have only one high and one low tide each day. The **tidal range** is the difference between the level of the ocean at high tide and low tide. Notice the tidal range in the photos in **Figure 15.**

Figure 16 The Bay of Fundy has the greatest tidal range in the world.
Infer *Was this picture taken at high tide or low tide?*

Extreme Tidal Ranges The shape of the seacoast and the shape of the ocean floor affect the ranges of tides. Along a smooth, wide beach, the incoming water can spread over a large area. There the water level might rise only a few centimeters at high tide. In a narrow gulf or bay, however, the water might rise many meters at high tide.

Most shorelines have tidal ranges between 1 m and 2 m. Some places, such as those on the Mediterranean Sea, have tidal ranges of only about 30 cm. Other places have large tidal ranges. Mont-Saint-Michel, shown in **Figure 15,** lies in the Gulf of Saint-Malo off the northwestern coast of France. There the tidal range reaches about 13.5 m.

The dock shown in **Figure 16** is in Digby, Nova Scotia in the Bay of Fundy. This bay is extremely narrow, which contributes to large tidal ranges. The difference between water levels at high tide and low tide can be as much as 15 m.

Science Journal

Grunion In the waters off southern California, the reproductive cycle of the grunion fish follows the tidal schedule. Have students research the egg-laying habits of the grunion and write about them in their Science Journals. On several nights after the full moon in spring and summer spawning season, grunion swim onto beaches. Females deposit their eggs in the sand and males spread their milt on top of the eggs. About ten nights later, as the tides are again increasing in height, the eggs are washed out of the sand and hatch. L2

Tidal Bores In some areas when a rising tide enters a shallow, narrow river from a wide area of the sea, a wave called a tidal bore forms. A tidal bore can have a breaking crest or it can be a smooth wave. Tidal bores tend to be found in places with large tidal ranges. The Amazon River in Brazil, the Tsientang River in China, and rivers that empty into the Bay of Fundy in Nova Scotia have tidal bores.

When a tidal bore enters a river, it causes water to reverse its flow. In the Amazon River, the tidal bore rushes 650 km upstream at speeds of 65 km/h, causing a wave more than 5 m in height. Four rivers that empty into the Bay of Fundy have tidal bores. In those rivers, bore rafting is a popular sport.

The Gravitational Effect of the Moon For the most part, tides are caused by the interaction of gravity in the Earth-Moon system. The Moon's gravity exerts a strong pull on Earth. Earth and the water in Earth's oceans respond to this pull. The water bulges outward as Earth and the Moon revolve around a common center of mass. These events are explained in **Figure 17.**

Two bulges of water form, one on the side of Earth closest to the Moon and one on the opposite side of Earth. The bulge on the side of Earth closest to the Moon is caused by the gravitational attraction of the Moon on Earth. The force of gravity here is greater than another, opposing force generated by the motion of Earth and the Moon. As a result, surface water is pulled in the direction of the Moon. The bulge on the opposite side of Earth is caused by the same opposing force that, here, is greater than the force of gravity. The imbalance in forces results in surface water being pulled away from the Moon. The ocean bulges are the high tides, and the areas of Earth's oceans that are not toward or away from the Moon are the low tides. As Earth rotates, different locations on its surface pass through high and low tide.

INTEGRATE Life Science

Life in the Tidal Zone
Limpets are sea snails that live on rocky shores. When the tide comes in, they glide over the rocks to graze on seaweed. When the tide goes out, they use strong muscles to pull their shells tight against the rocks. Find out how other organisms survive in the zone between high and low tides.

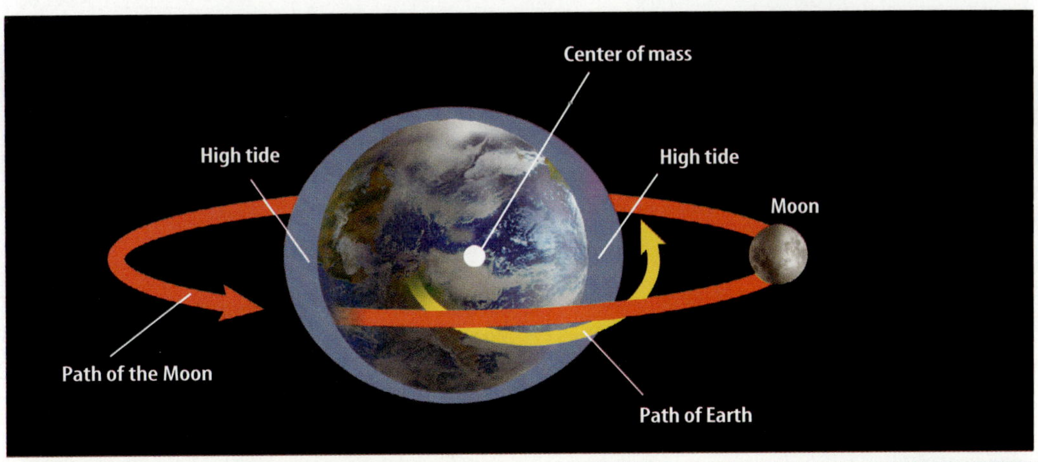

Figure 17 The Moon and Earth revolve around a common center of mass. Because the Moon's gravity pulls harder on parts of Earth closer to the Moon, a bulge of water forms on the side of Earth facing the Moon and the side of Earth opposite the Moon.

INTEGRATE Life Science

Life in the Tidal Zone On rocky coasts, barnacles produce a glue-like substance that prevents them from being washed off rocks. Mussels attach themselves to rocks by thin threads their bodies produce. Both barnacles and mussels feed during high tide and close their shells to hold in moisture during low tide. On sandy shores, worms burrow deep into moist sand to hide from predators at low tide. At high tide, they stick their tentacles above the sand to catch food floating in the water.

Research Have students research more information about how organisms that live in the intertidal zone are adapted to this unique environment. What mechanisms do they have or employ to guard against drying out? How do they handle extreme temperature changes or changes in salinity? L2

Virtual Labs

Tides *How are tides affected by the phases of the Moon?*

Teacher FYI

Slowing Rotation The action of the tides is slowing down Earth's rotation by about 0.002 second per century. Eventually, billions of years from now, Earth's rotation will be so slow that Earth will be "gravitationally locked" with the Moon. Just as the Moon keeps the same face always toward Earth, Earth will then keep one face always toward the Moon.

Differentiated Instruction

Visually Impaired Allow visually impaired students to hold a model of the Earth-Moon-Sun system. Have another student work with them to arrange the three bodies properly for a new moon and a full moon. Have the visually impaired student feel the model in each arrangement. Have the partner explain how the different arrangements affect the tides.

Visual Learning

Figure 18 What phase of the Moon would you expect to see during a neap tide and during a spring tide? first or third quarter; full or new Moon

Discussion

Gravitational Forces Why does the Moon have a greater effect on Earth's tides than the Sun, although the Sun is much larger? The Sun is much farther from Earth than the Moon is.

3 Assess

DAILY INTERVENTION

Check for Understanding

Visual-Spatial Provide students with an unlabeled diagram similar to **Figure 11.** Have them fill in the following labels: crest, trough, wavelength, wave height. L2

Reteach

Transfer of Energy Use a wave demonstration spring or coiled-spring toy to show how energy is transferred through a wave. Tie a piece of ribbon to the middle of the spring. Have two students create a wave while another student holds the ribbon. Have the class note that the ribbon does not move forward with the wave as energy moves through the spring. L2 LS **Interpersonal**

Assessment

Oral Have students write questions on the content of the chapter and quiz each other aloud. Use **Performance Assessment in the Science Classroom,** p. 91. L2

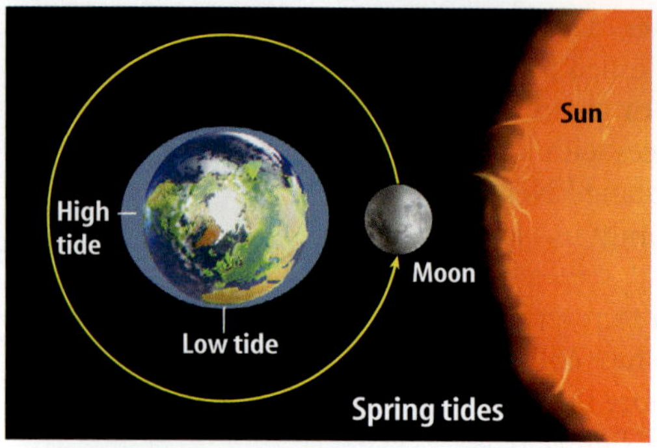

When the Sun, the Moon, and Earth are aligned, spring tides occur.

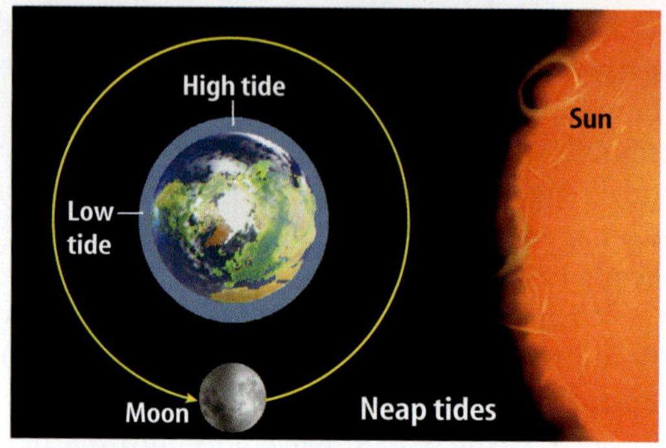

When the Sun, Earth, and the Moon form a right angle, neap tides occur.

Figure 18 The gravitational attraction of the Sun causes spring tides and neap tides.

The Gravitational Effect of the Sun The Sun also affects tides. The Sun can strengthen or weaken the Moon's effects. When the Moon, Earth, and the Sun are lined up together, the combined pull of the Sun and the Moon causes spring tides, shown in **Figure 18.** During spring tides, high tides are higher and low tides are lower than normal. The name *spring tide* has nothing to do with the season of spring. It comes from the German word *springen*, which means "to jump." When the Sun, Earth, and the Moon form a right angle, also shown in **Figure 18,** high tides are lower and low tides are higher than normal. These are called neap tides.

section 3 review

Summary

Waves

- A wave is a rhythmic movement that carries energy through matter or space.
- Water waves form as wind blows across a body of water.
- The height of a wave depends on the speed of the wind, the distance over which the wind blows, and the length of time the wind blows.

Tides

- Tides result from the gravitational pull of the Moon and the Sun on Earth.
- The shape of the seafloor and coast affect the range of tides in an area.
- Depending on the alignment of the Sun, the Moon, and Earth, spring tides or neap tides occur.

Self Check

1. **Identify** the parts of an ocean wave.
2. **Explain** how wind creates water waves.
3. **Describe** what causes high tides. Describe what causes spring tides.
4. **Summarize** the movement of water molecules in a wave and wave movement.
5. **Think Critically** At the ocean, you spot a wave about 200 m from shore. A few seconds later, the wave breaks on the beach. Explain why the water in the breaker is not the same water that was in the wave 200 m away.

Applying Skills

6. **Compare and contrast** the effects of the Sun and the Moon on Earth's tides.

Science Online bookh.msscience.com/self_check_quiz

section 3 review

1. crest: highest point, trough: lowest point, wave height: vertical distance between crest and trough, wavelength: horizontal distance between crests or troughs of two successive waves, amplitude: half of wave height
2. Friction pulls water along with wind.
3. Tides are caused by a giant wave formed as the result of the gravitational pull of the Sun and the Moon. As the wave's crest nears the shore, the water level appears to rise in a high tide. Spring tides are caused by the combined gravitational effect when the Moon, Earth, and the Sun are in line.
4. molecules of water are moved up, forward, down, and back
5. Although the energy in waves moves, the water molecules themselves do not.
6. Both affect tides because of gravitational attraction, but the effect of the Moon is greater because it's so much closer.

116 ◆ H CHAPTER 4 Ocean Motion

LAB

Wave Properties

Ocean wave energy impacts coastlines around the world. Understanding wave properties helps scientists predict the movement and effects of waves.

Real-World Question

How are wave characteristics related to each other and to the energy source that causes waves?

Goals

- **Test** statements about wave properties.
- **Summarize** the relationship between wave properties and the energy source which causes waves.

Materials

rectangular, clear-plastic box
water
straw
metric ruler
3-cm chalk piece
3-cm ball aluminum foil

Safety Precautions

Procedure

1. Copy the data table above.
2. Fill the clear, plastic box with water to a depth of about 5.5 cm.
3. Test statement 1. Hold the straw just above the water. Blow through the straw. Record your observations in your data table.
4. Test statement 2. Hold the straw just above the water at one end of the box. Blow gently and continuously. Use the metric ruler to compare the wavelengths close to the straw and on the other end of the box. Record your observations.

Statement	Observations
1. Wind causes waves.	
2. Wavelength increases as the distance from the energy sources increases.	Answers will vary.
3. The effects of wave motion are felt relatively close to the surface only.	
4. Wave energy is transferred through the water: the water itself does not move forward with the wave.	

5. Test statement 3. Sink the chalk piece in the middle of the box. Hold the straw just above the water at one end of the box. Blow gently and continuously. Observe any movement of the chalk and record your observations.
6. Test statement 4. Float the aluminum foil ball in the middle of the box. Hold the straw just above the water at one end of the box. Blow gently and continuously. Observe and record any movement of the aluminum foil ball.

Conclude and Apply

1. **Explain** how wind causes waves to form.
2. **Describe** Did your results support the statement *Wavelength increases as the distance from the energy source increases?* Why or why not?
3. **Explain** how you know the effects of wave motion are not felt below a certain depth.
4. **Infer** How did you prove that the water in a wave does not move forward with the wave? Did your observations surprise you? Why or why not?

BENCH TESTED

Real-World Question

Purpose Students will design and conduct experiments examining wave properties and the relationships among them.

Process Skills design experiments, manipulate variables, observe, describe relationships, interpret data, draw conclusions

Time Required 45 minutes

Procedure

Teaching Strategies

- To reinforce the concept that water does not move forward as the wave passes by, have two students hold each end of a length of rope with a piece of string tied in the middle. As they move the ends up and down, creating wave motion, emphasize that the string does not move forward or backwards as the energy passes through the rope.

Troubleshooting Have students hold the straw at roughly a 30 degree angle above the surface of the water as they blow through it.

Conclude and Apply

1. As wind blows across water, wind energy is transferred to the water, forming a wave.
2. Students should have been able to measure the wavelengths of the waves and show that wavelength increases as the wave gets farther away from the energy source.
3. The chalk did not move as waves passed over it.
4. The aluminum foil floating on the surface did not move forward as the wave passed by.

Assessment

Content Have students form new groups. Students share experimental designs and observations from their original groups for each statement, then create a chart which compares and contrasts the various methodologies and observations.

Differentiated Instruction

Visually Impaired Additional light above and to the sides of the box will help these students see wave motion more clearly. Clay models can be constructed which demonstrate most of the statements tested.

BENCH TESTED

Real-World Question

Purpose Students will design an experiment and data table to chart how increasing salinity affects the ability of a potato to float in water.

Process Skills design an experiment, form a hypothesis, compare, observe, and infer, communicate, interpret data, record data, analyze results, make and use tables

Time Required one class period

Form a Hypothesis

Possible Hypotheses Students may hypothesize that with increasing salinity, the potato rises, or the opposite—that it sinks with increasing salinity.

Test Your Hypothesis

Possible Procedures Fill a large glass bowl with approximately 960 mL of water. Using the balance, determine the mass of the potato. Record the mass in the data and carefully place the potato into the bowl. Use a metric ruler to record the displacement of the water. Remove the potato and add 1 teaspoon of salt to the water. Stir the water until the salt dissolves. Carefully place the potato into the water again and record the water displacements. Repeat the above steps until you notice a change in the potato's ability to float. For help figuring density, refer to the Applying Math in this chapter.

LAB Design Your Own

Sink or Float?

Goals

- **Design** an experiment to identify how increasing salinity affects the ability of a potato to float in water.

Possible Materials

small, uncooked potato
teaspoon
salt
large glass bowl
water
balance
large graduated cylinder
metric ruler

Safety Precautions

Real-World Question

As you know, ocean water contains many dissolved salts. How does this affect objects within the oceans? Why do certain objects float on top of the ocean's waves, while others sink directly to the bottom? Density is a measurement of mass per volume. You can use density to determine whether an object will float within a certain volume of water of a specific salinity. Based on what you know so far about salinity, why things float or sink, and the density of a potato, plus what it looks and feels like, formulate a hypothesis. Do you think the salinity of water has any effect on objects that are floating in water? What kind of effect? Will they float or sink? How would a dense object like a potato be different from a less dense object like a cork?

Test Your Hypothesis

Make a Plan

1. As a group, agree upon and write your hypothesis statement.
2. Devise a method to test how salinity affects whether a potato floats in water.
3. List the steps you need to take to test your hypothesis. Be specific, describing exactly what you will do at each step.
4. Read over your plan for testing your hypothesis.
5. How will you determine the densities of the potato and the different water samples? How will you measure the salinity of the water? How will you change the salinity of the water? Will you add teaspoons of salt one at a time?

Alternative Inquiry Lab

Exploring Further To extend this Lab into an Inquiry Lab, have students try the "Cartesian diver" experiment. Fill a 2 L bottle completely with water. There must be no air bubble in the bottle. Take in water with an eyedropper, but leave an air bubble in it. Put the dropper in the bottle. Adjust the air bubble's size until the "diver" dropper hovers somewhere in the middle. Squeezing the bottle increases the pressure on the air in the dropper, bringing more water into the dropper, and makes the diver sink. Releasing the pressure makes the diver hover higher up again. The effect is similar to what happens to skin divers' lungs as they dive deeper. This experiment may make students curious to research submarine raising and lowering functions, which use compressed air to expel water.

6. How will you measure the ability of an object to float? Could you somehow measure the displacement of the water? Perhaps you could draw a line somewhere on your bowl and see how the position of the potato changes.
7. **Design** a data table where you can record your results. Include columns/rows for the salinity and float/sink measurements. What else should you include?

Follow Your Plan

1. Make sure your teacher approves your plan before you start.
2. Carry out the experiment.
3. While conducting the experiment, record your data and any observations that you or other group members make in your Science Journal.

Analyze Your Data

1. **Compare** how the potato floated in water with different salinities.
2. How does the ability of an object to float change with changing salinity?

Conclude and Apply

1. Did your experiment support the hypothesis you made?
2. A heavily loaded ship barely floats in the Gulf of Mexico. Based on what you learned, infer what might happen to the ship if it travels into the freshwater of the Mississippi River.

Communicating Your Data

Prepare a large copy of your data table and share the results of your experiment with members of your class. For more help, refer to the Science Skill Handbook.

Teaching Strategy Before students create their own hypotheses, show three different beakers with different salt concentrations and float an object in each.

Expected Outcome The potato will float better in higher salt concentrations. It will take approximately 12-14 teaspoons of salt to float a 100 g potato.

Analyze Your Data

Answers to Questions

1. As the salinity of the water increased, the potato became more buoyant.
2. If the salinity is lowered, denser objects will not float as well. The density of the object affects whether or not it will float. For example, a bobber will float in freshwater or salt water.

Error Analysis If students are unable to get their potato to float in the time allowed, have them review the procedure steps. How much water did they use? How much salt?

Conclude and Apply

1. Answers will vary.
2. Students might indicate that the ship will sink if it travels into freshwater.

Assessment

Process Have students evaluate the data from other classmates. If their data is significantly different, have them discuss why it might be different. Use **Performance Assessment in the Science Classroom,** p. 99. L2

Communicating Your Data

Student charts should show how much salt was added each time and whether or not the potato changed position with the addition of salt.

Understanding Literature

Imagery Answers may vary but might include that the author uses the word "explosion" when referring to the tops of palm trees to convey a contrast between the trees and the gentle sea. He also compares the silence of the jungle to that of a library as a way of emphasizing the jungle's stillness and solitude.

Respond to the Reading

1. It was a gentle and happy place; he describes it as felicitous.
2. rippling ruff of feathers and foam
3. **Linking Science and Writing** Have students research conditions for a specific area for their reports.

INTEGRATE Earth Science **Monsoons** Currents in the Atlantic Ocean and the Pacific Ocean are nearly the same throughout the year. However, surface currents in the northern part of the Indian Ocean change with the monsoon winds. This is referred to as the Monsoon Drift. When the northeast monsoons blow over Sri Lanka, the Northeast Monsoon Drift flows toward the west. When the southwest monsoon blows, the current flow reverses.

Science and Language Arts

"The Jungle of Ceylon"

from Passions and Impressions

by Pablo Neruda

The following passage is part of a travel chronicle describing the Chilean poet Pablo Neruda's visit to the island of Ceylon, now called Sri Lanka, which is located southeast of India. The author considered himself so connected to Earth that he wrote in green ink.

Felicitous[1] shore! A coral reef stretches parallel to the beach; there the ocean interposes in its blues the perpetual white of a rippling ruff[2] of feathers and foam; the triangular red sails of sampans[3]; the unmarred line of the coast on which the straight trunks of the coconut palms rise like explosions, their brilliant green Spanish combs nearly touching the sky.

… In the deep jungle, there is a silence like that of libraries: abstract and humid.

1 Happy
2 round collar made of layers of lace
3 East Asian boats

Linking Science and Writing

Imagery Imagery is a series of words that evoke pictures to the reader. Poets use imagery to connect images to abstract concepts. The poet, here, wants to capture a particular feature of the reef and does so by describing it as a "ruff of feathers and foam," invoking the image of a gentle place, without the author saying so.
Where else in the poem does the poet use imagery to convey a mood or feeling?

Respond to the Reading

1. What were his impressions of the island on arrival?
2. What words does the author choose to describe waves?
3. **Linking Science and Writing** Write a weather report for fishers and others who work at sea.

INTEGRATE Earth Science Sri Lanka often is plagued by monsoons, which affect ocean conditions and local climate. Monsoons are seasonal reversals of the regional winds. During the wet season, moist winds blow in from the sea, causing storms and producing waves. During the dry season, winds blow from the land and sunny days are common.

Resources for Teachers and Students

Currents of Change: Impacts of El Nino and La Nina on Climate and Society, Cambridge University Press, February 2001

Residence on Earth, by Pablo Neruda, New Directions, March 1973

The Book of Questions, by Pablo Neruda, Copper Canyon Press, September 1991

Reviewing Main Ideas

Section 1 Ocean Water

1. Earth's ocean water might have originated from water vapor released from volcanoes. Over millions of years, the water condensed and rain fell, filling basins.
2. The oceans are a mixture of water, dissolved salts, and dissolved gases.
3. Ions are added to ocean water by rivers, volcanic eruptions, and the atmosphere. When seawater is evaporated, these ions combine to form salts.

Section 2 Ocean Currents

1. Wind causes surface currents. Surface currents are affected by the Coriolis effect.
2. Cool currents off western coasts originate far from the equator. Warmer currents along eastern coasts begin near the equator.
3. Differences in temperature and salinity between water masses in the oceans set up circulation patterns called density currents.

Section 3 Ocean Waves and Tides

1. A wave is a rhythmic movement that carries energy.
2. In a wave, energy moves forward while water molecules move around in small circles.
3. Wind causes water to pile up and form waves. Tides are caused by gravitational forces.

Visualizing Main Ideas

Copy and complete the following concept map on ocean motions.

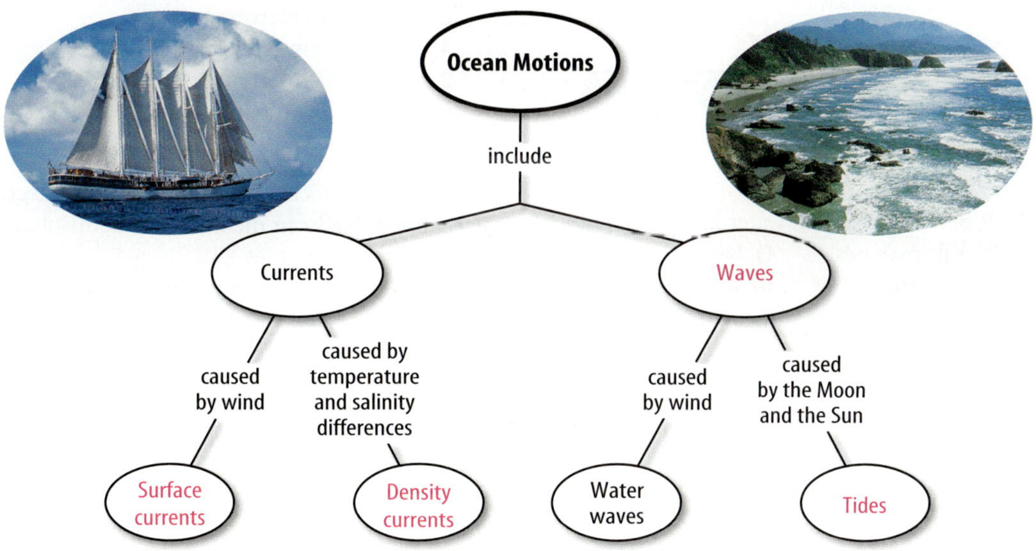

Science Online bookh.msscience.com/interactive_tutor

chapter 4 Study Guide

Reviewing Main Ideas

Summary statements can be used by students to review the major concepts of the chapter.

Visualizing Main Ideas

See student page.

Science Online

Visit bookh.msscience.com
/self_check_quiz
/interactive_tutor
/vocabulary_puzzlemaker
/chapter_review
/standardized_test

Assessment Transparency

For additional assessment questions, use the *Assessment Transparency* located in the transparency book.

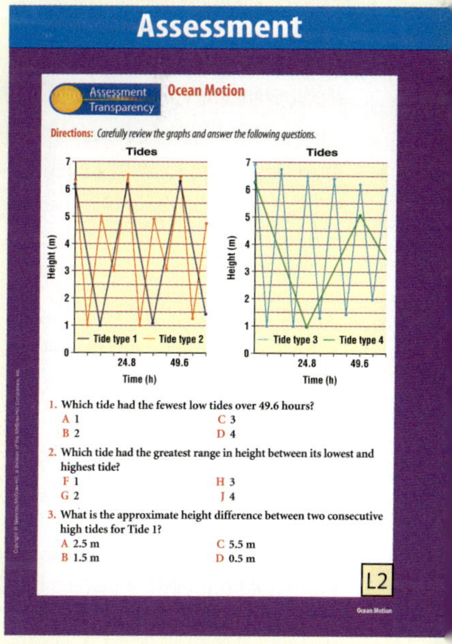

Assessment

Assessment Transparency **Ocean Motion**

Directions: *Carefully review the graphs and answer the following questions.*

1. Which tide had the fewest low tides over 49.6 hours?
   A 1 C 3
   B 2 D 4
2. Which tide had the greatest range in height between its lowest and highest tide?
   F 1 H 3
   G 2 J 4
3. What is the approximate height difference between two consecutive high tides for Tide 1?
   A 2.5 m C 5.5 m
   B 1.5 m D 0.5 m

L2

Ocean Motion

CHAPTER STUDY GUIDE H ◆ 121

Using Vocabulary

1. salinity
2. upwelling
3. density current
4. tide
5. tidal range

Checking Concepts

6. B
7. A
8. A
9. D
10. C
11. B
12. A
13. D
14. B

Thinking Critically

15. It could wash up onto a beach on the coast of Japan or somewhere in the Pacific Islands.
16. At the mouth of the Mississippi River the water has low salinity, while it is very salty in the Mediterranean Sea. The water is therefore denser in the Mediterranean.
17. Surface water is pushed away from an area by wind, allowing underlying cold water to rise to the surface, and sometimes fish numbers rise as well. If upwelling stops, good fishing areas can become barren.
18. Wind causes surface currents and waves. But surface currents move water from one place to another. Energy moves laterally in waves, but the water does not.
19. North from A, following shorelines around in clockwise pattern, moving south from B. Movement would be caused by surface currents.

chapter 4 Review

Using Vocabulary

basin p. 101	surface current p. 104
breaker p. 111	tidal range p. 114
Coriolis effect p. 105	tide p. 113
crest p. 110	trough p. 110
density current p. 107	upwelling p. 107
salinity p. 102	wave p. 110

Fill in the blanks with the correct vocabulary word or words.

1. The ________ of seawater has stayed about the same for hundreds of millions of years.
2. An area of ________ is a good place to catch fish.
3. A(n) ________ is created when a mass of more dense water sinks beneath less dense water.
4. Along most ocean beaches, a rise and fall of the ocean related to gravitational pull, or a(n) ________, is easy to see.
5. The difference between the level of the ocean at high tide and low tide is ________.

Checking Concepts

Choose the word or phrase that best answers the question.

6. Where might ocean water have originated?
   A) salt marshes **C)** basins
   B) volcanoes **D)** surface currents
7. How does chloride enter the oceans?
   A) volcanoes **C)** density currents
   B) rivers **D)** groundwater
8. What is the most common ion found in ocean water?
   A) chloride **C)** boron
   B) calcium **D)** sulfate
9. What causes most surface currents?
   A) density differences
   B) the Gulf Stream
   C) salinity
   D) wind

Use the illustration below to answer question 10.

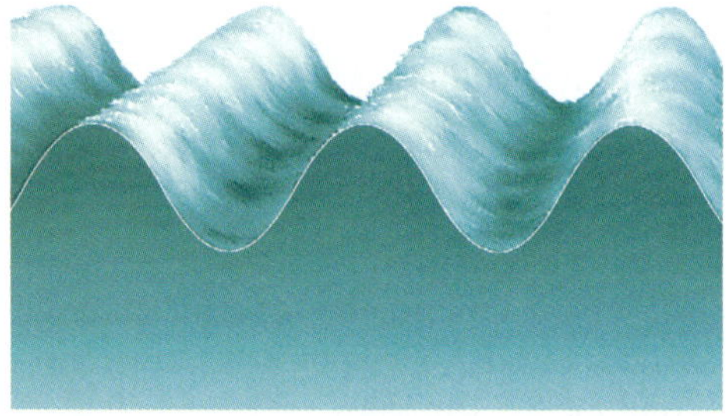

10. What is the highest point on a wave called?
    A) wave height **C)** crest
    B) trough **D)** wavelength
11. In the ocean, what is the rhythmic movement that carries energy through seawater?
    A) current **C)** crest
    B) wave **D)** upwelling
12. Which of the following causes the density of seawater to increase?
    A) a decrease in temperature
    B) a decrease in salinity
    C) an increase in temperature
    D) a decrease in pressure
13. In which direction does the Coriolis effect cause currents in the northern hemisphere to turn?
    A) east **C)** counterclockwise
    B) south **D)** clockwise
14. Tides are affected by the positions of which celestial bodies?
    A) Earth and the Moon
    B) Earth, the Moon, and the Sun
    C) Venus, Earth, and Mars
    D) the Sun, Earth, and Mars

Science Online bookh.msscience.com/vocabulary_puzzlemaker

Use the ExamView® Pro Testmaker CD-ROM to:
- create multiple versions of tests
- create modified tests with one mouse click for inclusion students
- edit existing questions and add your own questions
- build tests aligned with state standards using built-in State Curriculum Tags
- change English tests to Spanish with one mouse click and vice versa

Thinking Critically

15. Infer If a sealed bottle is dropped into the ocean off the coast of California, where do you think it might wash up?

16. Compare and contrast the density of seawater at the mouth of the Mississippi River and in the Mediterranean Sea.

17. Recognize Cause and Effect What causes upwelling? What effect does it have? What can happen when upwelling stops?

18. Compare and contrast ocean waves and ocean currents.

Use the figure below to answer question 19.

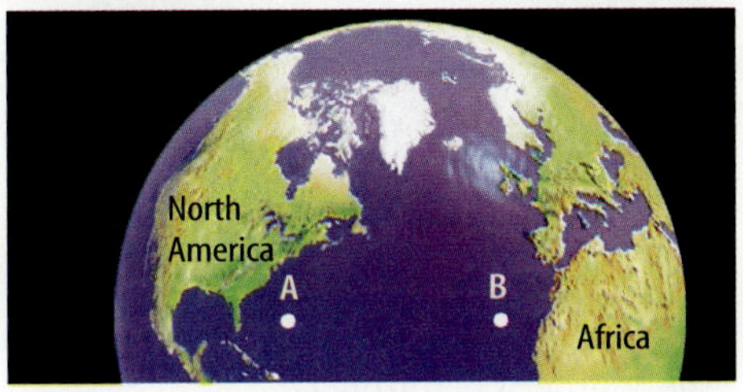

19. Predict how drift bottles dropped into the ocean at points A and B will move. Explain.

20. Recognize Cause and Effect In the Mediterranean Sea, a density current forms because of the high rate of evaporation of water from the surface. How can evaporation cause a density current?

21. Evaluate One water mass has a temperature of 5°C and a salinity of 37 parts per thousand. Another water mass has a temperature of 10°C and a salinity of 35 parts per thousand. Which water mass will sit on top of the other? Why?

22. Infer In some areas tidal energy is used as an alternative energy source. What are some advantages and disadvantages of doing this?

Performance Activities

23. Invention Design a method for desalinating water that does not use solar energy. Draw it, and display it for your class.

24. Design and Perform an Experiment Create an experiment to test the density of water at different temperatures.

Applying Math

25. Wave Speed Wave speed of deep water waves is calculated using the formula $S = L/T$, where S represents wave speed, L represents the wavelength, and T represents the period of the wave. What is the speed of a wave if $L = 100$ m and $T = 11$ s?

26. Wave Steepness The steepness of a wave is represented by the formula Steepness $= H/L$, where $H =$ wave height and $L =$ wavelength. When the steepness of a wave reaches 1/7, the wave becomes unstable and breaks. If $L = 50$ m, at what height will the wave break?

Use the graph below to answer question 27.

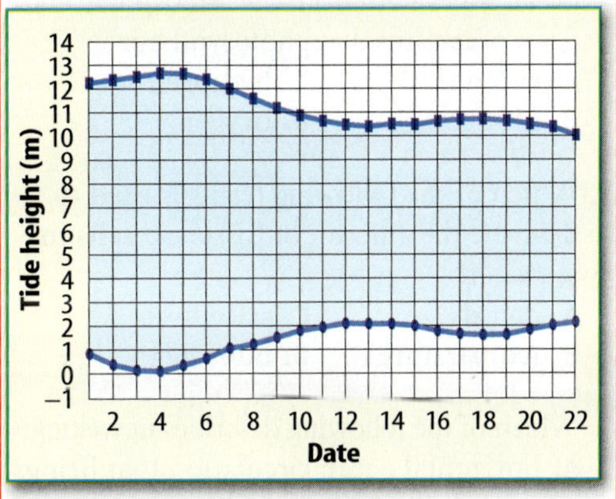

27. Tides The graph above shows the tidal ranges for each day. The maximum tidal range is called a spring tide. The minimum tidal range is referred to as a neap tide. Calculate the tidal range for the spring tide and the neap tide. On which date does each occur?

Science Online bookh.msscience.com/chapter_review

Assessment Resources

Reproducible Masters

Chapter *Fast File* Resources
Chapter Review, pp. 33–34
Chapter Tests, pp. 35–38
Assessment Transparency Activity, p. 45

Glencoe Science Web site
Chapter Review Test
Standardized Test Practice

Glencoe Technology
Assessment Transparency
Exam*View*® Pro Testmaker
MindJogger Videoquiz
Interactive Chalkboard

chapter 4 Review

20. When salt water evaporates, the water left behind has a higher concentration of salt, making it denser. The denser water sinks and flows out of the Mediterranean Sea into the Atlantic Ocean. Less dense surface water from the Atlantic flows in to replace it.

21. The colder, more saline water mass will sit beneath the warmer, less saline water mass. Density is affected by both salinity and temperature. As salinity increases, density increases. As temperature decreases, density increases.

22. Areas of the world that have extreme tidal ranges could use tidal energy. Advantages include that it is a renewable resource, predictable, and constant. Disadvantages include initial expense of setting up a suitable way of harnessing the energy and possible logistical problems due to location.

Performance Activities

23. Students can use some artificial source, such as a hot plate, to heat and evaporate water. Use **PASC,** p. 117.

24. Designs should include procedures for heating and cooling the water; ways to measure the temperature, volume, and mass of the water samples; and a method for calculating density. Hydrometers can be used to determine actual densities of samples. They can be made by students or purchased from aquarium supply stores. Use **PASC,** p. 95.

Applying Math

National Math Standards

1, 2, 5, 9

25. 9.09 m/s

26. 7.14 m

27. spring tide: about 12.5 m; neap tide: 8m; spring tide: day 4; neap tide: day 22

CHAPTER REVIEW H ◆ 123

FAST FILE

Answer Sheet A practice answer sheet can be found at bookh.msscience.com/answer_sheet.

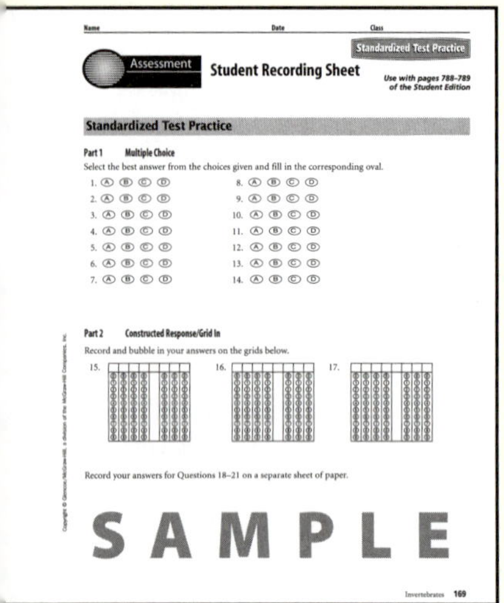
Name Date Class
Standardized Test Practice
Assessment Student Recording Sheet
Use with pages 788–789 of the Student Edition
Standardized Test Practice
Part 1 Multiple Choice
Select the best answer from the choices given and fill in the corresponding oval.
Part 2 Constructed Response/Grid In
Record and bubble in your answers on the grids below.
Record your answers for Questions 18–21 on a separate sheet of paper.
SAMPLE

Part 1 Multiple Choice

1. B
2. D
3. C
4. D
5. A
6. A
7. B
8. C

Part 2 Short Response

9. Surface currents move water horizontally and occur in the top 100 meters of the ocean. They are caused by wind. Density currents move water vertically and occur in deep water. They are caused when more dense water masses sink beneath less dense water masses. Density is affected by the temperature and salinity of the water.
10. spring tide

chapter 4 Standardized Test Practice

Part 1 Multiple Choice

Record your answers on the answer sheet provided by your teacher or on a sheet of paper.

Use the table below to answer question 1.

Ions in Seawater	
Ion	**Percentage**
Chloride	55.0
Sodium	30.6
Sulfate	7.7
Magnesium	3.7
Calcium	1.2
Potassium	1.1
Other	0.7

1. Which ion makes up 7.7 percent of the ions in seawater?
A. calcium **C.** chloride
B. sulfate **D.** sodium

2. Which of the following dissolved gases enters ocean water both from the atmosphere and from organisms that photosynthesize?
A. carbon **C.** hydrogen
B. nitrogen **D.** oxygen

3. Which of the following terms is used to describe the amount of dissolved salts in seawater?
A. density **C.** salinity
B. temperature **D.** buoyancy

4. Which of the following describes upwelling?
A. horizontal ocean circulation that brings deep, cold water to the surface
B. vertical ocean circulation that brings deep, warm water to the surface
C. horizontal ocean circulation that brings deep, warm water to the surface
D. vertical ocean circulation that brings deep, cold water to the surface

5. What is the lowest point on a wave called?
A. trough **C.** crest
B. wavelength **D.** wave height

Use the illustration below to answer questions 6 and 7.

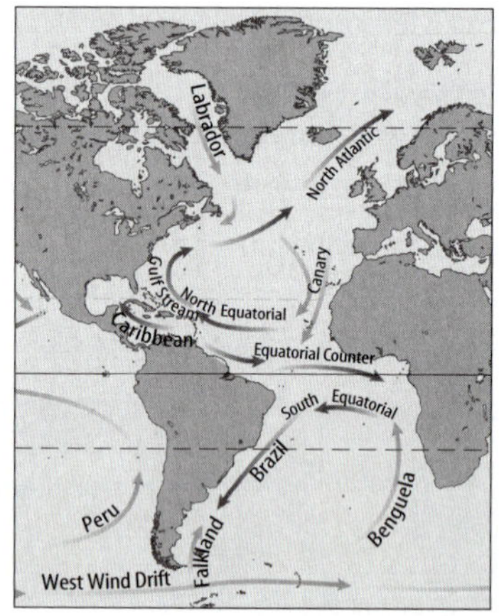

6. What is the direction of ocean currents in the southern hemisphere?
A. counterclockwise
B. north to south only
C. clockwise
D. east to west only

7. Which of the following is a reasonable conclusion based on the information in the figure?
A. The oceans' currents only flow in one direction.
B. The oceans' waters are constantly in motion.
C. The Gulf Stream flows east to west.
D. The Atlantic Ocean is deep.

8. What affects surface currents?
A. crests **C.** the Coriolis effect
B. upwellings **D.** tides

Part 2 Short Response/Grid In

11. The alignment of the Sun, Earth, and the Moon form a right angle.
12. Tidal range is the difference between the level of the ocean at high tide and low tide.
13. 1.485 m
14. Elements are added to the oceans at about the same rate they are removed. Dissolved salts are added to ocean water by rivers and volcanic eruptions. They are removed when they precipitate out of ocean water and become part of the sediment. Dissolved salts are also removed by biological processes such as marine organisms using certain salts to form hard body parts including bones and shells.
15. Surface currents that begin at the equator release heat as they flow away from the equator. This warms the atmosphere and influences climate.

Part 2 Short Response/Grid In

Record your answers on the answer sheet provided by your teacher or on a sheet of paper.

9. Explain the difference between surface currents and density currents in the ocean.

Use the illustration below to answer questions 10 and 11.

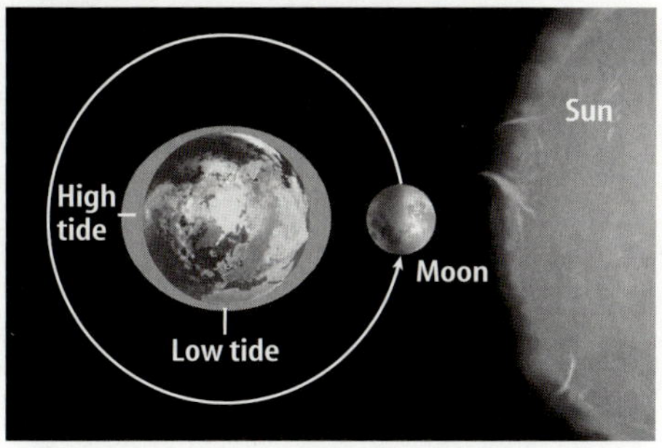

10. Which type of tide occurs when the Sun, the Moon, and Earth are aligned?

11. Describe how the Sun, the Moon, and Earth are positioned relative to each other during a neap tide.

12. What is tidal range?

13. On June 17th 2003, in Santa Barbara, California, the morning low tide was measured at −0.365 m. High tide was measured at 1.12 m. Calculate the tidal range between these tides.

14. Explain what the term *steady state* means in relation to ocean salinity. What processes keep ocean salinity in a steady state?

15. Explain how the ocean can influence the climate of an area.

Part 3 Open Ended

Record your answers on a sheet of paper.

16. Draw a diagram that explains the process of upwelling. An area of upwelling exists off of the western coast of South America. During El Niño events, upwelling does not occur and surface water is warm and nutrient-poor. What effect could this change have on the marine organisms in this area?

Use the illustration below to answer question 17.

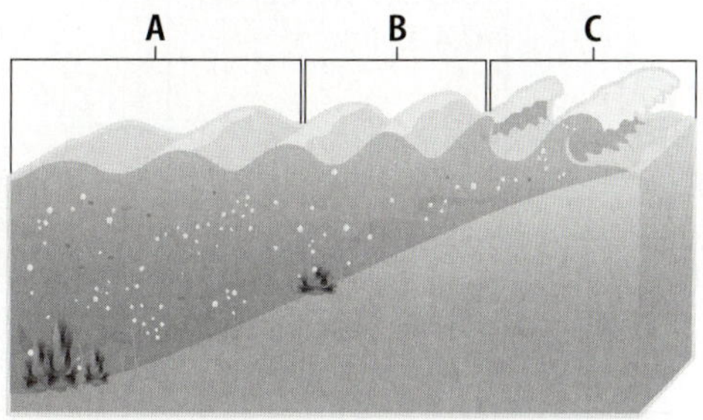

17. Describe the changes that occur as a wave approaches shore. Explain how wavelength is affected at each stage—A, B, and C—on the diagram.

18. What is the Coriolis effect? Explain how it affects ocean surface currents.

19. Compare and contrast the formation of North Atlantic Deep Water and Mediterranean Intermediate Water.

Test-Taking Tip

Answer All Parts Make sure each part of the question is answered when listing discussion points. For example, if the question asks you to compare and contrast, make sure you list both similarities and differences.

Question 19 Be sure to list the similarities and differences between the formation of the two masses of water.

Part 3 Open Ended

16. Diagrams will vary but should show the idea that upwelling occurs to replace surface water that has been carried away from an area. Upwelling usually provides an area of ocean with high concentrations of nutrients which promote plankton growth. This attracts small fish, larger fish and birds. During El Niño the source of nutrients decreases or disappears entirely. Without production occurring at the base of the food chain, each level of the food chain is affected.

17. As waves approach shore, friction with the ocean bottom slows water at the bottom of the wave. As the wave slows, the wavelength decreases and wave height increases. The top of the wave now moves faster than the bottom. The wave breaks when the top of the wave outruns the bottom. In section A the wavelengths are constant. In section B the wavelengths decrease as bottom drag increases. In section C the wave breaks.

18. The Coriolis effect is the shifting of winds and surface currents from their expected paths that is caused by Earth's rotation. The Coriolis effect causes currents north of the equator to turn to the right and currents south of the equator to turn to the left.

19. Both water masses are formed from density currents as surface water sinks beneath layers of less dense water. North Atlantic Deep Water, formed in the North Atlantic, becomes denser due to changes in temperature. It forms the bottom layer of ocean water. Mediterranean Intermediate Water forms because warm temperatures and dry air cause ocean water to evaporate. The surface water increases in salinity and becomes denser. It sinks to a middle depth of about 2000 m.

Rubrics

For more help evaluating open-ended assessment questions, see the rubrics on p. 10T.

chapter 5 Organizer

Section/Objectives	Standards		Labs/Features
	National	State/Local	
Chapter Opener	See p. 9T–10T for a Key to Standards.		**Launch Lab:** How deep is the ocean?, p. 127 **Foldables,** p. 127
Section 1 The Seafloor 2 sessions 1 block 1. **Differentiate** between a continental shelf and a continental slope. 2. **Describe** a mid-ocean ridge, an abyssal plain, and an ocean trench. 3. **Identify** the mineral resources found on the continental shelf and in the deep ocean.	National Content Standards: UCP.1–3, UCP.5, A.1, A.2, C.3, C.4, D.1		**Science Online,** p. 129 **Applying Math:** Calculating a Feature's Slope, p. 130 **MiniLAB:** Modeling the Mid-Atlantic Ridge, p. 131 **Lab:** Mapping the Ocean Floor, p. 134
Section 2 Life in the Ocean 2 sessions 1 block 4. **Describe** photosynthesis and chemosynthesis in the oceans. 5. **List** the key characteristics of plankton, nekton, and benthos. 6. **Compare** and contrast ocean margin habitats.	National Content Standards: UCP.1–3, UCP.5, A.1, A.2, B.1, B.3, C.1–C.5, D.1		**Integrate Career,** p. 136 **Integrate Chemistry,** p. 138 **MiniLAB:** Observing Plankton, p. 139 **Science Online,** p. 140 **Visualizing the Rocky Shore Habitat,** p. 141
Section 3 Ocean Pollution 4 sessions 2 blocks 7. **List** five types of ocean pollution. 8. **Explain** how ocean pollution affects the entire world. 9. **Describe** how ocean pollution can be controlled.	National Content Standards: UCP.1–3, UCP.5, A.1, A.2, F.2, G.3		**Lab:** Resources from the Ocean, p. 148 **Oops! Accidents in Science:** Strange Creatures from the Ocean Floor, p. 150

Glencoe Exclusive! TeacherWorks™ All-In-One Planner and Resource Center

Lab Materials	Reproducible Resources	Section Assessment	Technology
Launch Lab: spring, meterstick, stopwatch	**Chapter *Fast File* Resources** Foldables Worksheet, p. 17 Directed Reading Overview, p. 19 Note-taking Worksheets, pp. 33–35	Glencoe's Assessment Advantage	TeacherWorks includes: • Interactive Teacher Edition • Lesson Planner with calendar • Access to all program blacklines • Correlations to standards • Web links
MiniLAB: 2 tray tables, 10 paper towels, metric ruler **Lab:** graph paper, colored pencils	**Chapter *Fast File* Resources** Transparency Activity, p. 44 MiniLAB, p. 3 Lab Activity, pp. 9–11 Enrichment, p. 30 Reinforcement, p. 27 Directed Reading, p. 20 Transparency Activity, pp. 47–48 Lab Worksheets, pp. 5–6 **Cultural Diversity,** p. 27	**Portfolio** Science Journal, p. 132 **Performance** Applying, Math p. 130 MiniLAB, p. 131 Applying Math, p. 133 **Content** Section Review, p. 133	Section Focus Transparency Teaching Transparency Virtual Labs CD-ROM Guided Reading Audio Program Interactive Chalkboard CD-ROM Video Lab
MiniLAB: pond, lake, or ocean water; dropper; microscope slide; light microscope	**Chapter *Fast File* Resources** Transparency Activity, p. 45 MiniLAB, p. 4 Enrichment, p. 31 Reinforcement, p. 28 Directed Reading, p. 21 **Reading and Writing Skill Activities,** p. 7 **Home and Community Involvement,** p. 30	**Portfolio** Curriculum Connection, p. 137 **Performance** MiniLab, p. 139 Applying Skills p. 142 **Content** Section Review, p. 142	Section Focus Transparency Virtual Labs CD-ROM Guided Reading Audio Program Interactive Chalkboard CD-ROM
Lab: Internet or other resources describing products from the ocean *Need materials?* Contact Science Kit at 1-800-828-7777 or www.sciencekit.com on the Internet.	**Chapter *Fast File* Resources** Transparency Activity, p. 46 Lab Activity, pp. 13–16 Enrichment, p. 32 Reinforcement, p. 29 Directed Reading, pp. 21, 22 Lab Worksheets, pp. 7–8 **Lab Management and Safety,** p. 73 **Life Science Critical Thinking/ Problem Solving,** p. 10	**Portfolio** Curriculum Connection, p. 146 **Performance** Applying Skills, p. 147 **Content** Section Review, p. 147	Section Focus Transparency Virtual Labs CD-ROM Guided Reading Audio Program Interactive Chalkboard CD-ROM Video Lab

End of Chapter Assessment

Blackline Masters	Technology	Professional Series
Chapter *Fast File* Resources Chapter Review, pp. 37–38 Chapter Tests, pp. 39–42 **Standardized Test Practice**, pp. 23–26	MindJogger Videoquiz Virtual Labs CD-ROM Exam*View*® Pro Testmaker TeacherWorks CD-ROM Interactive Chalkboard CD-ROM	**Performance Assessment in the Science Classroom (PASC)**

Transparencies

Section Focus

Section 1 Section Focus Transparency — Let Go of Pressure

Areas called vents spew heated water and other materials from Earth's interior into the ocean. Water enters Earth's interior through cracks in ridges found on the ocean floor. The water is heated and later released through the vents. The materials that come from the vents include gases and various minerals.

1. Describe what you see in this picture. How is it different from the way you imagined the seafloor to be? How is it similar?
2. What resources come from the oceans?

L2

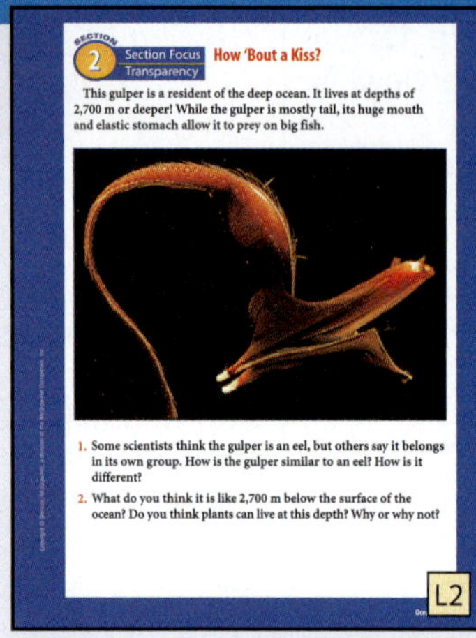
Section 2 Section Focus Transparency — How 'Bout a Kiss?

This gulper is a resident of the deep ocean. It lives at depths of 2,700 m or deeper! While the gulper is mostly tail, its huge mouth and elastic stomach allow it to prey on big fish.

1. Some scientists think the gulper is an eel, but others say it belongs in its own group. How is the gulper similar to an eel? How is it different?
2. What do you think it is like 2,700 m below the surface of the ocean? Do you think plants can live at this depth? Why or why not?

L2

Section 3 Section Focus Transparency — The Dead Zone

While most blooms are beautiful, algal blooms are an ugly exception. The result of pollution, they can devastate marine life. Below you see two photographs—one from a healthy marine environment and one from a similar area that experienced repeated algal blooms.

1. Describe the differences between the two images.
2. The lower photograph is from an area called the dead zone. How is this name appropriate?
3. What are some activities that might result in ocean pollution?

L2

This is a representation of key blackline masters available in the Teacher Classroom Resources. See Resource Manager boxes within the chapter for additional information.

Assessment

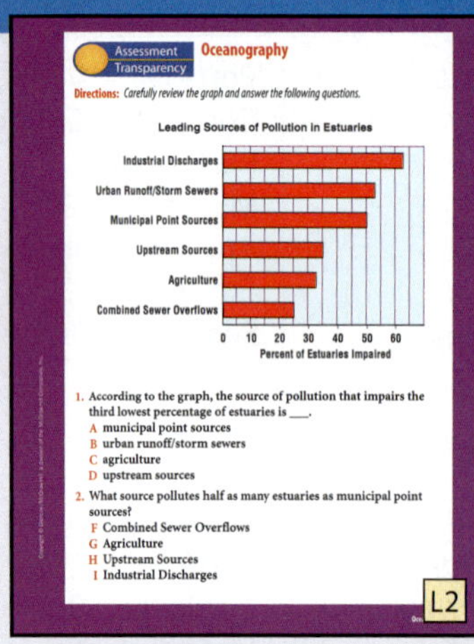
Assessment Transparency — Oceanography

Directions: Carefully review the graph and answer the following questions.

1. According to the graph, the source of pollution that impairs the third lowest percentage of estuaries is ___.
   - A municipal point sources
   - B urban runoff/storm sewers
   - C agriculture
   - D upstream sources
2. What source pollutes half as many estuaries as municipal point sources?
   - F Combined Sewer Overflows
   - G Agriculture
   - H Upstream Sources
   - I Industrial Discharges

L2

Teaching

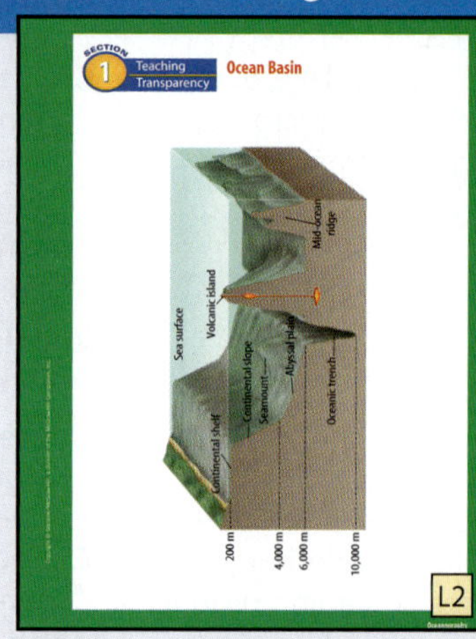
Section 1 Teaching Transparency — Ocean Basin

L2

Key to Teaching Strategies

The following designations will help you decide which activities are appropriate for your students.

- L1 Level 1 activities should be appropriate for students with learning difficulties.
- L2 Level 2 activities should be within the ability range of all students.
- L3 Level 3 activities are designed for above-average students.
- ELL ELL activities should be within the ability range of English Language Learners.
- COOP LEARN Cooperative Learning activities are designed for small group work.
- LS Multiple Learning Styles logos, as described on page 6T, are used throughout to indicate strategies that address different learning styles.
- P These strategies represent student products that can be placed into a best-work portfolio.
- PBL Problem-Based Learning activities apply real-world situations to learning.

Hands-on Activities

Student Text Lab Worksheet

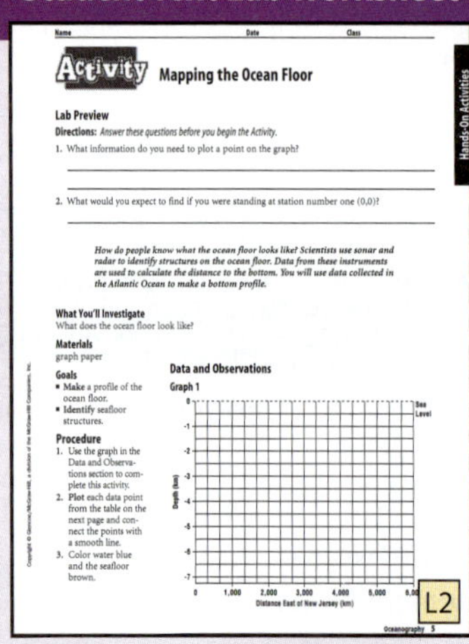
Activity — Mapping the Ocean Floor

Lab Preview

Directions: Answer these questions before you begin the Activity.

1. What information do you need to plot a point on the graph?
2. What would you expect to find if you were standing at station number one (0,0)?

How do people know what the ocean floor looks like? Scientists use sonar and radar to identify structures on the ocean floor. Data from these instruments are used to calculate the distance to the bottom. You will use data collected in the Atlantic Ocean to make a bottom profile.

What You'll Investigate

What does the ocean floor look like?

Materials

graph paper

Goals

- Make a profile of the ocean floor.
- Identify seafloor structures.

Procedure

1. Use the graph in the Data and Observations section to complete this activity.
2. Plot each data point from the table on the next page and connect the points with a smooth line.
3. Color water blue and the seafloor brown.

Data and Observations

Graph 1

L2

Laboratory Activities

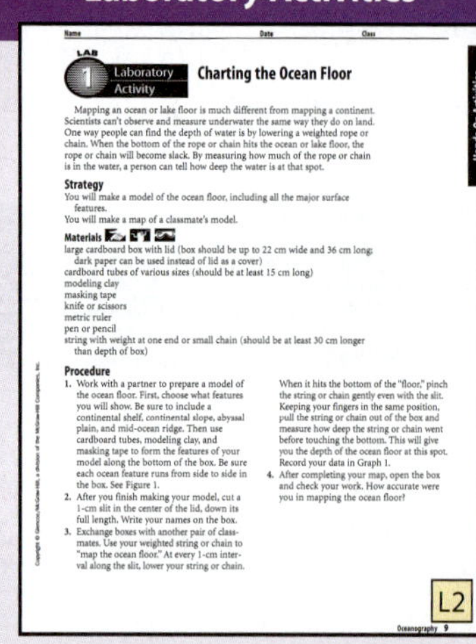
Laboratory Activity 1 — Charting the Ocean Floor

Mapping an ocean or lake floor is much different from mapping a continent. Scientists can't observe and measure underwater the same way they do on land. One way people can find the depth of water is by lowering a weighted rope or chain. When the bottom of the rope or chain hits the ocean or lake floor, the rope or chain will become slack. By measuring how much of the rope or chain is in the water, a person can tell how deep the water is at that spot.

Strategy

You will make a model of the ocean floor, including all the major surface features.

You will make a map of a classmate's model.

Materials

large cardboard box with lid (box should be up to 22 cm wide and 36 cm long; dark paper can be used instead of lid as a cover)
cardboard tubes of various sizes (should be at least 15 cm long)
modeling clay
masking tape
knife or scissors
metric ruler
pen or pencil
string with weight at one end or small chain (should be at least 30 cm longer than depth of box)

Procedure

1. Work with a partner to prepare a model of the ocean floor. First, choose what features you will show. Be sure to include a continental shelf, continental slope, abyssal plain, and mid-ocean ridge. Then use cardboard tubes, modeling clay, and masking tape to form the features of your model along the bottom of the box. Be sure each ocean feature runs from side to side in the box. See Figure 1.
2. After you finish making your model, cut a 1-cm slit in the center of the lid, down its full length. Write your names on the box.
3. Exchange boxes with another pair of classmates. Use your weighted string or chain to "map the ocean floor." At every 1-cm interval along the slit, lower your string or chain. When it hits the bottom of the "floor," pinch the string or chain gently even with the slit. Keeping your fingers in the same position, pull the string or chain out of the box and measure how deep the string or chain went before touching the bottom. This will give you the depth of the ocean floor at this spot. Record your data in Graph 1.
4. After completing your map, open the box and check your work. How accurate were you in mapping the ocean floor?

L2

Meeting Different Ability Levels

Content Outline

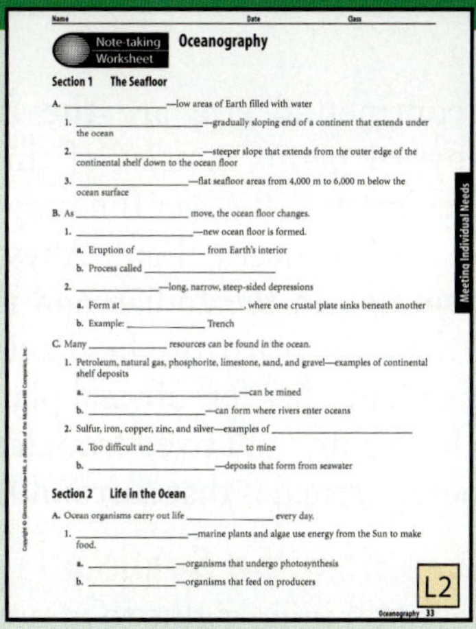

Reinforcement

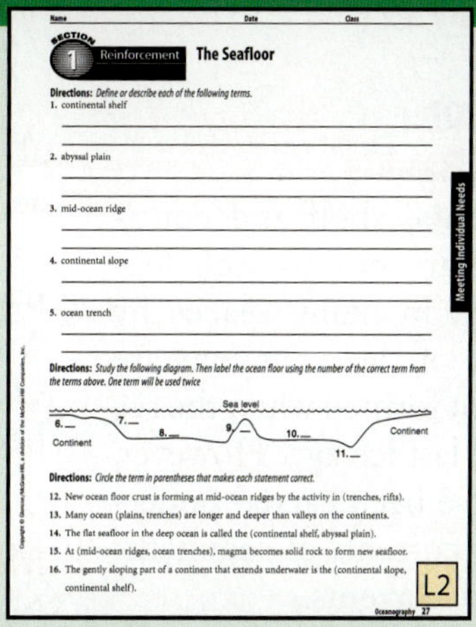

Enrichment

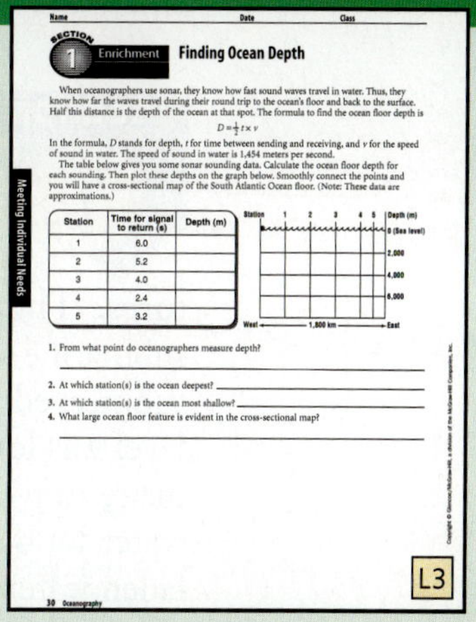

Directed Reading (English/Spanish)

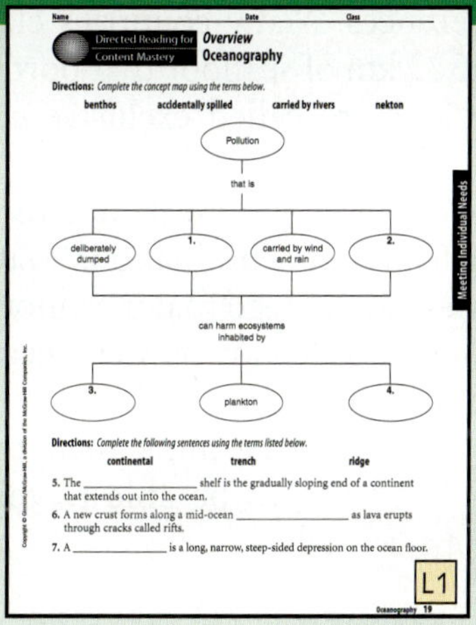

Study Guide

Reading Essentials

Assessment

Test Practice Workbook

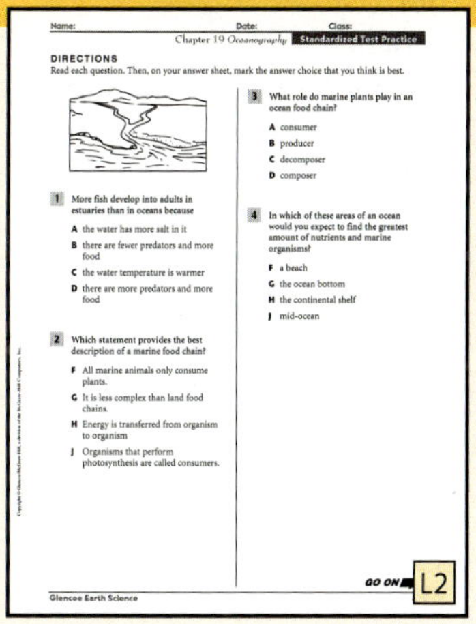

Chapter Review

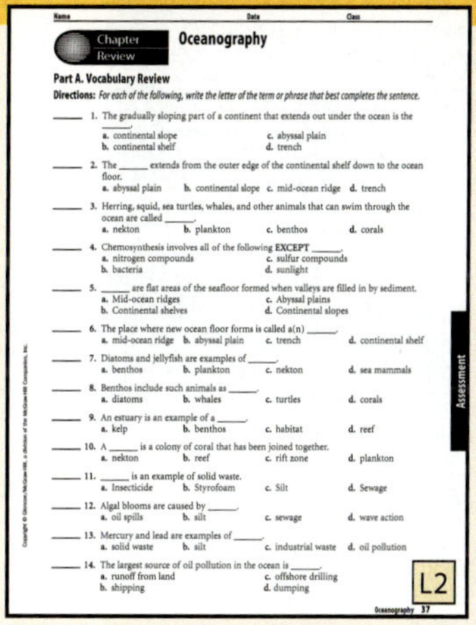

Chapter Tests

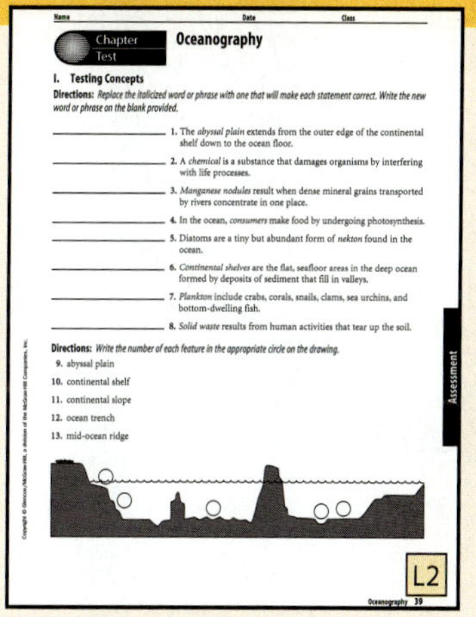

Science Content Background

section 1

The Seafloor

The Ocean Basins

The continental shelf and continental slope are not smooth features. They are dissected in many places by submarine canyons. Some of these canyons are submerged river valleys that formed when the sea level was lower during the last ice age. However, many canyons seem to have been carved underwater by bottom-hugging currents of sediment-laden water called turbidity currents.

At the base of the continental slope, sediment that is transported through submarine canyons accumulates to form large, underwater fans. These fans can coalesce to form a continental rise—a wedge-shaped body of sediment that separates the continental slope from the deep seafloor. Deep water currents flowing along the continental margin also can shape the continental rise.

Beyond the continental rise are the flat abyssal plains. Abyssal plains consist of thick accumulations of sediment that cover the rugged topography of the rock beneath them. Abyssal plains are more common in the Indian and the Atlantic Oceans than they are in the Pacific Ocean. The Pacific Ocean has few abyssal plains because much of the sediment derived from land is trapped in the ocean trenches that surround it.

Mineral Resources from the Seafloor

Nations disagree about mining rights to mineral deposits. Since 1958, nations have met to decide international policy at the United Nations Law of the Sea Conferences. Many countries claim mining rights to 322 km of seafloor that borders them. These regions are called exclusive economic zones.

Water beyond the economic zones of all countries is considered to be international water, and many countries have agreed that it cannot be mined. Some countries do not border oceans, but want the right to mine them. Other countries bordering the ocean can mine their own economic zones but want to be able to mine international waters as well.

chapter content resources

Internet Resources

For additional content background, visit bookh.msscience.com to:

- access your book online
- find references to related articles in popular science magazines
- access Web links with related content background
- access current events with science journal topics

Print Resources

The Oceans, by Ellen J. Prager, McGraw-Hill, 2000

Marine Biology, by Peter Castro and Michael Huber, McGraw-Hill, 2000

Guide to Marine Mammals of the World, by Randall Reeves, Brent Stewart, Phillip Clapham, and James Powell; Alfred A. Knopf, Inc., 2002

Chuck Davis/Stone

Section 2 Life in the Ocean

Life Processes

Photosynthetic marine organisms need nutrients and sunlight to survive. Water above continental shelves is shallow enough for sunlight to penetrate and is rich in nutrients because of runoff from the land. Factors such as temperature, light, nutrients, oxygen, substrate, and pollution limit where particular marine organisms can live. These are called limiting factors.

Ocean Life

Ocean organisms are adapted with a variety of structures and behaviors that help them to survive in their environment.

- Lanternfish, which are small deep-water fish, have photophores (light organs) on the sides of their bodies. By flashing their lights in a certain pattern, they can attract mates.
- When a flounder or flatfish is young, it swims like a normal fish and has one eye on either side of its head. As it matures, one of its eyes migrates to the other side of its head. The fish begins swimming along the ocean bottom on its side with both eyes on the top side, looking up.
- Camouflage helps keep many organisms from becoming a meal. For example, a decorator crab attaches sponges to the top of its shell so that it appears to be part of the ocean bottom.

Drugs from the Sea

Special adaptations of some organisms have attracted the attention of pharmaceutical companies. Because some organisms appear to have no natural enemies, scientists are interested in their chemical makeup.

When looking at developing new drugs from the sea, many different disciplines of scientists work together. Biologists and ecologists collect and separate the species. Chemists extract substances from the organisms. Biologists and pharmaceutical specialists test and purify the substances. Finally, researchers perform clinical trials to see whether these substances are effective in the treatment of certain diseases.

Section 3 Ocean Pollution

Sources of Pollution

The Environmental Protection Agency conducted a survey in conjunction with the coastal states in 1996 to determine the pollutants in their estuaries. The most common pollutants were excessive nutrients, which were found in 22 percent of the estuaries surveyed. Bacteria such as *E. coli* were the second most prevalent pollutants, indicating the presence of sewage. Bacteria were found to pollute 16 percent of the estuaries surveyed. Organic chemicals (15 percent), oxygen-depleting chemicals (12 percent), and petroleum products (8 percent) were the other major pollutants.

Teacher to Teacher

Rachel Shively, Junior High Science Teacher
Aptakisic Junior High School
Buffalo Grove, Illinois

"To help the students understand how and why our oceans are polluted, I have them create informational brochures. They research what the pollution is; what animal and plant life it affects; and give a couple of solutions as to how to fix the problem. They should include a bibliography, additional resources for concerned citizens, and pictures and graphs."

Rachel Shively

Art Brewer/International Stock

chapter 5

Chapter Vocabulary

continental shelf, p. 128
continental slope, p. 129
abyssal plain, p. 129
mid-ocean ridge, p. 130
trench, p. 131
photosynthesis, p. 135
chemosynthesis, p. 137
plankton, p. 138
nekton, p. 138
benthos, p. 139
estuary, p. 142
reef, p. 143
pollution, p. 144

Science Journal Student answers will vary but should include a description of each organism, its specific ocean habitat, and aspects of its behavior such as feeding or reproducing.

INTERACTIVE CHALKBOARD with Image Bank

PowerPoint® Presentations

This CD-ROM is an editable Microsoft® PowerPoint® presentation that includes:

- a pre-made presentation for every chapter
- interactive graphics
- animations
- audio clips
- image bank
- all new section and chapter questions
- Standardized Test Practice
- transparencies
- pre-lab questions for all labs
- Foldables directions
- links to bookh.msscience.com

chapter 5

Oceanography

chapter preview

sections

1 The Seafloor
Lab Mapping the Ocean Floor
2 Life in the Ocean
3 Ocean Pollution
Lab Resources from the Oceans
Virtual Lab What are some characteristics of the ocean and the ocean floor?

Marine Life

This inch-long jelly drifts with ocean currents and uses stinging tentacles to capture small prey. Through a chemical process called bioluminescence it is able to produce light and glow in the dark.

Science Journal Describe characteristics of three marine organisms you are familiar with.

Theme Connection

Scale and Structure The scale and structure of ocean floor topography are presented.

About the Photo

Marine Life This jelly is found in the frigid waters of the Southern Ocean. It is a member of the order Trachymedusae in Class Hydrozoa. This photo was taken at McMurdo Sound, Antarctica.

Start-Up Activities

Launch LAB

How deep is the ocean?

Sonar is used to measure ocean depth. You will model sonar in this lab.

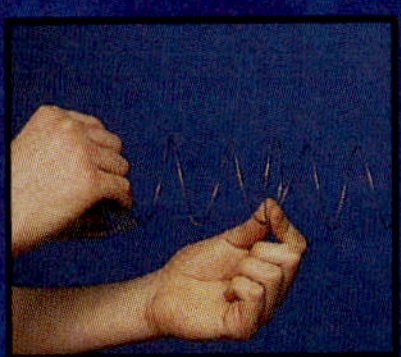

1. With one person holding each end, stretch a spring until it is taut. Measure the distance between the ends.
2. Pinch two coils together. When the spring is steady, release the coils to create a wave.
3. Record the time it takes the wave to travel back and forth five times. Divide this number by five to calculate the time of one round trip.
4. Calculate the speed of the wave by multiplying the distance by two and dividing this number by the time.
5. Move closer to your partner. Take in coils to keep the spring at the same tension. Repeat steps 2 and 3.
6. Calculate the new distance by multiplying the new time by the speed from step 4, and then dividing this number by two.
7. **Think Critically** Write a paragraph in your Science Journal that describes how this lab models sonar.

FOLDABLES™ Study Organizer

The Seafloor Make the following Foldable to help you identify the features of the seafloor.

STEP 1 **Fold** a sheet of paper in half lengthwise. Make the back edge about 1.25 cm longer than the front edge.

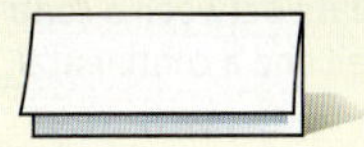

STEP 2 **Fold** in half, then fold in half again to make three folds.

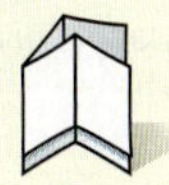

STEP 3 **Unfold and cut** only the top layer along the three folds to make four tabs.

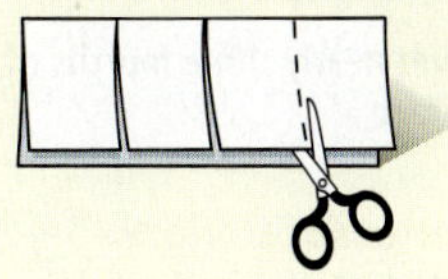

STEP 4 **Label** the tabs as shown.

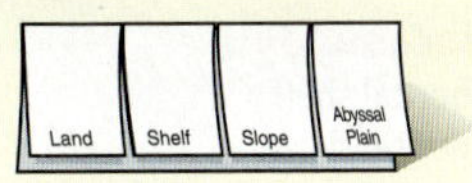

Find Main Ideas As you read the chapter, draw seafloor features on the front of the tabs and write information about them under the tabs.

Science Online

Preview this chapter's content and activities at bookh.msscience.com

Launch LAB

Purpose Students will model sonar by observing waves traveling on a spring.

Preparation Before beginning the lab, remind students that distance can be calculated from the travel time and the speed.

Materials large spring, timer, meterstick

Teaching Strategy

Troubleshooting Keeping the spring at the same tension in step 5 is important for obtaining accurate measurements. Before moving closer together, have students notice the distance between the coils and try to maintain that distance when they move. L2

Think Critically

Distance is calculated by knowing the speed of the wave and the travel time using the formula:

Distance = 1/2 travel time × wave speed

Assessment

Oral Ask students to compare and contrast the measurements in this activity with real sonar. In both this activity and in real sonar, the speed of compressional wave travel is used to calculate distance. In this activity, the waves travel along a spring. In real sonar, the waves are sound waves. Use **Performance Assessment in the Science Classroom**, p. 89. L2

FOLDABLES™ Study Organizer — Dinah Zike Study Fold

Student preparation materials for this Foldable are available in the Chapter *FAST FILE* Resources.

1 Motivate

Bellringer

Section Focus Transparencies also are available on the Interactive Chalkboard CD-ROM. L2 ELL

Section 1 Section Focus Transparency **Let Go of Pressure**

Areas called vents spew heated water and other materials from Earth's interior into the ocean. Water enters Earth's interior through cracks in ridges found on the ocean floor. The water is heated and later released through the vents. The materials that come from the vents include gases and various minerals.

1. Describe what you see in this picture. How is it different from the way you imagined the seafloor to be? How is it similar?
2. What resources come from the oceans?

L2

Tie to Prior Knowledge

Familiar Landforms Have students list landforms with which they are familiar. Explain that throughout their lives, students have observed geological features of Earth's surface. In this section, they will learn about similar structures on the ocean floor. L2

Text Answer Question

seamounts and abyssal plains

section 1

The Seafloor

as you read

What You'll Learn

- **Differentiate** between a continental shelf and a continental slope.
- **Describe** a mid-ocean ridge, an abyssal plain, and an ocean trench.
- **Identify** the mineral resources found on the continental shelf and in the deep ocean.

Why It's Important

Oceans cover nearly three fourths of Earth's surface.

Review Vocabulary
magma: hot, melted rock beneath Earth's surface

New Vocabulary
- continental shelf
- continental slope
- abyssal plain
- mid-ocean ridge
- trench

The Ocean Basins

Imagine yourself driving a deep-sea submersible along the ocean floor. Surrounded by cold, black water, the lights of your vessel reflect off of what looks like a mountain range ahead. As you continue, you find a huge opening in the seafloor—so deep you can't even see the bottom. What other ocean floor features can you find in **Figure 1?**

Ocean basins, which are low areas of Earth that are filled with water, have many different features. Beginning at the ocean shoreline is the continental shelf. The **continental shelf** is the gradually sloping end of a continent that extends under the ocean. On some coasts, the continental shelf extends a long distance. For instance, on North America's Atlantic and Gulf coasts, it extends 100 km to 350 km into the sea. On the Pacific Coast, where the coastal range mountains are close to the shore, the shelf is only 10 km to 30 km wide. The ocean covering the continental shelf can be as deep as 350 m.

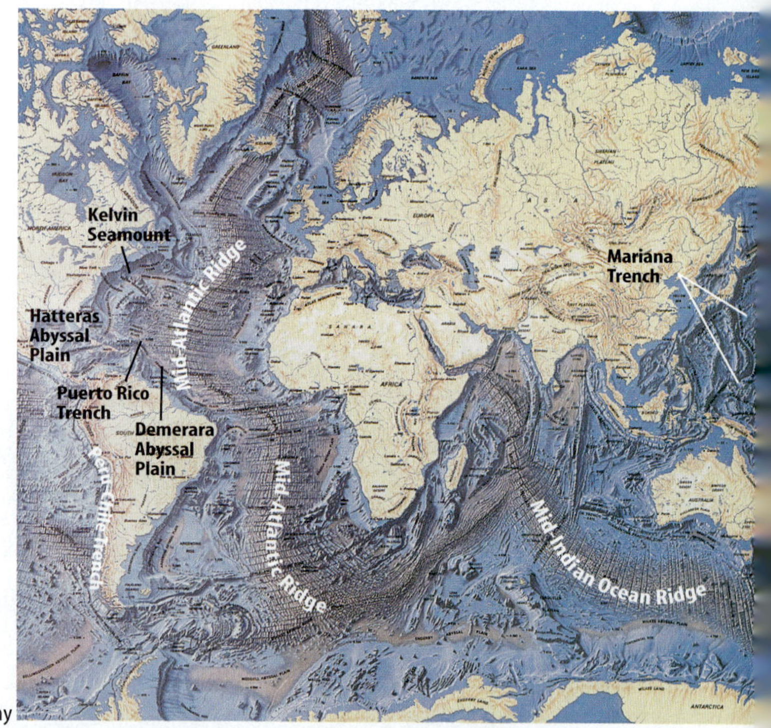

Figure 1 This map shows features of the ocean basins. Locate a trench and a mid-ocean ridge.

Section 1 Resource Manager

Chapter *Fast File* Resources
- Transparency Activity, pp. 44, 47–48
- Directed Reading for Content Mastery, pp. 19, 20
- Note-taking Worksheets, pp. 33–35
- MiniLAB, p. 3
- Enrichment, p. 30
- Reinforcement, p. 27
- Lab Worksheets, pp. 5–6
- Lab Activity, pp. 9–11

Mathematics Skill Activities, p. 61

Earth Science Critical Thinking/Problem Solving, p. 1

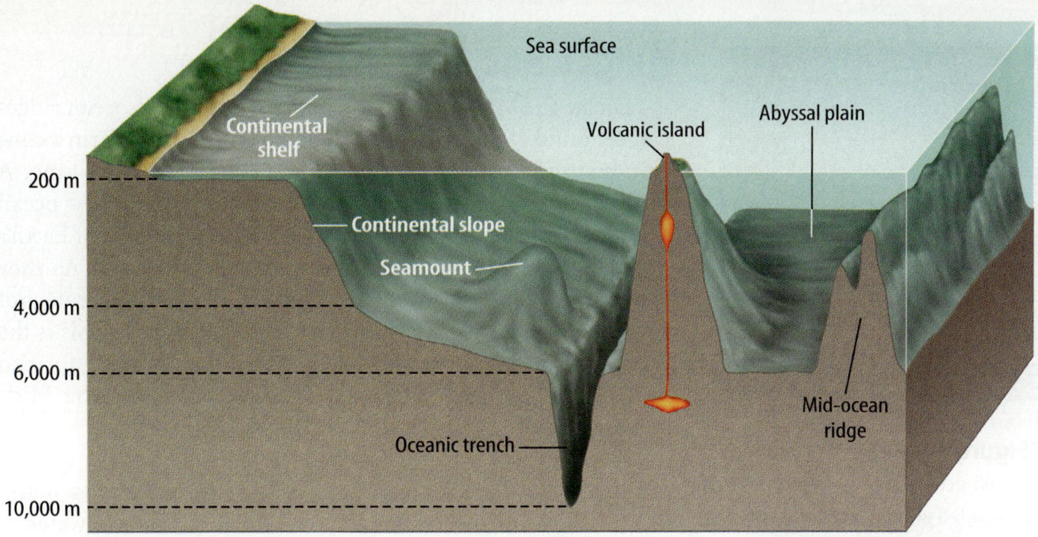

Figure 2 shows that beyond the shelf, the ocean floor drops more steeply, forming the continental slope. The **continental slope** extends from the outer edge of the continental shelf down to the ocean floor. Beyond the continental slope lie the trenches, valleys, plains, mountains, and ridges of the ocean basin.

In the deep ocean, sediment, derived mostly from land, settles constantly on the ocean floor. These deposits fill in valleys and create flat seafloor areas called **abyssal** (uh BIH sul) **plains.** Abyssal plains are from 4,000 m to 6,000 m below the ocean surface. Can you locate the abyssal plain shown in **Figure 2?**

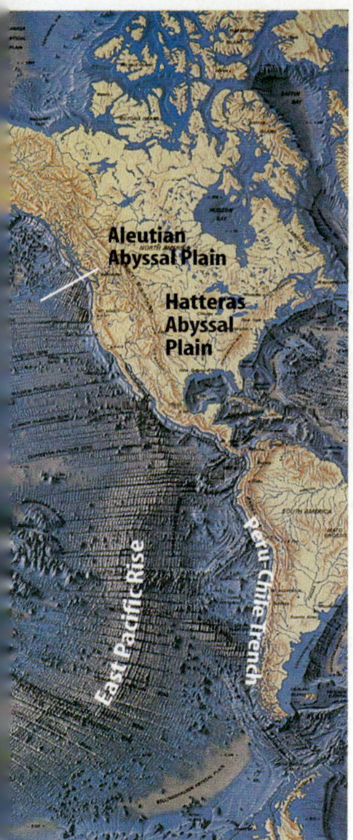

In the Atlantic Ocean, areas of extremely flat abyssal plains can be large. One example is the Canary Abyssal Plain, which has an area of approximately 900,000 km^2. Other abyssal plains found in the Atlantic Ocean include the Hatteras and Demerara Abyssal Plains, both shown in **Figure 1.** Some areas of abyssal plains have small hills and seamounts. Seamounts are under-water, inactive volcanic peaks. They most commonly are found in the Pacific Ocean. Can you locate a seamount in **Figure 1?**

Reading Check *What are seamounts?*

Figure 2 Ocean basin features are continuous from shore to shore. (Features in this diagram are not to scale.)
Describe *where the continental shelf ends and the continental slope begins.*

Science Online

Topic: Ocean Basins

Visit bookh.msscience.com for Web links to information about ocean basins.

Activity Find five fun facts about ocean basins, such as the location of the deepest known point in the ocean. Present your facts to the class.

2 Teach

Activity

Bathymetric Maps Obtain bathymetric maps (topographic maps of the bottom of the ocean). Refer to the maps frequently throughout this section. Have students contrast the width of the continental shelf off the east coast of the United States with that off the west coast. The shelf off the east coast is much wider. L2 LS **Visual-Spatial**

Caption Answer

Figure 2 The shelf ends and the slope begins where the elevation drops more steeply.

Visual Learning

Figure 2 Have students describe the structure of trenches and islands. Trenches are like very deep valleys. Islands look like the tops of mountains. L2 LS **Linguistic**

Use Science Words

Word Meaning Have students find out why plains from 4,000 to 6,000 m below the ocean's surface are called *abyssal plains.* *Abyssal* relates to bottom waters in the ocean. L2 LS **Linguistic**

Reading Check

Answer submerged, inactive volcanic peaks

Differentiated Instruction

English-Language Learners Have students learning English as their second language make index cards with this section's vocabulary words on one side and their correct pronunciation and definition on the other side. Students can work in pairs to review pronunciations. L2

Fun Fact

Many of the rocks that form the mountains of the world were formed underwater. Evidence of this comes from fossils of marine life found within rocks on mountain slopes and peaks.

Discussion

Crust Ages Where is the oldest crust on the Atlantic Ocean bottom? Where is the youngest crust? The oldest is at its margins; the youngest is in its middle. L2

Reading Check

Answer Lava erupting from mid-ocean ridges forms new ocean floor.

Applying Math

National Math Standards

Correlation to Mathematics Objectives

1, 2, 9

Teaching Strategy

Follow the steps in the example problem. The slope can be found dividing 2000 m by 40 km.

Answers to Practice Problems

1. 50 m/km

2. 50 km

Figure 3 New seafloor forms at mid-ocean ridges. A type of lava called pillow lava lies newly formed at this ridge on the ocean floor.

Ridges and Trenches

Locate the Mid-Atlantic Ridge in **Figure 1.** Mid-ocean ridges can be found at the bottom of all ocean basins. They form a continuous underwater ridge approximately 70,000 km long. A **mid-ocean ridge** is the area in an ocean basin where new ocean floor is formed. Crustal plates, which are large sections of Earth's uppermost mantle and crust, are moving constantly. As they move, the ocean floor changes. When ocean plates separate, hot magma from Earth's interior forms new ocean crust. This is the process of seafloor spreading. New ocean floor is being formed at a rate of approximately 2.5 cm per year along the Mid-Atlantic Ridge.

New ocean floor forms along mid-ocean ridges as lava erupts through cracks in Earth's crust. **Figure 3** shows newly erupted lava on the seafloor. When the lava hits the water, it cools quickly into solid rock, forming new seafloor. While seafloor is being formed in some parts of the oceans, it is being destroyed in others. Areas where old ocean floor slides beneath another plate and descends into Earth's mantle are called subduction zones.

Reading Check *How does new ocean floor form?*

Applying Math — Find the Slope

CALCULATING A FEATURE'S SLOPE If the width of a continental shelf is 320 km and it increases in depth a total of 300 m in that distance, what is its slope?

Solution

1. *This is what you know:*
   - width = 320 km
   - increase in depth = 300 m
2. *This is what you need to find:* slope: s
3. *This is the equation you need to use:* s = increase in depth ÷ width
4. *Solve the equation by substituting in known values:* $s = 300 \text{ m} \div 320 \text{ km} = 0.94 \text{ m/km}$

Practice Problems

1. The width of a continental slope is 40 km. It increases in depth by 2,000 m. What is the slope of the continental slope?
2. If the depth of a continental slope increases by 3,700 m and the slope is 74 m/km, what is the width of the slope?

Science Online For more practice, visit bookh.msscience.com/math_practice

Science Journal

Ocean Features Ask each student to write a paragraph that begins, "Many landforms were revealed the day the oceans suddenly disappeared. I saw. . ." L2 LS **Linguistic**

Teacher FYI

Ocean Depths Four general ocean depths are recognized. The neritic depth is 0–200 m below the surface, and the bathyal depth is 200–4,000 m. The abyssal depth is 4,000–6,000 m below the surface. Eighty-five percent of the ocean floor is at abyssal depth. The hadal depth is anything deeper than 6,000 m.

Figure 4 Located at subduction zones, trenches are important ocean basin features.

In 1960, the world's deepest dive was made in the Mariana Trench. The *Trieste* carried Jacque Piccard and Donald Walsh to a depth of almost 11 km.

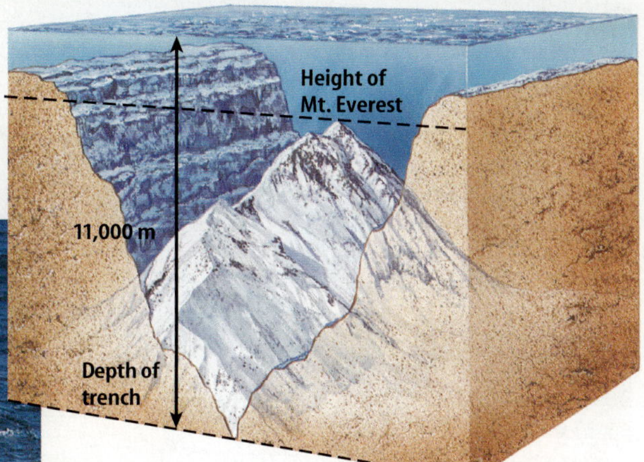

If Earth's tallest mountain, Mount Everest, were set in the bottom of the Mariana Trench of the Pacific Basin, it would be covered with more than 2,000 m of water.

Subduction Zones On the ocean floor, subduction zones are marked by deep ocean trenches. A **trench** is a long, narrow, steep-sided depression where one crustal plate sinks beneath another. Most trenches are found in the Pacific Basin. Ocean trenches are usually longer and deeper than any valley on any continent. One trench, famous for its depth, is the Mariana Trench. It is located to the south and east of Japan in the Pacific Basin. This trench reaches 11 km below the surface of the water, and it is the deepest place in the Pacific. The photo in **Figure 4** shows the deep-sea vessel, the *Trieste,* that descended into the trench in 1960. **Figure 4** also illustrates that the Mariana Trench is so deep that Mount Everest could easily fit into it.

Mineral Resources from the Seafloor

Resources can be found in many places in the ocean. Some deposits on the continental shelf are relatively easy to extract. Others can be found only in the deep abyssal regions on the ocean floor. People still are trying to figure out how to get these valuable resources to the surface. As you read, suggest some methods that could be used to retrieve hard-to-reach resources.

Mini LAB

Modeling the Mid-Atlantic Ridge

Procedure

1. Set two **tray tables** 2 cm apart.
2. Gather ten **paper towels** that are still connected. Lay one end of the paper towels on each table so the towels hang down into the space between the tables.
3. Slowly pull each end of the paper towels away from each other.

Analysis

1. Explain how this models the Mid-Atlantic Ridge.
2. How long does it take for 2.5 cm of new ocean crust to form at the Mid-Atlantic Ridge? How long does it take 25 cm to form?

Try at Home

Curriculum Connection

History In 1960, Auguste Piccard's bathyscaphe submersible, the *Trieste*, descended to the deepest point a person has ever been, the Mariana Trench (–10,912 meters). Have students research to find out more about the *Trieste* expedition. L3 LS **Linguistic**

Use an Analogy

Recycling Crust Relate what happens at trenches and rifts to recycling. At trenches, crustal material disappears into the mantle. At rifts, material from the mantle emerges to form new crust. In this way, Earth's crust is recycled. L2

Mini LAB

Purpose Students will create a model of the Mid-Atlantic Ridge.

Materials two tray tables, ten paper towels that are still connected

Teaching Strategy Before beginning the activity, use colored markers to draw parallel lines on the paper towels to help students visualize the spreading of the ocean floor.

Analysis

1. Paper towel emerges at the "rift" between the tables and spreads in both directions. This is similar to the way new oceanic crust forms in a rift valley and spreads away from the Mid-Atlantic Ridge.
2. one year; ten years

Assessment

Process Show students a map of the ocean floor and point out that the Mid-Atlantic Ridge runs through the country of Iceland. Ask students to infer how this affects Iceland today and how it will affect the geography of Iceland in the future. Today, Iceland experiences many earthquakes and volcanic eruptions. The country is widening at about 2 cm per year as the North American and Eurasian Plates separate. Use **Performance Assessment in the Science Classroom,** p. 89. L2

Try at Home

Caption Answer

Figure 5 phosphorite: Pacific coast of North America and South America, among others; diamonds: Southern Africa, Eastern India, Southeast Asia, Northern Australia

Quick Demo

Setting Rates

Materials gravel, sand, clay, water, glass jar with lid

Estimated Time 10 minutes

Procedure To demonstrate the settling rates of sediments where rivers flow into oceans, pour a mixture of gravel, sand, and clay into a jar of water. Put a lid on the jar and slowly swirl the contents until they are well mixed. Allow the sediments to settle. Students will observe that the gravel will settle to the bottom first, followed by the sand, and then the clay. Why does the gravel settle first? It has the largest particle size and is the heaviest of the three sediments. L2 LS **Visual-Spatial**

Virtual Labs

Ocean Characteristics

What are some characteristics of the ocean and the ocean floor?

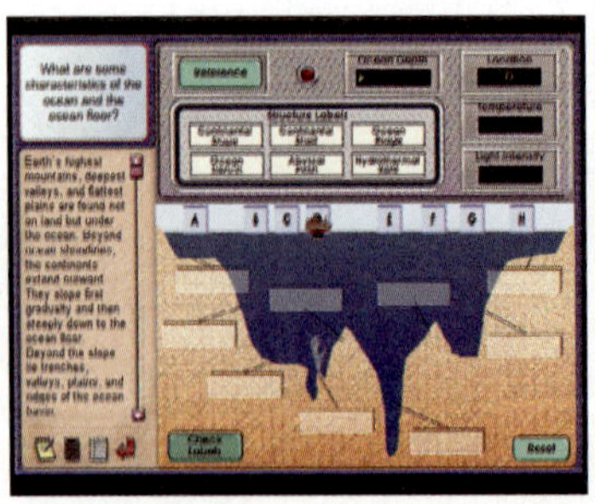

Figure 5 The ocean is rich with mineral resources.
Determine *In which areas of the world can phosphorite be found? Where can diamonds be found?*

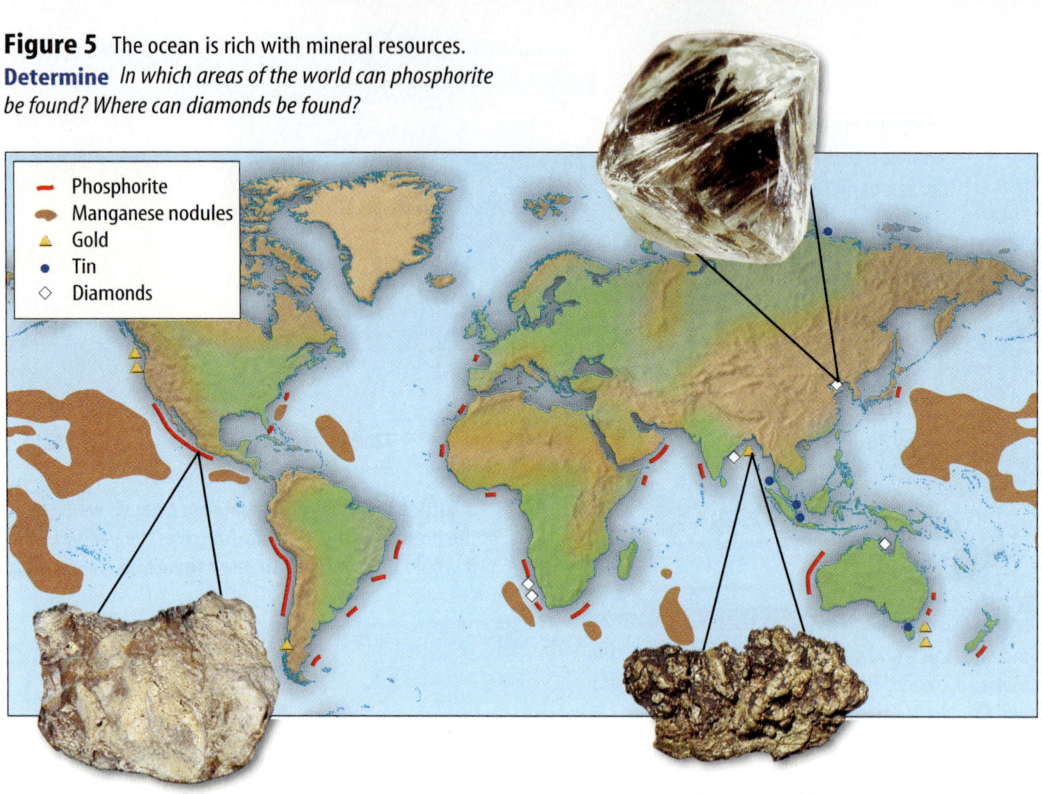

Continental Shelf Deposits A high amount of organic activity occurs in the waters above the continental shelf, and sediment accumulates to great thickness on the ocean floor. This is why many different kinds of resources can be found there, such as petroleum and natural gas deposits. Approximately 20 percent of the world's oil comes from under the seabed. To extract these substances, wells are drilled into the seafloor from floating vessels and fixed platforms.

Other deposits on the continental shelf include phosphorite, which is used to make fertilizer, and limestone, which is used to make cement. Sand and gravel, both economically important, also can be dredged from the continental shelf.

Rivers that flow into oceans transport important minerals to the continental shelf from land. Sometimes the energy of ocean waves and currents can cause denser mineral grains that have been brought in by rivers to concentrate in one place. These deposits, called placer (PLAHS ur) deposits, can occur in coastal regions where rivers entering the ocean suddenly lose energy, slow down, and drop their sediment. Metals such as gold and titanium and gemstones such as diamonds are mined from placer deposits in some coastal regions. **Figure 5** shows where some resources in the ocean can be found.

Differentiated Instruction

Challenge Have students write research reports about how placer deposits are mined. Reports should describe panning and sluicing. L3 LS **Linguistic**

Science Journal

Law of the Sea Many nations disagree about mining rights to mineral deposits on the ocean floor. Ask students to research United Nations Law of the Sea Conferences, find out why nations disagree, and summarize the information in a brief report. Students may find nations disagree on definition of international waters. There are also environmental questions. L3 LS **Linguistic** P

132 ◆ H CHAPTER 5 Oceanography

Deep-Water Deposits Through the holes and cracks along mid-ocean ridges, plumes of hot water billow out into surrounding seawater. As the superheated water cools, mineral deposits sometimes form. As a result, elements such as sulfur and metals like iron, copper, zinc, and silver can be concentrated in these areas. Today, no one is mining these valuable materials from the depths because it would be too expensive to recover them. However, in the future, these deposits could become important.

Other mineral deposits can precipitate from seawater. In this process, minerals that are dissolved in ocean water come out of solution and form solids on the ocean floor. Manganese nodules are small, darkly colored lumps strewn across large areas of the ocean basins. **Figure 6** shows these nodules. Manganese nodules form by a chemical process that is not fully understood. They form around nuclei such as discarded sharks' teeth, growing slowly, perhaps as little as 1 mm to 10 mm per million years. These nodules are rich in manganese, copper, iron, nickel, and cobalt, which are used in the manufacture of steel, paint, and batteries. Most of the nodules lie thousands of meters deep in the ocean and are not currently being mined, although suction devices similar to huge vacuum cleaners have been tested to collect them.

Figure 6 These manganese nodules were found on the floor of the Pacific Ocean.
Think Critically *Can you think of an efficient way to gather the nodules from a depth of 4 km?*

section 1 review

Summary

The Ocean Basins
- Ocean basins have many different features, including the continental shelf, continental slope, and abyssal plains.

Ridges and Trenches
- New ocean floor forms along mid-ocean ridges.
- Trenches mark areas of ocean floor where one crustal plate is sinking beneath another.

Mineral Resources from the Seafloor
- Many mineral deposits, such as petroleum and natural gas, can be found on the continental shelf.
- Other mineral deposits, such as manganese nodules, can be found in deep water.

Self Check

1. **Compare and contrast** continental shelves and continental slopes.
2. **Contrast** mid-ocean ridges and trenches.
3. **Describe** how an abyssal plain looks and how it forms.
4. **Think Critically** Why is the formation of continental shelf deposits different from that of deep-water deposits? Name two examples of each type of deposit.

Applying Skills

5. **Infer** Depth soundings, taken as a ship moves across an ocean, are consistently between 4,000 m and 4,500 m. Over which area of seafloor is the ship passing?

Caption Answer

Figure 6 Possible answers: robots with suction devices, scoops, nets

3 Assess

DAILY INTERVENTION

Check for Understanding

Visual-Spatial Provide students with an unlabeled diagram of **Figure 2**. Have them identify the following structures: continental shelf, continental slope, seamount, oceanic trench, volcanic island, abyssal plain, mid-ocean ridge. L2

Reteach

Quiz Organize the class into groups of five. Have each student in the group write one of the terms learned in this section on one side of a note card and its meaning on the opposite side. Then, have students quiz one another on the terms. L1 COOP LEARN IN Interpersonal

Assessment

Oral Which ocean basins are getting larger? Which is getting smaller? What would cause these changes? The Atlantic and Indian Basins are getting larger; the plates are diverging at the mid-ocean ridges. The Pacific Basin is getting smaller; crust is sinking into its trenches as plates converge. Use **Performance Assessment in the Science Classroom**, p. 89. L2

section 1 review

1. They are adjacent to each other. Shelves are relatively flat; slopes are steeper.
2. Mid-ocean ridges are mountains formed where plates diverge. Trenches are valleys formed where plates converge and one plate moves under the other.
3. An abyssal plain is very flat, and it forms where deposits fill in valleys on the ocean basin floor.
4. Continental shelf deposits form as a result of sediment accumulation from organic activity and river input. Deep-water deposits form as superheated water along mid-ocean ridges cools. They also precipitate from seawater. Continental shelf deposits include phosphorite, limestone, sand, and gravel. Deep-water deposits include iron, copper, and manganese nodules.
5. the abyssal plain

BENCH TESTED

Real-World Question

Purpose Students make and analyze a profile map of the ocean floor. L3 COOP LEARN LS **Visual-Spatial and Kinesthetic**

Process Skills Use numbers, observe and infer, interpret data, form operational definitions, interpret scientific illustrations, make models, measuring in SI

Time Required 30 to 40 minutes

Procedure

Teaching Strategy Be sure that students realize that the graph is set up in a form that is different from what they may be used to. Usually, zero is at the bottom of the vertical axis. However, on this graph, it is at the top of the vertical axis because the data represent depths below sea level. Students may become confused and reverse the data as they begin to plot the points.

Conclude and Apply

1. A continental slope occurs between 160 and 1,050 km. The Mid-Atlantic Ridge occurs between 2,000 and 4,500 km from New Jersey. A continental slope lies between 5,300 and 5,500 km from New Jersey.
2. No; the profile has extensive vertical exaggeration. To make an accurate profile, the vertical and horizontal scales must be the same.

LAB

Mapping the Ocean Floor

In this lab you will use sonar data from the Atlantic Ocean to make a profile of the ocean bottom.

Real-World Question

What does the ocean floor look like?

Goals

- **Make** a profile of the ocean floor.
- **Identify** seafloor structures.

Materials

graph paper

Procedure

1. Copy and complete a graph like the one shown.
2. **Plot** each data point and connect the points with a smooth line.

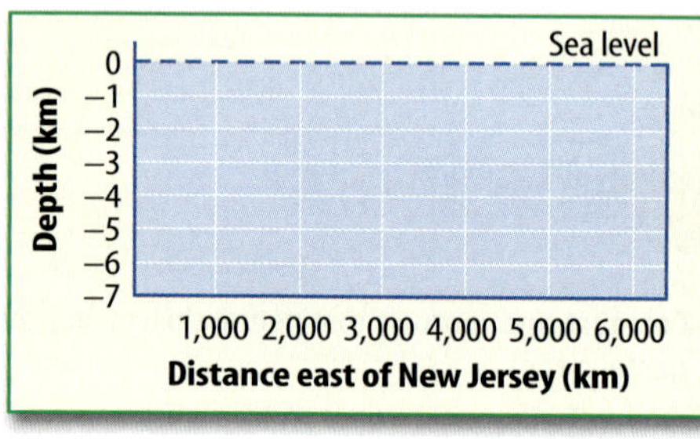

3. Color water blue and the seafloor brown.

Conclude and Apply

1. What ocean floor structures occur between 160 km and 1,050 km east of New Jersey? Between 2,000 km and 4,500 km? Between 5,300 km and 5,500 km?
2. When a profile of a feature is drawn to scale, the horizontal and vertical scales must be the same. Does your profile give an accurate picture of the ocean floor? Explain.

Ocean Floor Data

Station Number	Distance from New Jersey (km)	Depth to Ocean Floor (m)
1	0	0
2	160	165
3	200	1,800
4	500	3,500
5	1,050	5,450
6	1,450	5,100
7	1,800	5,300
8	2,000	5,600
9	2,300	4,750
10	2,400	3,500
11	2,600	3,100
12	3,000	4,300
13	3,200	3,900
14	3,450	3,400
15	3,550	2,100
16	3,700	1,275
17	3,950	1,000
18	4,000	0
19	4,100	1,800
20	4,350	3,650
21	4,500	5,100
22	5,000	5,000
23	5,300	4,200
24	5,450	1,800
25	5,500	920
26	5,650	0

Assessment

Performance To further assess students' understanding of the ocean floor profile, have them draw a profile of an abyssal plain and a seamount. Use **Performance Assessment in the Science Classroom,** p. 127. L3

section 2

Life in the Ocean

Life Processes

Life processes such as breathing oxygen, digesting food, making new cells, and growing take place in your body every day. It takes energy to do this, plus walk between classrooms or play soccer. Organisms that live in the ocean also carry out life processes every day. The octopus shown in **Figure 7** will get the oxygen it needs from the water. It will have to eat, and it will use energy to capture prey and to escape predators. It will make new cells and eventually reproduce. Like other marine organisms, it is adapted to accomplish these processes in the salty water of the ocean.

One of the most important processes in the ocean, as it is on land, is that organisms obtain food to use for energy. Obtaining the food necessary to survive can be done in several ways.

INTEGRATE Life Science **Photosynthesis** Nearly all of the energy used by organisms in the ocean ultimately comes from the Sun. Radiant energy from the Sun penetrates seawater to an average depth of 100 m. Marine organisms such as plants and algae use energy from the Sun to build their tissues and produce their own food. This process of making food is called **photosynthesis.** During photosynthesis, carbon dioxide and water are changed to sugar and oxygen in the presence of sunlight. Organisms that undergo photosynthesis are called producers. Producers also need nutrients, such as nitrogen and phosphorus, in order to produce organic matter. These and other nutrients are obtained from the surrounding water. Marine producers include sea grasses, seaweeds, and microscopic algae. Although they might seem unimportant because they are small, microscopic algae are responsible for approximately 90 percent of all marine production. Organisms that feed on producers are called consumers. Consumers in the marine environment include shrimp, fish, dolphins, whales, and sharks.

as you read

What You'll Learn

- **Describe** photosynthesis and chemosynthesis in the oceans.
- **List** the key characteristics of plankton, nekton, and benthos.
- **Compare and contrast** ocean margin habitats.

Why It's Important

The ocean environment is fragile, and many organisms, including humans, depend on it for their survival.

Review Vocabulary

nutrient: a substance needed for the production of organic matter

New Vocabulary

- photosynthesis
- chemosynthesis
- plankton
- nekton
- benthos
- estuary
- reef

Figure 7 Hunting at night, this Pacific octopus feeds on snails and crabs. It uses camouflage, ink, and speed to avoid predators.

1 Motivate

Bellringer

INTERACTIVE CHALKBOARD PowerPoint® Presentations

Section Focus Transparencies also are available on the Interactive Chalkboard CD-ROM. L2 ELL

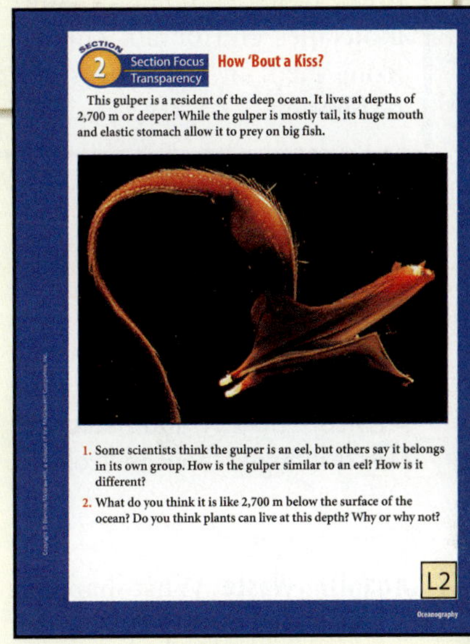

Tie to Prior Knowledge

Food Chain Review the structure of a simple food chain, such as grass ➡ cow ➡ humans. Remind students that they have observed different kinds of organisms interacting with one another. In this section, they will learn about life processes in the ocean, including the transfer of energy through food webs.

Section 2 Resource Manager

Chapter *FAST FILE* Resources

Transparency Activity, p. 45

Directed Reading for Content Mastery, p. 21

MiniLAB, p. 4

Enrichment, p. 31

Reinforcement, p. 28

Life Science Critical Thinking/Problem Solving, p. 1

Earth Science Critical Thinking/Problem Solving, p. 16

2 Teach

IDENTIFYING Misconceptions

Respiration Students may have the misconception that only animals respire. Plants and algae produce food through photosynthesis. They also break down complex organic molecules and obtain energy from them in the process of respiration.

INTEGRATE Career

Ecologist Students interested in ecology as a career should take science, math, and statistics classes in high school and college.

Discussion

Organism Waste What happens to the wastes of organisms that live in the oceans? The wastes decay. They are then used as nutrients by other organisms. L2

Reading Check

Answer energy

Make a Model

Ocean Food Chains Have student teams make mobiles of food chains that exist in the ocean. Hang the mobiles from the classroom ceiling. L2 ELL

COOP LEARN LS Kinesthetic

INTEGRATE Career

Ecologist An ecologist is a scientist who studies the interactions between organisms and their environment. Ecologists may specialize in areas such as marine ecosystems or tropical rain forests. They may also study how energy is transferred from one organism to another, such as through a food web.

Energy Relationships Energy from the Sun is transferred through food chains. Although the organisms of the ocean capture only a small part of the Sun's energy, this energy is passed from producer to consumer, then to other consumers. In **Figure 8,** notice that in one food chain, a large whale shark consumes small, shrimplike organisms as its basic food. In the other chain, microscopic algae found in water are eaten by microscopic animals called copepods (KOH puh pahdz). The copepods are, in turn, eaten by herring. Cod eat the herring, seals eat the cod, and eventually great white sharks eat the seals. At each stage in the food chain, energy obtained by one organism is used by other organisms to move, grow, repair cells, reproduce, and eliminate waste.

Reading Check *What is passed on at each stage in a food chain?*

In an ecosystem—a community of organisms and their environment—many complex feeding relationships exist. Most organisms depend on more than one species for food. For example, herring eat more than copepods, cod eat more than herring, seals eat more than cod, and white sharks eat more than seals. In an ecosystem, food chains overlap and are connected much like the threads of a spider's web. These highly complex systems are called food webs.

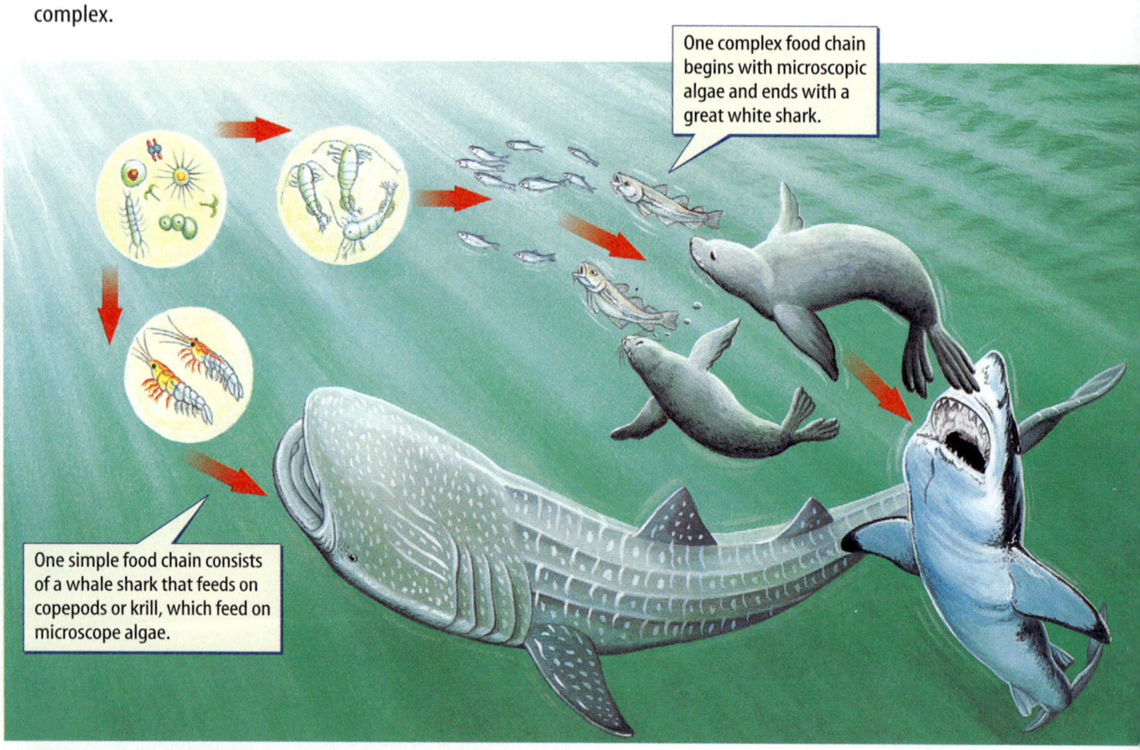

Figure 8 Numerous food chains exist in the ocean. Some food chains are simple and some are complex.

Visual Learning

Figure 8 Have a student volunteer compare and contrast the food chains that include great white sharks and whale sharks. L2

Active Reading

Learning Journal Have students draw a vertical line down each page of the Learning Journal. Have them enter research notes, lecture notes, or vocabulary terms on the left. On the right, have them respond to, interpret, question, or analyze left column entries. Suggest students write a Learning Journal about energy relationships in the ocean. L2

Curriculum Connection

Math Blue whales can consume 8 metric tons of krill daily. Adults average 26 m in length and have a mass of 137 metric tons. Have students calculate the number of grams of krill a blue whale consumes each day and the length in centimeters and mass in kilograms of an average adult blue whale. 8 million g of krill each day, 2,600 cm long, mass of 137,000 kg L3 LS **Logical-Mathematical**

Chemosynthesis Other types of food webs do not depend on the Sun and photosynthesis. These food webs depend on bacteria that perform chemosynthesis. **Chemosynthesis** (kee moh SIHN thuh sus) involves using sulfur or nitrogen compounds as an energy source, instead of light from the Sun, to produce food. Bacteria that perform chemosynthesis using sulfur compounds live along mid-ocean ridges near hydrothermal vents where no light is available. Recall that superheated water from the crust contains high amounts of sulfur. The bacteria found here form the base of a food chain and support a host of highly specialized organisms such as giant tube worms, clams, crabs, and shrimp.

Figure 9 Because sponges live attached to the ocean bottom, they depend on currents to carry their reproductive cells to nearby sponges.
Infer *what would happen if a sponge settled in an area without strong currents.*

Other Life Processes Reproduction also is a vital life process. Some organisms, such as corals and sponges, depend on ocean currents for successful reproduction. Shown in **Figure 9,** these organisms release reproductive cells into the water where they unite to form more organisms of the same type. Other organisms, such as salmon and the Atlantic eel, travel long distances across the ocean in order to reproduce in a specific location. One important aspect of successful reproduction is finding a safe place for eggs and newly hatched larvae to develop. You will learn later in this section that some places in the ocean are used by marine organisms for this purpose.

Ocean Life

Many varieties of plants and animals live in the ocean. Although some organisms live in the open ocean or on the deep ocean floor, most marine organisms live in the waters above or on the floor of the continental shelf. In this relatively shallow water, the Sun penetrates to the bottom, allowing for photosynthesis. Because light is available for photosynthesis, large numbers of producers live in the waters above the continental shelf. These waters also contain many nutrients that producers use to carry out life processes. As a result, the greatest source of food is located in the waters of the continental shelf.

Caption Answer

Figure 9 Their reproductive cells could not be widely dispersed.

Inquiry Lab

Comparing Freshwater and Saltwater Organisms

Purpose Students will deepen their understanding of ocean organisms and ecosystems by comparing and contrasting them with freshwater organisms and ecosystems.

Possible Materials two fish tanks with all necessary accessories and supplies (consult a pet store)

Estimated Time one period—3 weeks

Teaching Strategies

- This could be done as a one-class brainstorming activity, an extended research project, or a lengthy experiment, setting up both a saltwater aquarium and a freshwater aquarium.
- Allow students to generate questions and design experiments.
- Investigation topics may include: plankton, nekton, and benthos species (plant and animal); light adaptations; available nutrients; pollution issues
- An alternative: have students experiment with buoyancy to compare sharks' oil-filled livers (oil-filled balloons in a water tank) with fishes' swim bladders (air-filled balloon with weights in a water tank) with fishes with no swim bladders (water-filled balloon in a water tank).

For additional inquiry activities, see *Science Inquiry Labs*.

Teacher FYI

Phytoplankton The producers at the base of most food webs in the ocean are phytoplankton. During photosynthesis, water is chemically combined with carbon dioxide in the presence of sunlight, and sugar and oxygen are produced. Phytoplankton store the sugars in their cells and release the oxygen into the water.

Curriculum Connection

Literature Suggest that students read a poem about the ocean. Ask them to write a paragraph on how the ocean and ocean life were used in the poem. Paragraphs may say fish from the ocean were used for food, or products were made from ocean life. L2 LS **Linguistic** P

Activity

Ocean Pressure For every 10 m depth in the ocean, water pressure increases 100,000 N/m^2. Have students calculate the pressure that a deep-sea fish experiences at 4,000 m depth. To complete the calculation, divide 4,000 m by 10 m to get 400. Then multiply 100,000 N/m^2 by 400 to get 40,000,000 N/m^2. Add 100,000 N/m^2, which is atmospheric pressure to get the final answer of 40,100,000 N/m^2. L3 LS **Logical-Mathematical**

INTEGRATE Chemistry

Bioluminescence Students may be familiar with the bioluminence of fireflies. Fireflies produce light by the oxidation of luciferin. In fireflies, the flashing is probably a mating response. In marine organisms, flashing is used to attract food or a mate or to temporarily blind predators.

Research Have students research which organisms, both on land and in water, are bioluminencent. Students can choose one organism and give a detailed report to the class. L2

Reading Check

Answer animals that actively swim rather than drift with the current

Diatoms are phytoplankton that live in freshwater and ocean water.

The zooplankton shown here is a copepod. Although it has reached its adult size, it is still microscopic.

Figure 10 Some plankton are producers, others are consumers.

Plankton Marine organisms that drift with the currents are called **plankton.** Plankton range from microscopic algae and animals to organisms as large as jellyfish. Most phytoplankton—plankton that are producers—are one-celled organisms that float in the upper layers of the ocean where light needed for photosynthesis is available. One abundant form of phytoplankton is a one-celled organism called a diatom. Diatoms are shown in **Figure 10.** Diatoms and other phytoplankton are the source of food for zooplankton, animals that drift with ocean currents.

Zooplankton includes newly hatched fish and crabs, jellies, and tiny adults of some organisms like the one shown in **Figure 10,** which feed on phytoplankton and are usually the second step in ocean food chains. Most animal plankton depend on surface currents to move them, but some can swim short distances.

Nekton Animals that actively swim, rather than drift with the currents in the ocean, are called **nekton.** Nekton include all swimming forms of fish and other animals, from tiny herring to huge whales. Nekton can be found from polar regions to the tropics and from shallow water to the deepest parts of the ocean. In **Figure 11,** the Greenland shark, the manatee, and the deep-ocean fish are all nekton. As nekton move throughout the oceans, it is important that they are able to control their buoyancy, or how easily they float or sink. What happens when you hold your breath underwater, then let all of the air out of your lungs at once? The air held in your lungs provides buoyancy and helps you float. As the air is released, you sink. Many fish have a special organ filled with gas that helps them control their buoyancy. By changing their buoyancy, organisms can change their depth in the ocean. The ability to move between different depths allows animals to search more areas for food.

Reading Check *What are nekton?*

Some deep-dwelling nekton are adapted with special light-generating organs. The light has several uses for these organisms. The deep-sea fish, shown in **Figure 11,** dangles a luminous lure from beneath its jaw. When prey attracted by the lure are close enough, they are swallowed quickly. Some deep-sea organisms use this light to momentarily blind predators so they can escape. Others use it to attract mates.

INTEGRATE Chemistry

Bioluminescence Some marine organisms, including types of bacteria, one-celled algae, and fish, can make their own light through a process called bioluminescence. The main molecule involved in producing light is luciferin. In the process of a chemical reaction involving luciferin, a burst of light is produced.

LAB DEMONSTRATION

Purpose to demonstrate buoyancy

Materials 2-L plastic bottle with cap, dropper, water, 1,000-mL beaker

Preparation Fill the beaker with water. Fill the bottle three-quarters full with water.

Procedure Draw water into the dropper from the beaker until the bottom of its bulb is level with the water in the beaker. Place the dropper in the bottle. Cap the bottle. Then squeeze and release the bottle.

Expected Outcome The dropper sank when the bottle was squeezed and rose when pressure was released.

Assessment

Why did the dropper move up and down? When water was pushed into the dropper, it gained mass, became less buoyant, and sank. When pressure was released, air in the dropper expanded and pushed water out, mass was lost, buoyancy increased, and the dropper rose. L2

Figure 11 Nekton are found living in all areas of the ocean, warm or cold, shallow or deep.

This Greenland shark lives in the cold waters of the North Atlantic.

Manatees are found in tropical regions around the world.

This deep-sea fish is adapted to living under pressure at a depth of 4 km.

Bottom Dwellers The plants and animals living on or in the seafloor are the **benthos** (BEN thahs). Benthic animals include crabs, snails, sea urchins, and bottom-dwelling fish such as flounder. These organisms move or swim across the bottom to find food. Other benthic animals that live permanently attached to the bottom, such as sea anemones and sponges, filter out food particles from seawater. Certain types of worms live burrowed in the sediment of the ocean floor. Bottom-dwelling animals can be found living from the shallow water of the continental shelf to the deepest areas of the ocean. Benthic plants and algae, however, are limited to the shallow areas of the ocean where enough sunlight penetrates the water to allow for photosynthesis. One example of a benthic algae is kelp, which is anchored to the bottom and grows toward the surface from depths of up to 30 m.

Ocean Margin Habitats

The area of the environment where a plant or animal normally lives is called a habitat. Along the near-shore areas of the continental shelf, called ocean margins, a variety of habitats exist. Beaches, rocky shores, estuaries, and coral reefs are some examples of the different habitats found along ocean margins.

Mini LAB

Observing Plankton

Procedure

1. Place one or two drops of **pond, lake, or ocean water** onto a **microscope slide.**
2. Use a **microscope** to observe your sample. Look for microscopic life.
3. Find at least three different types of plankton.

Analysis

1. Draw detailed pictures of three types of plankton.
2. Classify the plankton as phytoplankton or zooplankton.

Mini LAB

Purpose Students observe plankton with a microscope. L2 LS **Visual-Spatial**

Materials ocean, lake, river, or pond water containing microscopic organisms; microscope; glass depression slide; dropper

Teaching Strategy To enhance the number of plankton in a sample, use a plankton net and collect in an area of fast-moving water, or drag the net through the water for about five minutes. Leave the storage container uncovered. If possible, use a small aerator to keep oxygen in the water. The best time to collect plankton is in late spring or early fall. Dispose of samples so as to prevent the transfer of microorganisms from one body of water to another.

Troubleshooting Examine the sample for type and abundance of microorganisms before students start this activity.

Analysis

1. Drawings should show as much detail as possible.
2. Plantlike plankton will be green from chlorophyll in their cells. Animal-like plankton will not have chlorophyll and will most likely be moving.

Assessment

Performance To further assess students' knowledge of what plankton look like, have them make three-dimensional models of some of the specimens they observed. Use **Performance Assessment in the Science Classroom,** p. 123.

Differentiated Instruction

Learning Disabled Provide students with a list of various kinds of seafood. Have each student select one organism from the list and research some fun facts about the organism. Students could also depict its appearance in a drawing. L2

Activity

Underwater Cities Explain that some scientists predict future human societies will live in underwater cities. Have students design such a city. Have them plan their cities so that waves and currents will not threaten the city; the rough terrain and tremendous water pressure will not be problems; and people will have air, food, and drinking water. L2 ELL COOP LEARN IN **Kinesthetic and Interpersonal**

Caption Answer

Figure 12 Encourage students to completely describe the habitat in which they would like to live.

Quick Demo

Particle Size

Materials softball, tennis ball, golf ball, fine- or coarse-grained sand, metric ruler

Estimated Time 10 minutes

Procedure To demonstrate to students that not all beaches consist of sand, discuss the Wentworth Scale of Particle Size. The Wentworth scale classifies the size of beach particles ranging from cobble (64 mm–256 mm) to pebble (4 mm–64 mm) to different coarsenesses of sand (0.063 mm–1 mm). Have a volunteer measure the size of the softball, tennis ball, golf ball, and sand. What would each of these be classified as? softball: cobble; tennis ball: pebble; golf ball: pebble; sand: will vary with sample Have students compare how it would feel to walk on a beach with cobble-sized particles to a beach with fine-grained sand.

Beaches At the edge of a sandy beach where the waves splash, you can find some microscopic organisms and worms that spend their entire lives between moist grains of sand. Burrowing animals such as small clams and mole crabs make holes in the sand. When water covers the holes, these animals rise to the surface to filter food from the water. Where sand is covered constantly by water, larger animals like horseshoe crabs, snails, fish, turtles, and sand dollars reside. **Figure 12** shows some of the organisms that are found living on sandy beaches.

Although the beach is great fun for people, it is a very stressful environment for the plants and animals that live there. They constantly deal with waves, changing tides, and storms, all of which redistribute large amounts of sand. Large waves produced by storms, such as hurricanes, can cause damage to beaches as they crash onto shore. These organisms must adapt to natural changes as well as changes created by humans. Damming rivers, building harbors, and constructing homes and hotels near the shoreline disrupts natural processes on the beach.

Science Online

Topic: Beach Erosion
Visit bookh.msscience.com for Web links to information about beach erosion.

Activity Find two examples of beaches that are having problems with erosion. What is being done to help solve the problem?

Rocky Shore Areas In some regions the shoreline is rocky, as shown in **Figure 13.** Algae, sea anemones, mussels, and barnacles encrust submerged rocks. Sea stars, sea urchins, octopuses, and hermit crabs crawl along the rock surfaces, looking for food.

Tide pools are formed when water remains onshore, trapped by the rocks during low tide. Tide pools are an important habitat for many marine organisms. They serve as protected areas where many animals, such as octopuses and fish, can develop safely from juveniles to adults. Tide pools contain an abundance of food and offer protection from larger predators.

Figure 12 Organisms inhabit many different shore areas.
Describe *where you would want to live if you were a marine organism.*

Differentiated Instruction

Challenge Have students research different symbiotic relationships in the oceans, including competition, predation, parasitism, and mutualism. Students can use pictures or their own drawings to present their findings to the class. L3 IN **Linguistic and Visual-Spatial**

Curriculum Connection

Art Have students make a mural showing a cross section of the ocean from the shore to the deep ocean. The mural should show the diversity of life that occurs from the ocean floor to the surface and from the shore to the deep ocean. L2 ELL COOP LEARN IN **Visual-Spatial**

NATIONAL GEOGRAPHIC **VISUALIZING THE ROCKY SHORE HABITAT**

Figure 13

Life is tough in the intertidal zone—the coastal area between the highest high tide and the lowest low tide. Organisms here are pounded by waves and alternately covered and uncovered by water as tides rise and fall. These organisms tend to cluster into three general zones along the shore. Where they live depends on how well they tolerate being washed by waves, submerged at high tide, or exposed to air and sunlight when the tide is low.

UPPER INTERTIDAL ZONE This part of the intertidal zone is splashed by high waves and is usually covered by water only during the highest tides each month. It is home to crabs that scuttle among periwinkle snails, limpets, and a few kinds of algae that can withstand long periods of dryness.

MID-INTERTIDAL ZONE Submerged at most high tides and exposed at most low tides, this zone is populated by brown algae, sponges, barnacles, mussels, chitons, snails, and sea stars. These creatures are resistant to drying out and good at clinging to slippery surfaces.

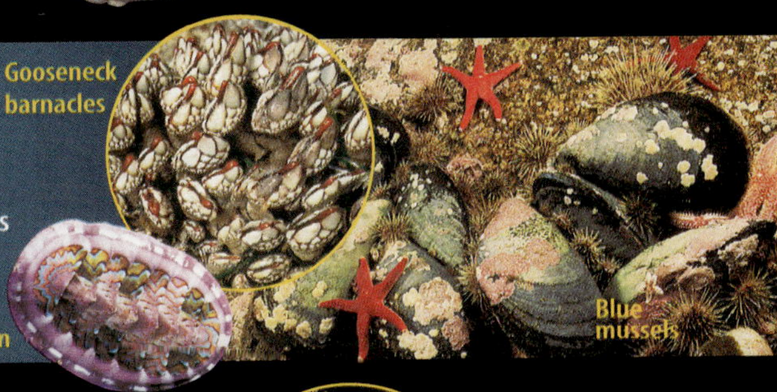

LOWER INTERTIDAL ZONE This section of the intertidal zone is exposed only during the lowest tides each month. It contains the most diverse collection of living things. Here you find sea urchins, large sea stars, brittle stars, nudibranchs, sea cucumbers, anemones, and many kinds of fish.

NATIONAL GEOGRAPHIC

Visualizing the Rocky Shore Habitat

Have students examine the pictures and read the captions. Then ask the following question.

Mussels feed by straining tiny bits of food from ocean water. Why do they live in the mid-intertidal zone rather than the upper intertidal zone? In the upper intertidal zone, they would be covered by water only during the highest tides each month and would not be able to get enough food, but in the mid-intertidal zone, they are submerged at most high tides so they are able to get enough food.

Activity

Intertidal Zones Organize the class into three groups. Assign each group one of the three intertidal zones. Have them research information about how the organisms that live there are specifically adapted to life in that zone. Have each group make a multimedia presentation of its information to the class. L2 COOP LEARN LS **Naturalist and Linguistic**

Differentiated Instruction

Challenge Have students locate an area where people could visit to experience a rocky shore habitat. Have students prepare a travel brochure on the area. L3 LS **Linguistic**

Discussion

Reef Life Why might nekton live within a reef? Reefs provide shelter from predators and readily available sources of food. L2

3 Assess

DAILY INTERVENTION

Check for Understanding

Linguistic Why is a fish considered plankton at one stage of its life and nekton at another stage? When a fish is still inside an egg or is a young hatchling, it drifts with the surface currents and can be considered plankton. When it matures, a fish can move from one depth and place to another and is considered nekton. L3

Reteach

Marine Organisms Have each student write the names of three marine organisms on separate index cards. Collect and mix the cards. Organize the class into teams. Each team will try to identify the organisms as producers or consumers, and then as plankton, nekton, or benthos. For example, kelp can be classified as producer and benthos. L2

COOP LEARN

Assessment

Performance To assess students' abilities to use variables, constants, and controls, have them describe how they could test the effects of temperature on marine organisms. Use **Performance Assessment in the Science Classroom,** p. 95. L2

Figure 14 Estuaries are called the nurseries of the oceans because many creatures spend their early lives there.

Estuaries An **estuary** is an area where the mouth of a river opens into an ocean. Because estuaries receive freshwater from rivers, they are not as salty as the ocean. Rivers also bring nutrients to estuaries. Areas with many nutrients usually have many phytoplankton, which form the base of the food chain. Shown in **Figure 14,** estuaries are full of life from salt-tolerant grasses to oysters, clams, shrimps, fish, and even manatees.

Estuaries are an important habitat to many marine organisms. Newly hatched fish, shrimps, crabs, and other animals enter estuaries as microscopic organisms and remain there until adulthood. For these vulnerable animals, fewer predators and more food are found in estuaries.

Coral Reefs Corals thrive in clear, warm water that receives a lot of sunlight. This means that they generally live in warm latitudes, between 30°N and 30°S, and in water that is no deeper than 40 m. Each coral animal builds a hard capsule around its body from the calcium it removes from seawater. Each capsule is cemented to others to form a large colony called a reef. A **reef** is a rigid, wave-resistant structure built by corals from skeletal material. As a coral reef forms, other benthos such as sea stars and sponges and nekton such as fish and turtles begin living on it.

In all ocean margin habitats, nutrients, food, and energy are cycled among organisms in complex food webs. Plankton, nekton, and benthos depend on each other for survival.

section 2 review

Summary

Life Processes

- Organisms in the ocean obtain food to use for energy in several ways.
- Photosynthesis and chemosynthesis are processes used by producers to make food. Other organisms are consumers.
- Reproduction is also a vital life process.

Ocean Life

- Organisms in the ocean can be classified as plankton, nekton, or benthos depending on where they live.

Ocean Margin Habitats

- Ocean margin habitats such as beaches, rocky shore areas, estuaries, and coral reefs exist along the near-shore areas of the continental shelf.

Self Check

1. **Describe** the processes of photosynthesis and chemosynthesis.
2. **Identify** the key characteristics of plankton, nekton, and benthos.
3. **Compare and contrast** the characteristics of coral reef and estuary habitats.
4. **Think Critically** The amount of nutrients in the water decreases as the distance from the continental shelf increases. What effect does this have on open-ocean food chains?

Applying Skills

5. **Use Graphics Software** Design a creative poster that shows energy relationships in a food chain. Begin with photosynthesis. Use clip art, scanned photographs, or computer graphics.

Science Online bookh.msscience.com/self_check_quiz

section 2 review

1. Producers use CO_2, water, and sunlight to make food by photosynthesis, and sulfur or nitrogen to make food by chemosynthesis.
2. Plankton drift in surface currents. Nekton move from one depth to another. Benthos live on or near the bottom.
3. Both are important habitats to many shallow-water organisms. Coral reefs: clear, warm, salt water; estuary: freshwater and salt water are mixed.
4. There aren't as many organisms in open-ocean food chains.
5. Posters should be colorful, describe photosynthesis, and explain how energy moves through a food chain.

section 3

Ocean Pollution

Sources of Pollution

How would you feel if someone came into your bedroom; spilled oil on your carpet; littered your room with plastic bags, cans, bottles, and newspapers; then sprayed insect killer and scattered sand all over? Organisms in the ocean experience these things when people pollute seawater.

Pollution is the introduction of harmful waste products, chemicals, and other substances not native to an environment. A pollutant is a substance that causes damage to organisms by interfering with life processes.

Pollutants from land eventually will reach the ocean in one of four main ways. They can be dumped deliberately and directly into the ocean. Material can be lost overboard accidentally during storms or shipwrecks. Some pollutants begin in the air and enter the ocean through rain. Other pollutants will reach the ocean by being carried in rivers that empty into the ocean. **Figure 15** illustrates how pollutants from land enter the oceans.

as you read

What You'll Learn

- **List** five types of ocean pollution.
- **Explain** how ocean pollution affects the entire world.
- **Describe** how ocean pollution can be controlled.

Why It's Important

Earth's health depends on the oceans being unpolluted.

Review Vocabulary

runoff: water that does not soak into the ground or evaporate but instead flows over Earth's surface, eventually entering streams, lakes, or oceans

New Vocabulary

- pollution

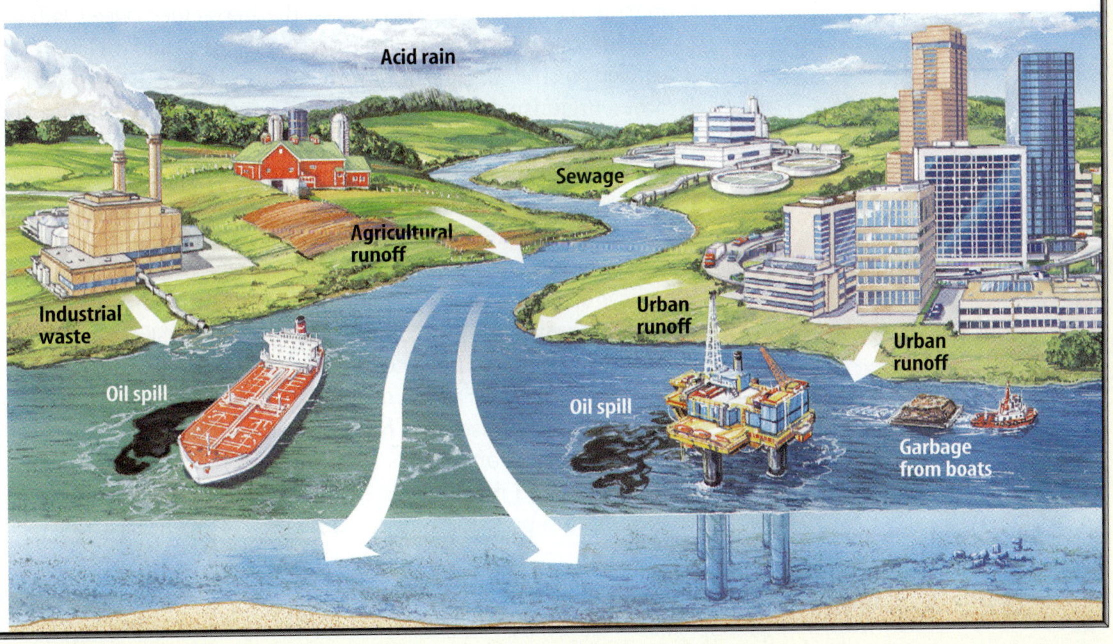

Figure 15 Ocean pollution comes from many sources.

1 Motivate

Bellringer

INTERACTIVE CHALKBOARD PowerPoint® Presentations

Section Focus Transparencies also are available on the Interactive Chalkboard CD-ROM. L2 ELL

SECTION 3 Section Focus Transparency **The Dead Zone**

While most blooms are beautiful, algal blooms are an ugly exception. The result of pollution, they can devastate marine life. Below you see two photographs—one from a healthy marine environment and one from a similar area that experienced repeated algal blooms.

1. Describe the differences between the two images.
2. The lower photograph is from an area called the dead zone. How is this name appropriate?
3. What are some activities that might result in ocean pollution?

L2

Oceanography

Tie to Prior Knowledge

Ocean Life Ask volunteers to tell the class about one thing they have learned about life in the ocean. Explain that they will learn how marine life is affected by pollution. In this section they will learn how pollution affects these processes.

Section 3 Resource Manager

Chapter *FAST FILE* Resources

Transparency Activity, p. 46

Directed Reading for Content Mastery, pp. 21, 22

Enrichment, p. 32

Reinforcement, p. 29

Lab Worksheets, pp. 7–8

Lab Activity, pp. 13–16

Cultural Diversity, p. 7

Reading and Writing Skill Activities, p. 29

Lab Management and Safety, p. 73

2 Teach

Reading Check

Answer rapid reproduction of algae

Caption Answer

Figure 16 The organisms that feed on the fish no longer have food. Also, the decomposition of many fish at one time may cause chemical changes in the water.

Discussion

Ocean Pollution Stress that ocean pollution is an international problem. Have student teams discuss who should pay for the cleanup of ocean pollution. L2 LS **Interpersonal**

Quick Demo

Ocean Pollution

Materials Photos of oil slicks in the ocean created by tanker leaks or collisions

Estimated Time 10 minutes

Procedure Have students examine the photos then discuss methods of how it can be contained and cleaned up. Ask students to consider how the presence of large amounts of oil in an area of ocean or shore can affect marine life. Eggs of fish and other organisms can be killed. Intertidal and subtidal animals can be killed. If birds' feathers are saturated with oil, they lose their ability to stay insulated from the cold.

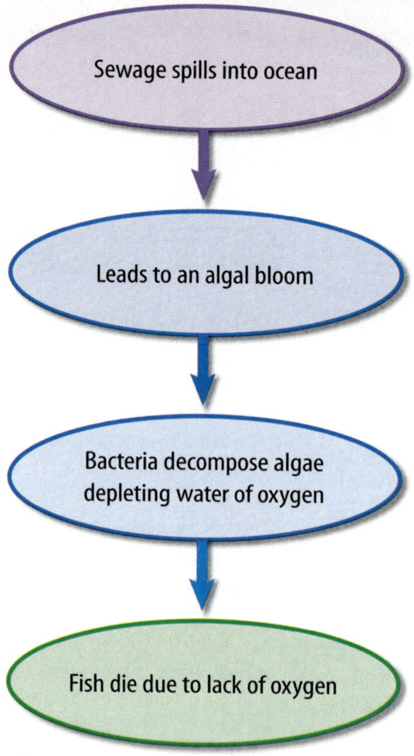

Figure 16 Fish kills occur when the oxygen supply is low.
Infer *how fish kills affect the food web.*

Sewage In some regions, human sewage leaks from septic tanks or is pumped directly into oceans or into rivers leading to an ocean. The introduction of sewage to an area of the ocean can cause immediate changes in the ecosystem, as shown by the following example. Sewage is a pollutant that acts like fertilizer. It is rich in nutrients that cause some types of algae to reproduce rapidly, creating what is called a bloom. The problem occurs when the algae die. As huge numbers of bacteria reproduce and decompose the algae, much of the oxygen in the water is used up. Other organisms, such as fish, cannot get enough oxygen. As a result, fish die in a phenomenon called a fish kill, as illustrated in **Figure 16.**

Reading Check *What is an algal bloom?*

When sewage is dumped routinely into the same area year after year, changes take place. Entire ecosystems have been altered drastically as a result of long-term, repeated exposure to sewage and fertilizer runoff. In some areas of the world, sewage is dumped directly onto coral reefs. When this happens algae can outgrow the coral because the sewage acts like a fertilizer. Eventually, the coral organisms die. If this occurs, other organisms that depend on the reef for food and shelter also can be affected.

Chemical Pollution Industrial wastes from land can harm marine organisms. When it rains, the herbicides (weed killers) and insecticides (insect killers) used in farming and on lawns are carried to streams. Eventually, they can reach the ocean and kill other organisms far from where they were applied originally. Sometimes industrial wastes are released directly into streams that eventually empty into oceans. Other chemicals are released into the air, where they later settle into the ocean. Industrial chemicals include metals like mercury and lead and chemicals like polychlorinated biphenyls (PCBs). In a process called biological amplification (am plah fah KAY shun), harmful chemicals can build up in the tissues of organisms that are at the top of the food chain. Higher consumers like dolphins and seabirds accumulate greater amounts of a toxin as they continue to feed on smaller organisms. At high concentrations, some chemicals can damage an organism's immune and reproductive systems. Explosives and nuclear wastes also have been dumped, by accident and on purpose, into some regions of the oceans.

Cultural Diversity

Reef Destruction In Belize, the 180-mile-long barrier reef is dying. Possible causes include agricultural pollutants, overfishing, and abuse by scuba divers and snorkelers. Fertilizers, pesticides, and sediments wash into the ocean, killing the coral animals. Overfishing disrupts the food web. Scuba divers and snorkelers step on the coral and take away pieces of the reef for souvenirs. To protect the Belize reef, a Coastal Zone Management Project has recently designated certain areas of the reef as commercial and recreational zones. Other areas have been designated as marine preserves and wilderness areas. This project also regulates fishing and recreational use. Ask students what effect they think the Coastal Zone Management Project will have on the health of the reef.

Oil Pollution Although oil spills from tankers that have collided or are leaking are usually highly publicized, they are not the biggest source of oil pollution in the ocean. As much as 44 percent of oil that reaches the ocean comes from land. Oil that washes from cars and streets, or that is poured down drains or into soil, flows into streams. Eventually, this oil reaches the ocean. Other sources of oil pollution are leaks at offshore oil wells and oil mixed with wastewater that is pumped out of ships. **Figure 17** shows the percentage of different sources of oil entering the oceans each year.

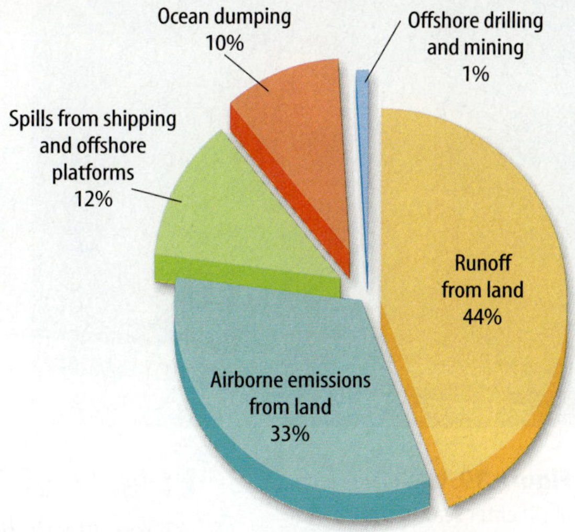

Figure 17 Although oil spills are highly publicized and tragic, the same harmful oil enters the ocean every day from many other sources.

Think Critically *What can be done to reduce the amount of oil entering the oceans?*

Solid-Waste Pollution Even in the most remote areas of the world, such as uninhabited islands that are thousands of miles from any major city, large amounts of trash wash up on the beach. **Figure 18** shows the amount of debris collected by a scientist on an island in the Pacific Ocean, 8,000 km east of Australia in just one day. The presence of trash ruins a beautiful beach, and solid wastes, such as plastic bags and fishing line, can entangle animals. Animals such as sea turtles mistakenly eat plastic bags, because they look so much like their normal prey, floating jellyfish. Illegally dumped medical waste such as needles, plastic tubing, and bags also are a threat to humans and other animals.

Figure 18 These items are like the ones found washed ashore on one of the Pitcairn Islands in the South Pacific. The number of each item found is shown below the figure. Also among the rubble were broken toys, two pairs of gloves, and an asthma inhaler.

Caption Answer

Figure 17 Possible answers: fix leaky cars, treat runoff from roadways before water enters rivers and oceans, reduce the likelihood of tanker collisions and oil spills

Activity

Oil Spill Have students use olive oil, a pan, gravel, water, toothpicks, a sponge, paper towels, cotton balls, feathers, string, liquid soap, a sheet of cardboard, and a dropper to make a model of an oil spill along a beach. Then have students clean up the spill. Students should analyze which materials were the most effective in cleaning up the oil from the water's surface and from the rocks. At the conclusion of the activity, ask students why oil spills are difficult to clean up. The oil spreads over both the water and the beach. Different techniques must be used to clean up these different areas. L2 LS **Kinesthetic**

Fun Fact

In one survey, 86 percent of the solid waste observed floating in the northern Pacific Ocean was plastic.

Differentiated Instruction

Challenge Have students research oil spills that have occurred over the last ten years. On a world map, have them pinpoint the location of these spills. L3 LS **Linguistic and Visual-Spatial**

Visual Learning

Figure 17 What is the greatest source of oil pollution? Runoff from the land. L2

SECTION 3 Ocean Pollution H ◆ 145

Discussion

Sediment How does cutting down forests increase the amount of sediment in the oceans? When trees are removed, soil erosion increases. Runoff carries the soil to streams and eventually to an ocean. Why is sediment pollution a special problem for benthos that are attached to the bottom? Their filter-feeding mechanisms clog with sediments and they cannot obtain food. L2

Teacher FYI

Coral Bleaching Corals exist symbiotically with algae. The algae provide nutrients and rid the coral animals of wastes. The algae are protected by the coral's hard skeleton. In recent years, many corals have lost their algae. The reefs look bleached. Many coral animals are dying. No one knows for certain what causes coral bleaching. Some possible explanations are variations in the water temperature, increased deposition of silt and sediments, increased ultraviolet radiation, and pollution.

Sediment Silt also can pollute the ocean. Human activities such as agriculture, deforestation, and construction tear up the soil. Rain washes soil into streams and eventually into an ocean nearby, as shown in **Figure 19.** This causes huge amounts of silt to accumulate in many coastal areas. Coral reefs and saltwater marshes are safe, protected places where young marine organisms grow to adults. When large amounts of silt cover coral reefs and fill marshes, these habitats are destroyed. Without a safe place to grow larger, many organisms will not survive.

Figure 19 When large amounts of silt enter seawater, the filter-feeding systems of animals such as oysters and clams can be clogged.

Effects of Pollution

You already have learned some examples of how pollution affects the ocean and the organisms that live there. Today, there is not a single area of the ocean that is not polluted in some way. As pollution from land continues to reach the ocean, scientists are recording dramatic changes in this environment.

Estuaries and the rivers that feed into them from Delaware to North Carolina have suffered from toxic blooms of *Pfiesteria* since the late 1980s. These blooms have killed billions of fish. *Pfiesteria,* a type of plankton, also has caused rashes, nausea, memory loss, and the weakening of the immune system in humans. The cause of these blooms is thought to be runoff contaminated by fertilizers and other waste materials. In Florida, toxic red tides kill fish and manatees. Some people also blame these red tides on sewage releases and fertilizer runoff. **Figure 20** shows an increase in the number of harmful algal blooms since the early 1970s.

Figure 20 Some scientists hypothesize that a relationship exists between increased pollution in the ocean and the number of harmful algal blooms in the last 30 years.

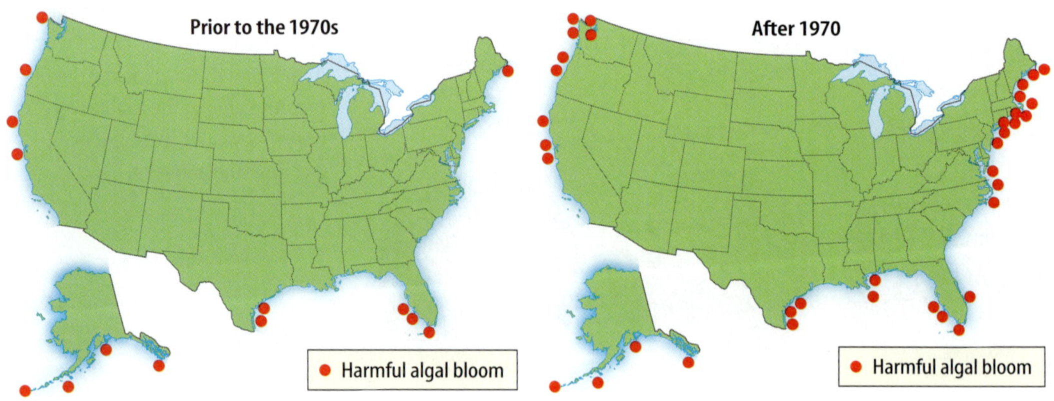

Differentiated Instruction

Visually Impaired Enlarge **Figure 20** for students who are visually impaired. Have them count the dots that represent a harmful algal bloom on each part of the figure. How many more harmful algal blooms can they find after 1970 compared to before 1970? 22 L1

Curriculum Connection

Health In the 1950s, mercury was discharged into Minamata Bay, Japan by a plastics manufacturer. When people ingested seafood caught in the bay, they became very ill. Have students research the health effects on these people and write a report on their findings. Some effects were cerebral palsy and mental retardation in children; deformities in newborns; tremors, slurred speech, delirium, convulsions, and coma in people of all ages. Some people died. L3 P

Controlling Pollution

Some people believe that oceans take care of themselves because they are large. However, other people view ocean pollution as a serious problem. Many international organizations have met to develop ways of reducing ocean pollution. Treaties prohibit the dumping of some kinds of hazardous wastes from vessels, aircraft, and platforms. One treaty requires that some ships and operators of offshore platforms have oil pollution emergency plans. This includes having the proper equipment to combat oil spills and practicing what to do if a spill takes place. Recall that a large amount of pollution enters the ocean from land. Although the idea of reducing land pollution to better protect the ocean has been discussed, no international agreement exists to prevent and control land-based activities that affect the oceans.

Reading Check *What has been done to help control ocean pollution?*

What You Can Do Current international and U.S. laws aren't effective enough. Further cooperation is needed to reduce ocean pollution. You can help by disposing of wastes properly and volunteering for beach or community cleanups, like the one shown in **Figure 21.** You can recycle materials such as newspapers, glass, and plastics and never dump chemicals like oil or paint onto soil or into water. One of the best things you can do is continue to learn about marine pollution and how people affect the oceans. What other things will help reduce ocean pollution?

Figure 21 Under careful supervision, picking up trash is an easy way to help reduce ocean pollution.

section 3 review

Summary

Sources of Pollution

- Sewage, Industrial wastes, oil, solid waste, and sediment are the main types of pollution entering ocean water.

Effects of Pollution

- As pollution from land continues to reach the ocean, ecosystems and organisms are negatively affected.

Controlling Pollution

- International treaties and U.S. laws have been made to help reduce ocean pollution.
- Everyone can help reduce ocean pollution.

Self Check

1. **Identify** five human activities that pollute the oceans. Suggest a solution to each.
2. **Explain** how pollution of the oceans affects the world.
3. **Describe** the ways international treaties have helped reduce pollution.
4. **Think Critically** To widen beaches, some cities pump offshore sediment onto them. How might this affect organisms that live in coastal waters?

Applying Skills

5. **Concept Map** Make an events-chain concept map that describes how runoff can reach the ocean. Include examples of pollution that could be in the runoff.

Reading Check

Answer Treaties prohibit dumping of some types of wastes into the ocean.

3 Assess

DAILY INTERVENTION

Check for Understanding

Linguistic Why is ocean pollution an international problem? Earth's oceans are interconnected, making pollution a worldwide problem. Pollution disrupts food chains and oxygen supplies. L2

Reteach

Ocean Pollutants Have students discuss why locating the exact source of an ocean pollutant is sometimes difficult. Students should include the fact that there are many sources of pollution. Rivers, air, and ocean currents also move pollutants from place to place. L2 LS **Linguistic**

Assessment

Process Assess students' ability to relate cause and effect by asking them to explain why the health of our planet depends on the oceans being unpolluted. Use **Performance Assessment in the Science Classroom,** p. 89. L2

section 3 review

1. dumping untreated sewage—build treatment plants; using herbicides on lawns—use fewer chemicals; oil leaks from car—check and repair leaks; leaving trash on beach—provide trash cans/take waste home; construction—plant grass over bare soil
2. When organisms die as a result of pollution, food chains and oxygen supply are affected.
3. They prohibit hazardous waste dumping from vessels, aircraft, and platforms, and they require merchant fleets and operators of platforms to have oil pollution emergency plans.
4. Habitats are destroyed; organisms are covered by sediments and sucked into pumps.
5. Runoff flows into stream, stream water flows into river, river flows into ocean. Examples include fertilizer, oil and gasoline, and pesticides and herbicides.

BENCH TESTED

Real-World Question

Purpose Students will study ways that marine organisms are used in making products, and they will explain why it is important to keep oceans clean. L2

Process Skills Search, gather, and analyze information from the Internet and various other sources

Time Required one to two weeks

Make a Plan

Internet Students will gather data from the Internet sites that can be accessed by visiting bookh.msscience.com/internet_lab. Students can post their findings on the Web site and get information from other schools around the country.

Non-Internet Sources If you do not have access to the Internet, then the school and local libraries are a good place to start. Science television programs and video tapes dealing with the ocean are also reliable sources of information.

LAB Use the Internet

Resources from the Oceans

Goals
- **Research and identify** organisms that are used to make products.
- **Explain** why it is important to keep oceans clean.

Data Source

Science Online

Visit bookh.msscience.com/internet_lab for Web links to more information about resources from the oceans, hints on which products come from the oceans, and data from other students.

Real-World Question

Oceans cover most of Earth's surface. Humans get many things from oceans, such as seafood, medicines, oil, and diamonds. Humans also use oceans for recreation and to transport materials from place to place. What else comes from oceans? Scientists continue to discover and research new uses for ocean resources. You might not realize that you probably use many products every day that are made from organisms that live in oceans. Think about the plants and animals that live in the oceans. How could these organisms be used to make everyday products? Form a hypothesis about the types of products that could be manufactured from these organisms.

Make a Plan

1. **Identify** Web links shown in the Data Source section above and identify other resources that will help you complete the data table shown on the right.
2. **Observe** that to complete the table you must identify products made from marine organisms, where the organisms are collected or harvested, and alternative products.
3. **Plan** how and when you will locate the information.

Ocean Resources Data

Organism	Location Where Collected or Harvested	Product (name and use)	Alternatives

148 ◆ H CHAPTER 5 Oceanography

Alternative Inquiry Lab

Asking Additional Questions To extend this Lab into an Inquiry Lab, have students study technologies that are used for mining both organic and inorganic resources from the ocean. What is the history of harvesting resources (live and inert) from the sea? What risks are caused to the marine ecosystem? What measures are taken to reduce or prevent problems? What clean-up methods are available when problems do occur (such as oil spills)? What issues are associated with noise from ships, explosions, and rigs? Are there new technologies that have yet to be tried? Why have they not yet been used? If time permits, choose one or two of these questions to explore as a class.

Using Scientific Methods

Follow Your Plan

1. Make sure your teacher approves your plan and your resource list before you begin.
2. **Describe** at least three ocean organisms that are used to make products you use every day.
3. **Identify** the name and any uses of the product.
4. **Research** where each organism lives and the method by which it is collected or harvested.
5. **Identify** alternative products.

Analyze Your Data

1. **Describe** the different ways in which ocean organisms are useful to humans.
2. **Explain** Are there any substitutes or alternatives available for the ocean organisms in the products?

Conclude and Apply

1. **Infer** How might the activities of humans affect any of the ocean organisms you researched?
2. **Determine** Are the substitute or alternative products more or less expensive?
3. **Describe** Can you tell whether the ocean-made product is better than the substitute product?
4. **Explain** why it is important to conserve ocean resources and keep oceans clean.

Communicating Your Data

Find this lab using the link below. Post your data in the table provided. Compare your data to those of other students.

Science Online

bookh.msscience.com/internet_lab

Follow Your Plan

Teaching Strategies

- Motivate students by bringing in unusual products from the sea. For example, bring in Japanese nori (dried red algae).
- Present the idea of overharvesting ocean resources and the use of substitute products. Recount the story of whale oil in the late nineteenth century and the eventual substitution of petroleum products.

Analyze Your Data

1. They can be used for food and as a raw material for products and medicines. For example, seaweeds are a source of iodine; algin from kelp is used in making ice cream, pudding, and paint; diatomaceous earth is used to make toothpaste and car polish; horseshoe crab blood is used to detect dangerous bacteria in fluids used on human patients.
2. Answer depends on products chosen. Possible answers: using synthetic sponges instead of sponges, using plastic or glass instead of pearls.

Conclude and Apply

1. Activities that pollute the oceans kill organisms.
2. Answers will vary depending on the product chosen. Alternative products may be more expensive.
3. Answers will vary.
4. Marine organisms are a good source of foods, medicines, and oxygen. Polluting the oceans may endanger these organisms.

Assessment

Performance Have students form large groups to design and produce skits that demonstrate the key concepts they have learned in their investigation. Suggest that their presentation target an audience of upper elementary school students. Use **Performance Assessment in the Science Classroom**, p. 147. L2

Communicating Your Data

Have students use a computer program to prepare a display of the products they find that are made from marine organisms. Students should present their displays to the class.

Content Background

Just as photosynthetic organisms use the Sun's radiant energy to produce food, chemosynthetic bacteria produce food from sulphur compounds blasted from ocean vents. These bacteria form the base of the food chain. The bacteria can be found growing freely in large mats covering the hot rocks near the vents.

Some organisms feed directly on the bacteria. Many of the other organisms at the vents, such as the giant tube worms and the large white clams that are found scattered throughout the area around the vents, have a mutualistic relationship with the bacteria. They house the bacteria in their tissues and absorb hydrogen sulfide for the bacteria to use. In return they live off of the carbohydrates produced by the bacteria.

Discussion

Vent Closings Ask students to think about why a vent would "shut down" and why the organisms would either die or move to another vent. Possible answers: In the short term, vents may shut down as a result of the magma beneath the rift moving and no longer providing a heat source. In the long term, vents shut down as the section of crust they are on moves further from the area of spreading and its source of heat. Either way, the organisms can no longer survive if there is not a source of hydrogen sulfide from the black smoker. With no hydrogen sulfide, the bacteria can no longer fulfill their role as the primary producer in the food chain. L2

Activity

Vent Communities Have students find more information on the organisms of hydrothermal vent communities. Have students make a report or a poster to share their results. LS **Visual-Spatial** L2

Oops! Accidents in SCIENCE

SOMETIMES GREAT DISCOVERIES HAPPEN BY ACCIDENT!

Strange Creatures from the Ocean Floor

In 1977, the *Alvin*, a small submersible craft specially designed to explore the ocean depths, took three geologists down about 2,200 m below the sea surface. They wanted to be the first to observe and study the formations of the Galápagos Rift deep in the Pacific Ocean. What they saw was totally unexpected. Instead of barren rock, the geologists found life—a lot of life. And they had never even considered having a life scientist as part of the research team!

The crew of the *Alvin* discovered hydrothermal vents—underwater openings where hot water (400°C) spurts from cracks in the rocks on the ocean floor. Some organisms thrive there because of the hydrogen sulfide that exists at the vents. Many of these organisms are like nothing humans had ever seen before. They are organisms that live in extremely hot temperatures and use hydrogen sulfide as their food supply.

The discovery and study of hydrothermal vents almost has been overshadowed by the amazing variety of life that was found there. But scientists think these openings on the ocean floor (many located along the Mid-Atlantic Ridge) control the temperature and movement of nearby ocean waters, as well as have a significant effect on the ocean's chemical content. These vents also act as outlets for Earth's inner heat.

Scientists also have discovered that the vent communities are temporary. Each vent eventually shuts down and the organisms somehow disperse to other vents. Exactly how this happens is an area of ongoing research.

Blood-red tube worms live deep beneath the sea.

Creative Writing Imagine you were a passenger on the *Alvin*. Write about your adventure as you came upon the hydrothermal communities. Describe and draw in detail some of the unique creatures you saw.

Science Online
For more information, visit bookh.msscience.com/oops

Creative Writing Students' answers will vary. The scientists who were in the Alvin at the time, all geologists, had no idea what they were looking at. What they described as a field of dandelions turned out to be a new species of jellyfish.

Resources for Teachers and Students

Eternal Darkness, A Personal History of Deep-Sea Exploration, by Robert Ballard, Princeton University Press, 2000

"Explosive Earth," videotape, National Geographic and Salmon City Productions, 1996

Reviewing Main Ideas

Section 1 The Seafloor

1. The continental shelf is a gently sloping part of the continent that extends into the oceans. The continental slope extends from the continental shelf to the ocean floor. The abyssal plain is a flat area of the ocean floor.
2. Along mid-ocean ridges, new seafloor forms. Seafloor slips beneath another crustal plate at a trench.
3. Petroleum, natural gas, and placer deposits are mined from continental shelves. Manganese nodules and other mineral deposits can be found in deep water.

Section 2 Life in the Ocean

1. Marine organisms are specially adapted to live in salt water. They produce or consume food, and reproduce in the oceans.
2. Photosynthesis is the basis of most of the food chains in the ocean. Chemosynthesis is a process of making food using chemical energy. Energy is transferred through food webs.
3. Organisms that drift in ocean currents are called plankton. Nekton are marine organisms that swim. Benthos are plants and animals that live on or near the ocean floor.
4. Ocean margin habitats, found along the continental shelf, include sandy beaches, rocky shores, estuaries, and coral reefs.

Section 3 Ocean Pollution

1. Sources of pollution include sewage, chemical pollution, oil spills, solid waste pollution, and sediment.
2. Ocean pollution can disrupt food webs and threaten marine organisms.
3. International treaties and U.S. laws have been made to help reduce ocean pollution.

Visualizing Main Ideas

Copy and complete the following chart on ocean-margin organisms.

Ocean-Margin Organisms

Organism	Ocean-Margin Environments			
	Sandy Beach	Rocky Shore	Estuary	Coral Reef
Plankton	phytoplankton zooplankton	phytoplankton zooplankton	phytoplankton zooplankton	phytoplankton zooplankton
Nekton	fish	octopuses, fish	shrimps, fish	fish, turtles
Benthos	worms, horseshoe crabs	algae, sponges, mussels, sea urchins	grasses, snails, clams	corals, sea stars, sponges

Science Online bookh.msscience.com/interactive_tutor

chapter 5 Study Guide

Reviewing Main Ideas

Summary statements can be used by students to review the major concepts of the chapter.

Visualizing Main Ideas

See student page.

Science Online

Visit bookh.msscience.com
/self_check_quiz
/interactive_tutor
/vocabulary_puzzlemaker
/chapter_review
/standardized_test

Assessment Transparency

For additional assessment questions, use the *Assessment Transparency* located in the transparency book.

Assessment

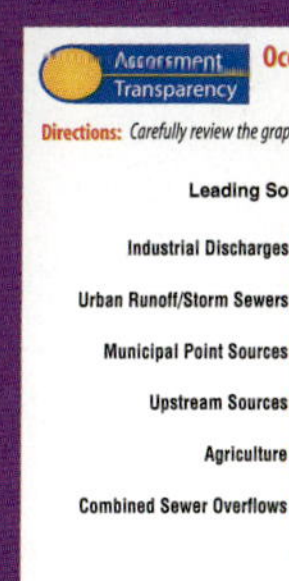

Assessment Transparency **Oceanography**

Directions: *Carefully review the graph and answer the following questions.*

1. According to the graph, the source of pollution that impairs the third lowest percentage of estuaries is ___.
   A municipal point sources
   B urban runoff/storm sewers
   C agriculture
   D upstream sources
2. What source pollutes half as many estuaries as municipal point sources?
   F Combined Sewer Overflows
   G Agriculture
   H Upstream Sources
   I Industrial Discharges

L2

Oceanography

Using Vocabulary

1. nekton
2. Chemosynthesis
3. continental slope
4. mid-ocean ridges
5. estuary

Checking Concepts

6. C
7. D
8. B
9. D
10. D
11. C
12. C
13. B
14. B

chapter 5 Review

Using Vocabulary

abyssal plain p. 129
benthos p. 139
chemosynthesis p. 137
continental shelf p. 128
continental slope p. 129
estuary p. 142
mid-ocean ridge p. 130
nekton p. 138
photosynthesis p. 135
plankton p. 138
pollution p. 143
reef p. 142
trench p. 131

Fill in the blanks with the correct words.

1. Animals such as whales, sea turtles, and fish are examples of _________.
2. _________ occurs in areas where organisms use sulfur as energy to produce their own food.
3. The _________ drops from the edge of a continent out to the deep abyssal plains of the ocean floor.
4. New ocean floor is formed at _________.
5. An area where the mouth of a river opens into an ocean is a(n) _________.

Checking Concepts

Choose the word or phrase that best answers the question.

6. What are formed along subduction zones?
   A) mid-ocean ridges
   B) continental shelves
   C) trenches
   D) density currents
7. What are ocean organisms that drift in ocean currents called?
   A) nekton **C)** benthos
   B) pollutants **D)** plankton
8. How do some deep-water bacteria in the ocean make food?
   A) photosynthesis **C)** respiration
   B) chemosynthesis **D)** rifting

Use the illustration below to answer question 9.

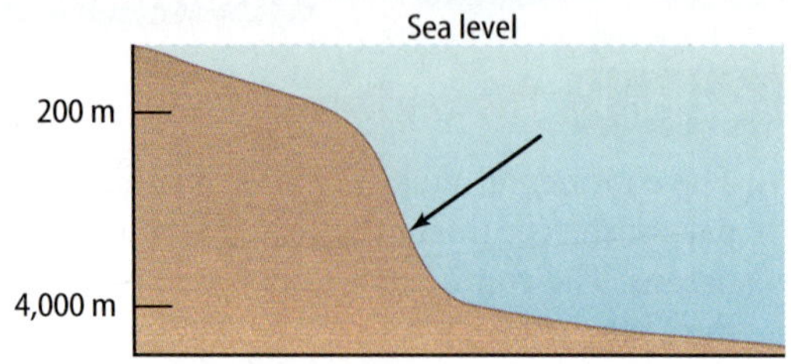

9. Which feature of the ocean floor is the arrow pointing to?
   A) rift valley **C)** abyssal plain
   B) seamount **D)** continental slope
10. What might be found in areas where rivers enter oceans?
    A) rift valleys
    B) manganese nodules
    C) abyssal plains
    D) placer deposits
11. Which organisms reproduce rapidly, resulting eventually in a lack of oxygen?
    A) fish **C)** algae
    B) corals **D)** animal plankton
12. In which area of the ocean is the greatest source of food found?
    A) on abyssal plains
    B) in trenches
    C) along continental shelves
    D) along the mid-ocean ridge
13. Where does most oil pollution originate?
    A) tanker collisions
    B) runoff from land
    C) leaks at offshore wells
    D) in wastewater pumped from ships
14. Where does new seafloor form?
    A) trenches
    B) mid-ocean ridges
    C) abyssal plains
    D) continental shelves

Science Online bookh.msscience.com/vocabulary_puzzlemaker

Use the Exam*View*® Pro Testmaker CD-ROM to:

- create multiple versions of tests
- create modified tests with one mouse click for inclusion students
- edit existing questions and add your own questions
- build tests aligned with state standards using built-in State Curriculum Tags
- change English tests to Spanish with one mouse click and vice versa

Thinking Critically

15. **Infer** why some industries might be interested in mining manganese nodules.
16. **Explain** why ocean pollution is considered to be a serious international problem.
17. **Summarize** How can agricultural chemicals kill marine organisms?
18. **Draw Conclusions** Would you expect coral reefs to grow around the bases of underwater volcanoes off the coast of Alaska?
19. **Think Critically** Scientists currently are researching the use of chemicals produced by marine organisms to help fight diseases including certain types of cancer. How would ocean pollution affect the ability to discover and research new drugs?
20. **Use Scientific Illustrations** Use **Figure 8** to determine which organisms would starve if phytoplankton became extinct.
21. **Classify** each of these sea creatures as plankton, nekton, or benthos: shrimps, dolphins, sea stars, krill, coral, manatees, and algae.

Use the illustration below to answer question 22.

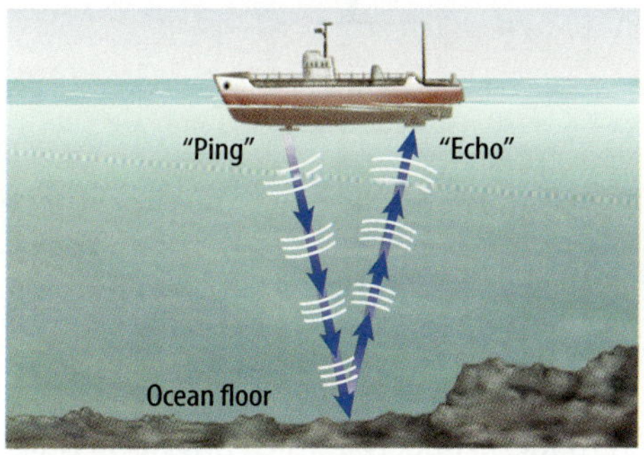

22. **Draw Conclusions** At point A an echo, a sound wave bounced off the ocean floor, took 2 s to reach the ship. It took 2.4 s at point B. Which point is deeper?

Performance Activities

23. **Graph a Profile** Previously, you made a profile of the ocean bottom along the 38° N latitude line, but it was not drawn to scale. Make a scale profile of the area between 3,600 km and 4,100 km from New Jersey. Use the scale 1 mm = 1,000 m.
24. **Poster** Choose a sea animal and research its life processes. Classify it as plankton, nekton, or benthos. Design a poster that includes all of this information.

Applying Math

Use the graph below to answer questions 25 and 26.

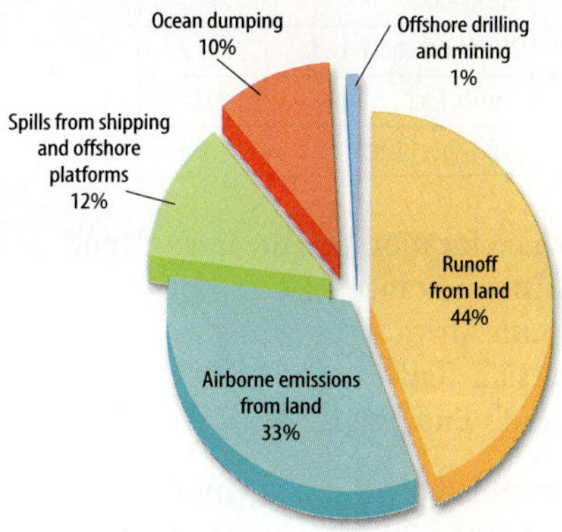

25. **Oil Production** If 6,600,000 tons of oil enter the world's oceans in one year, approximately what amount (in tons) is from runoff from land?
26. **Sources of Pollution** How many more times are airborne emissions from land a source of oil pollution than ocean dumping?
27. **Kelp Growth** If kelp grows at a steady rate of 30 cm per day, how long would it take to reach a length of 25 m?
28. **Seafloor Spreading** The distance between two locations across an ocean basin increases by 1.8 cm, 4.1 cm, and 3.2 cm, each year respectively. What is the average rate of separation of these locations during this time?

Thinking Critically

15. They are rich in manganese, copper, iron, nickel, and cobalt. These metals are used to manufacture steel, paint, and batteries.
16. Ocean pollution knows no boundaries. Currents carry pollution from one region to another.
17. Herbicides, pesticides, and fertilizers are carried in runoff to streams. From streams, they go to rivers, and then to the ocean. Herbicides and pesticides kill marine organisms. Fertilizers cause algal blooms and fish kills.
18. No; the waters in the region are too cold for coral reefs, which need warm, tropical waters.
19. Ocean pollution kills many organisms. As a result, we would have fewer organisms to analyze as possible drug sources.
20. copepods, krill, small and medium fish, whale sharks, seals, great white sharks
21. Shrimp, dolphins, and manatees are nekton. Coral and seastars are benthos. Krill and most algae are plankton.
22. The ocean is deeper at point B because the echo took longer to reach the ship from point B.

Performance Activities

23. Using the scale 1 mm = 1 km, the horizontal axis will be 5 m long; the vertical axis will be 3 mm high. Use **PASC**, p. 111.
24. Posters will vary but should be colorful and informative. Use **PASC**, p. 145.

Applying Math

National Math Standards

5, 9

25. 2,904,000 tons
26. 3.3
27. about 84 days
28. 3.03 cm per year

Assessment Resources

Reproducible Masters

Chapter *Fast File* Resources
- Chapter Review, pp. 37–38
- Chapter Tests, pp. 39–42
- Assessment Transparency Activity, p. 49

Glencoe Science Web site
- Chapter Review Test
- Standardized Test Practice

Glencoe Technology

- Assessment Transparency
- Exam*View*® Pro Testmaker
- MindJogger Videoquiz
- Interactive Chalkboard

FAST FILE

Answer Sheet A practice answer sheet can be found at bookh.msscience.com/answer_sheet.

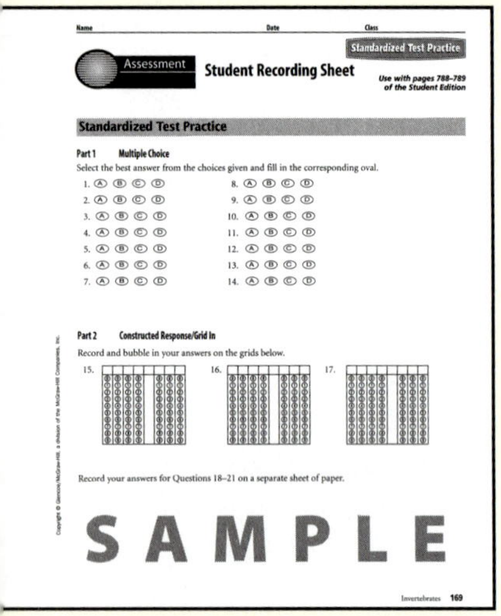

Part 1 Multiple Choice

1. C
2. C
3. A
4. B
5. A
6. B
7. D

Part 2 Short Response

8. A producer is an organism that can produce its own food using energy from the sun or a chemical source such as sulfur or nitrogen. A consumer is an organism that feeds on other organisms for energy. Producers include phytoplankton, seaweeds, and sea grasses. Consumers include fish, octopus, dolphins, and whales.
9. Both habitats are ocean margin habitats. Both can be harsh environments for organisms to live in because of the wave action in these areas. Smaller organisms may live between grains of sand on the beach, larger organisms such as horseshoe crabs and sand dollars live on top of the sand. Organisms on the rocky shore live attached to rocks or crawl along rock surfaces looking for food.
10. The number of harmful algal blooms has increased. Some scientists hypothesize that this is a result of increased runoff containing sewage, fertilizers and other waste materials.
11. Crustal plates are moving constantly. When ocean plate separate, hot magma from Earth's interior forms new ocean crust.
12. 253.3
13. There are fewer predators and more food sources in estuaries as compared to the open ocean. Tide pools and coral reefs are also nurseries for the same reasons.

chapter 5 Standardized Test Practice

Part 1 Multiple Choice

Record your answers on the answer sheet provided by your teacher or on a sheet of paper.

Use the table below to answer questions 1 and 2.

Oil Spills Around the World		
Year	Location	Spill Size (millions of liters)
1967	Land's End, England	144.7
1972	Gulf of Oman, Oman	143.5
1978	Brittany, France	260.2
1979	Bay of Campeche, Mexico	530.3
1983	South Africa	297.3
1988	Newfoundland, Canada	163.2
1991	Persian Gulf	909.0
2001	Galápagos Islands	0.6

1. At which location was the largest spill?
A. Brittany, France
B. South Africa
C. Persian Gulf
D. Land's End, England

2. Approximately how many more liters were spilled in the Persian Gulf than in the Bay of Campeche, Mexico?
A. 530 million liters
B. 470 million liters
C. 379 million liters
D. 279 million liters

3. Why is oil entering the ocean a concern?
A. The presence of oil can reduce water quality.
B. The presence of oil in the water is not harmful to marine life.
C. Large spills can be easy to clean up.
D. The presence of oil can improve water quality.

4. A rigid, wave-resistant structure built by corals from skeletal materials is
A. an estuary. **C.** a beach.
B. a reef. **D.** a rocky shore.

5. Some bacteria undergo chemosynthesis. Chemosynthesis is
A. a process that involves using sulfur or nitrogen compounds as an energy source to produce food.
B. a process by which other organisms are consumed as a source of energy.
C. a process by which reproductive cells are released into the water.
D. a process that involves using light from the Sun as an energy source to produce food.

Use the illustration of a simple food chain below to answer questions 6 and 7.

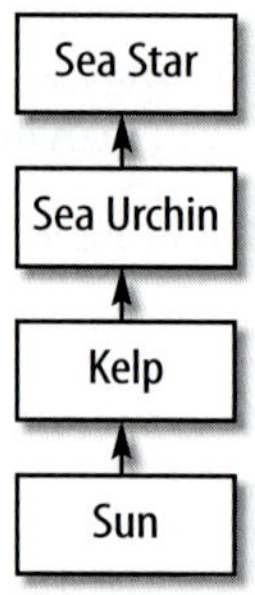

6. Which is a producer?
A. the Sun **C.** sea urchin
B. kelp **D.** sea star

7. Producers undergo which process in order to make food?
A. bioluminescence
B. respiration
C. reproduction
D. photosynthesis

Part 2 Short Response/Grid In

Record your answers on the answer sheet provided by your teacher or on a sheet of paper.

8. Define the words *producer* and *consumer*. Give two examples of each that can be found in the ocean.
9. Compare and contrast rocky shore and beach habitats.

Use the illustration below to answer question 10.

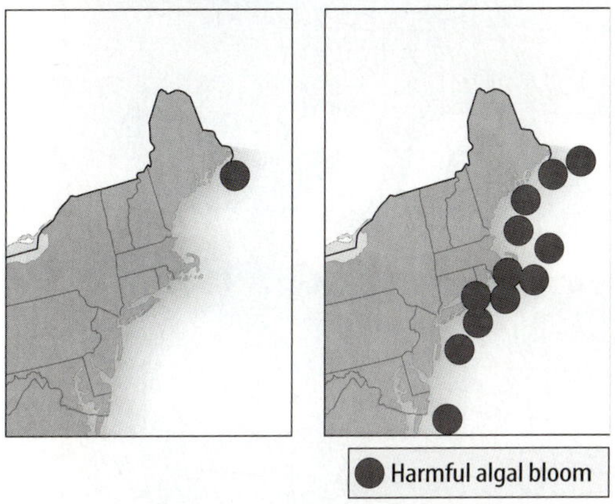

10. What changes have occurred in regard to harmful algal blooms since 1970? What do some scientists hypothesize is the cause of these changes?
11. Explain the process of seafloor spreading.
12. Ten percent of the total available energy is stored by a consumer at each level of the food chain. If 2,533 energy units are passed on to a salmon feeding on zooplankton, how many energy units will the salmon store?
13. Why are estuaries referred to as nurseries? What other marine habitat could also be referred to this way? Why?

Part 3 Open Ended

Record your answers on a sheet of paper.

14. Compare and contrast the Atlantic Ocean Basin with the Pacific Ocean Basin. Which basin contains many deep-sea trenches? Which basin is getting larger with time?
15. Describe how you could set up an experiment to test the effects of different amounts of light on marine producers.
16. Write a paragraph that explains why ocean pollution is a problem that people can help prevent. List examples of things people can do to help.

Use the illustration below to answer question 17.

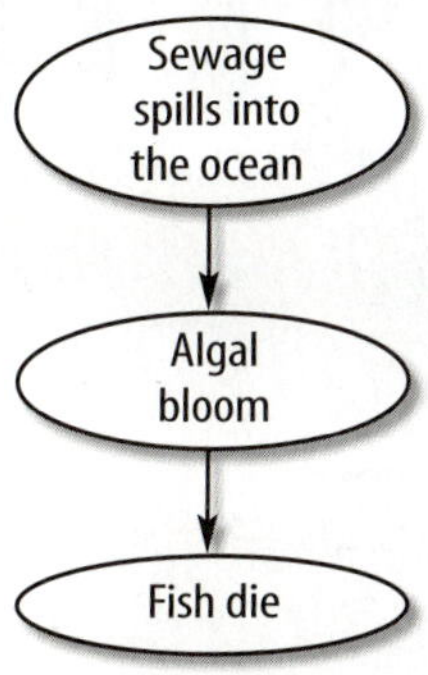

17. Explain in detail how the events are related.
18. Explain the relationship between hydrothermal vents and minerals on the ocean floor. Why are these minerals not being mined?

Test-Taking Tip

Organize Main Ideas For essay questions, spend a few minutes listing and organizing the main ideas on your scratch paper.

Question 14 Make two columns, titled *Atlantic Ocean* and *Pacific Ocean*, on your scratch paper. Fill in information about each topic in the columns.

Rubrics

The following rubrics are sample scoring devices for short response and open-ended questions.

Short Response

Points	Description
2	The student demonstrates a thorough understanding of the science of the task. The response may contain minor flaws that do not detract from the demonstration of a thorough understanding.
1	The student has provided a response that is only partially correct.
0	The student has provided a completely incorrect solution or no response at all.

Open Ended

Points	Description
4	The student demonstrates a thorough understanding of the science of the task. The response may contain minor flaws that do not detract from the demonstration of a thorough understanding.
3	The student demonstrates an understanding of the science of the task. The response is essentially correct and demonstrates an essential but less than thorough understanding of the science.
2	The student demonstrates only a partial understanding of the science of the task. Although the student may have used the correct approach to a solution or may have provided a correct solution, the work lacks an essential understanding of the underlying science concepts.
1	The student demonstrates a very limited understanding of the science of the task. The response is incomplete and exhibits many flaws.
0	The student provides a completely incorrect solution or no response at all.

Part 3 Open Ended

14. They are both large oceans. The Pacific Ocean Basin has the most trenches. The Atlantic Ocean Basin's plates are separating which means the basin is getting larger.
15. Test the same type of marine algae with three different light levels. The control has natural sunlight. Salinity and temperature are kept constant.
16. The health of the oceans is important to the ecology of other systems on Earth. People can properly dispose of wastes, recycle and educate themselves about marine pollution.
17. Sewage is a pollutant that acts like fertilizer, causing some types of algae to reproduce rapidly, creating a bloom. When the algae die, bacteria decompose the algae using much of the oxygen in the water. Fish and other organisms cannot get enough oxygen and die.
18. When the superheated water that billows out from vents cools, mineral deposits sometimes form. It is too expensive to mine the minerals because of the depth at which they are located.

Student Resources

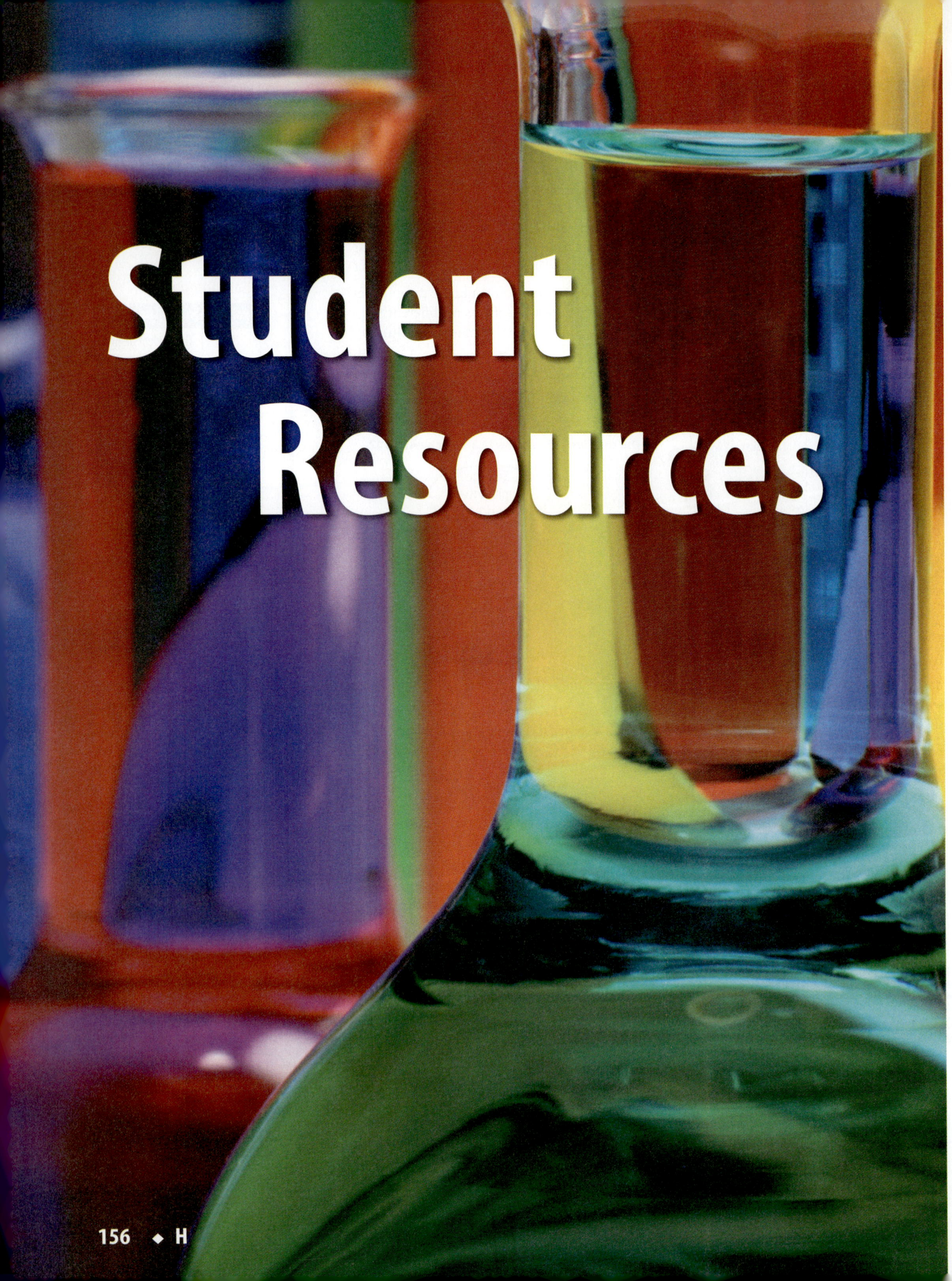

CONTENTS

Science Skill Handbook

Scientific Methods

Scientists use an orderly approach called scientific methods to solve problems. These include organizing and recording data so others can understand them. Scientists use many variations in these methods when they solve problems.

Identify a Question

The first step in a scientific investigation or experiment is to identify a question to be answered or a problem to be solved. For example, you might ask which gasoline is the most efficient.

Gather and Organize Information

After you have identified your question, begin gathering and organizing information. There are many ways to gather information, such as researching in a library, interviewing those knowledgeable about the subject, testing and working in the laboratory and field. Fieldwork is investigations and observations done outside of a laboratory.

Researching Information Before moving in a new direction, it is important to gather the information that already is known about the subject. Start by asking yourself questions to determine exactly what you need to know. Then you will look for the information in various reference sources, like the student is doing in **Figure 1.** Some sources may include textbooks, encyclopedias, government documents, professional journals, science magazines, and the Internet. Always list the sources of your information.

Figure 1 The Internet can be a valuable research tool.

Evaluate Sources of Information Not all sources of information are reliable. You should evaluate all of your sources of information, and use only those you know to be dependable. For example, if you are researching ways to make homes more energy efficient, a site written by the U.S. Department of Energy would be more reliable than a site written by a company that is trying to sell a new type of weatherproofing material. Also, remember that research always is changing. Consult the most current resources available to you. For example, a 1985 resource about saving energy would not reflect the most recent findings.

Sometimes scientists use data that they did not collect themselves, or conclusions drawn by other researchers. This data must be evaluated carefully. Ask questions about how the data were obtained, if the investigation was carried out properly, and if it has been duplicated exactly with the same results. Would you reach the same conclusion from the data? Only when you have confidence in the data can you believe it is true and feel comfortable using it.

Interpret Scientific Illustrations As you research a topic in science, you will see drawings, diagrams, and photographs to help you understand what you read. Some illustrations are included to help you understand an idea that you can't see easily by yourself, like the tiny particles in an atom in **Figure 2.** A drawing helps many people to remember details more easily and provides examples that clarify difficult concepts or give additional information about the topic you are studying. Most illustrations have labels or a caption to identify or to provide more information.

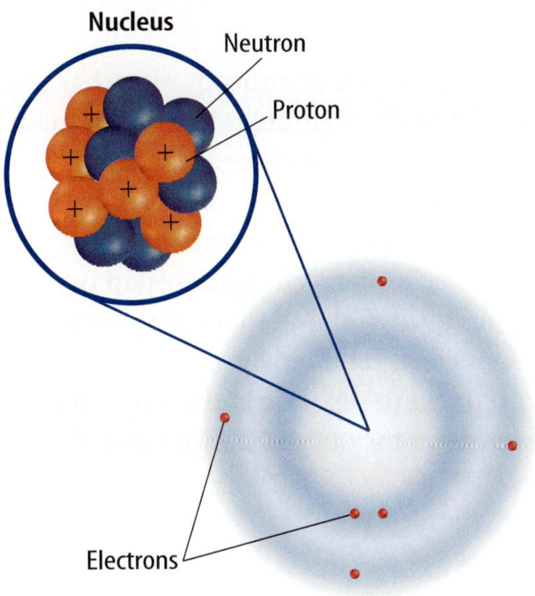

Figure 2 This drawing shows an atom of carbon with its six protons, six neutrons, and six electrons.

Concept Maps One way to organize data is to draw a diagram that shows relationships among ideas (or concepts). A concept map can help make the meanings of ideas and terms more clear, and help you understand and remember what you are studying. Concept maps are useful for breaking large concepts down into smaller parts, making learning easier.

Network Tree A type of concept map that not only shows a relationship, but how the concepts are related is a network tree, shown in **Figure 3.** In a network tree, the words are written in the ovals, while the description of the type of relationship is written across the connecting lines.

When constructing a network tree, write down the topic and all major topics on separate pieces of paper or notecards. Then arrange them in order from general to specific. Branch the related concepts from the major concept and describe the relationship on the connecting line. Continue to more specific concepts until finished.

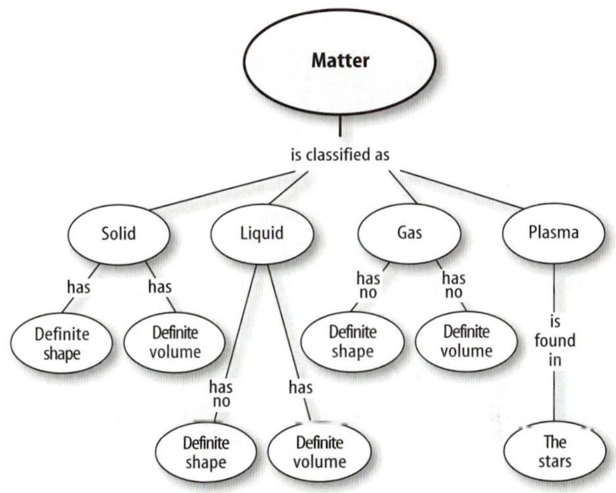

Figure 3 A network tree shows how concepts or objects are related.

Events Chain Another type of concept map is an events chain. Sometimes called a flow chart, it models the order or sequence of items. An events chain can be used to describe a sequence of events, the steps in a procedure, or the stages of a process.

When making an events chain, first find the one event that starts the chain. This event is called the initiating event. Then, find the next event and continue until the outcome is reached, as shown in **Figure 4.**

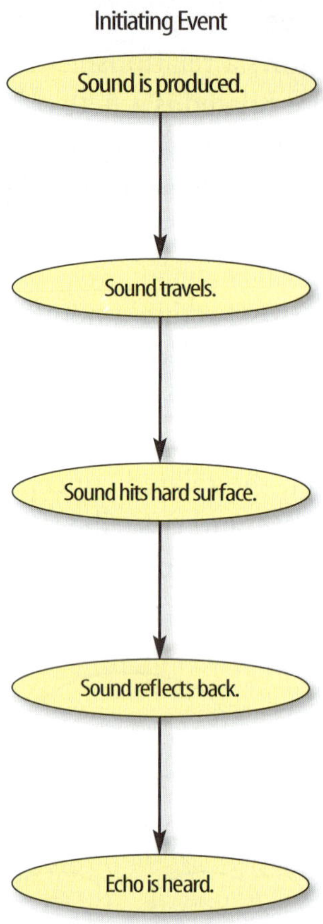

Figure 4 Events-chain concept maps show the order of steps in a process or event. This concept map shows how a sound makes an echo.

Cycle Map A specific type of events chain is a cycle map. It is used when the series of events do not produce a final outcome, but instead relate back to the beginning event, such as in **Figure 5.** Therefore, the cycle repeats itself.

To make a cycle map, first decide what event is the beginning event. This is also called the initiating event. Then list the next events in the order that they occur, with the last event relating back to the initiating event. Words can be written between the events that describe what happens from one event to the next. The number of events in a cycle map can vary, but usually contain three or more events.

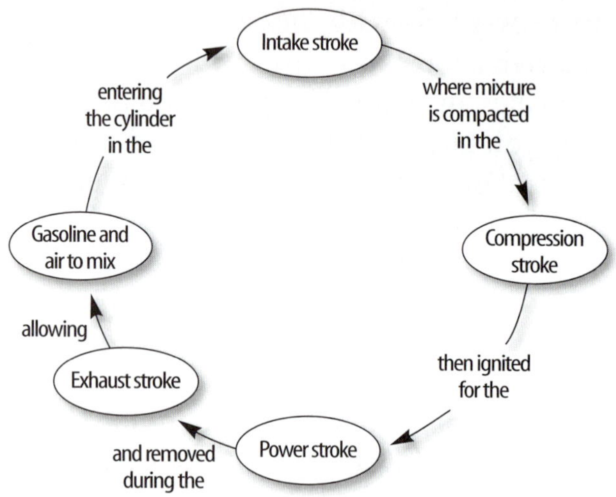

Figure 5 A cycle map shows events that occur in a cycle.

Spider Map A type of concept map that you can use for brainstorming is the spider map. When you have a central idea, you might find that you have a jumble of ideas that relate to it but are not necessarily clearly related to each other. The spider map on sound in **Figure 6** shows that if you write these ideas outside the main concept, then you can begin to separate and group unrelated terms so they become more useful.

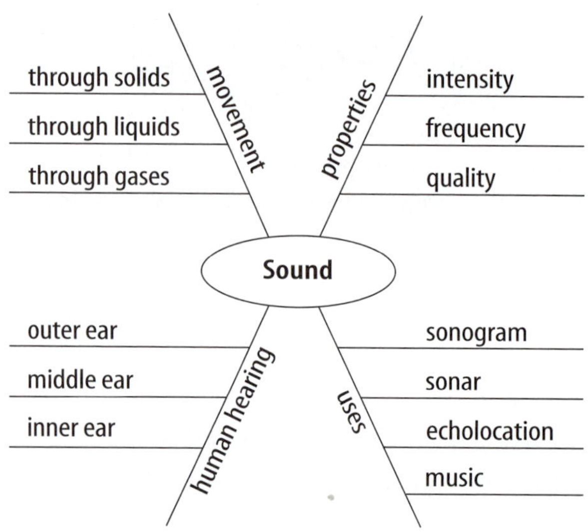

Figure 6 A spider map allows you to list ideas that relate to a central topic but not necessarily to one another.

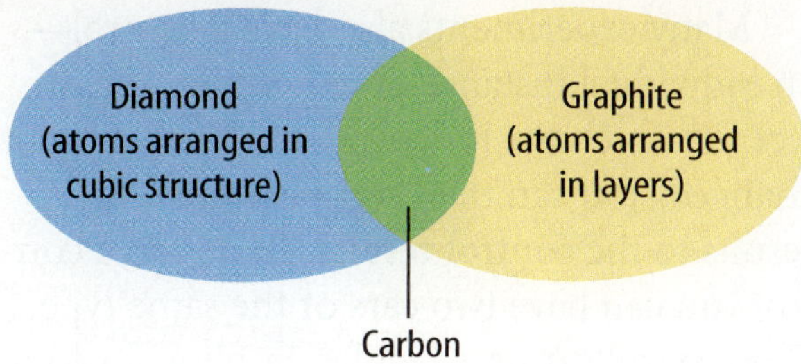

Figure 7 This Venn diagram compares and contrasts two substances made from carbon.

Venn Diagram To illustrate how two subjects compare and contrast you can use a Venn diagram. You can see the characteristics that the subjects have in common and those that they do not, shown in **Figure 7.**

To create a Venn diagram, draw two overlapping ovals that that are big enough to write in. List the characteristics unique to one subject in one oval, and the characteristics of the other subject in the other oval. The characteristics in common are listed in the overlapping section.

Make and Use Tables One way to organize information so it is easier to understand is to use a table. Tables can contain numbers, words, or both.

To make a table, list the items to be compared in the first column and the characteristics to be compared in the first row. The title should clearly indicate the content of the table, and the column or row heads should be clear. Notice that in **Table 1** the units are included.

Table 1 Recyclables Collected During Week

Day of Week	Paper (kg)	Aluminum (kg)	Glass (kg)
Monday	5.0	4.0	12.0
Wednesday	4.0	1.0	10.0
Friday	2.5	2.0	10.0

Make a Model One way to help you better understand the parts of a structure, the way a process works, or to show things too large or small for viewing is to make a model. For example, an atomic model made of a plastic-ball nucleus and pipe-cleaner electron shells can help you visualize how the parts of an atom relate to each other. Other types of models can by devised on a computer or represented by equations.

Form a Hypothesis

A possible explanation based on previous knowledge and observations is called a hypothesis. After researching gasoline types and recalling previous experiences in your family's car you form a hypothesis—our car runs more efficiently because we use premium gasoline. To be valid, a hypothesis has to be something you can test by using an investigation.

Predict When you apply a hypothesis to a specific situation, you predict something about that situation. A prediction makes a statement in advance, based on prior observation, experience, or scientific reasoning. People use predictions to make everyday decisions. Scientists test predictions by performing investigations. Based on previous observations and experiences, you might form a prediction that cars are more efficient with premium gasoline. The prediction can be tested in an investigation.

Design an Experiment A scientist needs to make many decisions before beginning an investigation. Some of these include: how to carry out the investigation, what steps to follow, how to record the data, and how the investigation will answer the question. It also is important to address any safety concerns.

Science Skill Handbook

Test the Hypothesis

Now that you have formed your hypothesis, you need to test it. Using an investigation, you will make observations and collect data, or information. This data might either support or not support your hypothesis. Scientists collect and organize data as numbers and descriptions.

Follow a Procedure In order to know what materials to use, as well as how and in what order to use them, you must follow a procedure. **Figure 8** shows a procedure you might follow to test your hypothesis.

Procedure

1. Use regular gasoline for two weeks.
2. Record the number of kilometers between fill-ups and the amount of gasoline used.
3. Switch to premium gasoline for two weeks.
4. Record the number of kilometers between fill-ups and the amount of gasoline used.

Figure 8 A procedure tells you what to do step by step.

Identify and Manipulate Variables and Controls In any experiment, it is important to keep everything the same except for the item you are testing. The one factor you change is called the independent variable. The change that results is the dependent variable. Make sure you have only one independent variable, to assure yourself of the cause of the changes you observe in the dependent variable. For example, in your gasoline experiment the type of fuel is the independent variable. The dependent variable is the efficiency.

Many experiments also have a control—an individual instance or experimental subject for which the independent variable is not changed. You can then compare the test results to the control results. To design a control you can have two cars of the same type. The control car uses regular gasoline for four weeks. After you are done with the test, you can compare the experimental results to the control results.

Collect Data

Whether you are carrying out an investigation or a short observational experiment, you will collect data, as shown in **Figure 9.** Scientists collect data as numbers and descriptions and organize it in specific ways.

Observe Scientists observe items and events, then record what they see. When they use only words to describe an observation, it is called qualitative data. Scientists' observations also can describe how much there is of something. These observations use numbers, as well as words, in the description and are called quantitative data. For example, if a sample of the element gold is described as being "shiny and very dense" the data are qualitative. Quantitative data on this sample of gold might include "a mass of 30 g and a density of 19.3 g/cm^3."

Figure 9 Collecting data is one way to gather information directly.

Figure 10 Record data neatly and clearly so it is easy to understand.

When you make observations you should examine the entire object or situation first, and then look carefully for details. It is important to record observations accurately and completely. Always record your notes immediately as you make them, so you do not miss details or make a mistake when recording results from memory. Never put unidentified observations on scraps of paper. Instead they should be recorded in a notebook, like the one in **Figure 10.** Write your data neatly so you can easily read it later. At each point in the experiment, record your observations and label them. That way, you will not have to determine what the figures mean when you look at your notes later. Set up any tables that you will need to use ahead of time, so you can record any observations right away. Remember to avoid bias when collecting data by not including personal thoughts when you record observations. Record only what you observe.

Estimate Scientific work also involves estimating. To estimate is to make a judgment about the size or the number of something without measuring or counting. This is important when the number or size of an object or population is too large or too difficult to accurately count or measure.

Sample Scientists may use a sample or a portion of the total number as a type of estimation. To sample is to take a small, representative portion of the objects or organisms of a population for research. By making careful observations or manipulating variables within that portion of the group, information is discovered and conclusions are drawn that might apply to the whole population. A poorly chosen sample can be unrepresentative of the whole. If you were trying to determine the rainfall in an area, it would not be best to take a rainfall sample from under a tree.

Measure You use measurements everyday. Scientists also take measurements when collecting data. When taking measurements, it is important to know how to use measuring tools properly. Accuracy also is important.

Length To measure length, the distance between two points, scientists use meters. Smaller measurements might be measured in centimeters or millimeters.

Length is measured using a metric ruler or meter stick. When using a metric ruler, line up the 0-cm mark with the end of the object being measured and read the number of the unit where the object ends. Look at the metric ruler shown in **Figure 11.** The centimeter lines are the long, numbered lines, and the shorter lines are millimeter lines. In this instance, the length would be 4.50 cm.

Figure 11 This metric ruler has centimeter and millimeter divisions.

Mass The SI unit for mass is the kilogram (kg). Scientists can measure mass using units formed by adding metric prefixes to the unit gram (g), such as milligram (mg). To measure mass, you might use a triple-beam balance similar to the one shown in **Figure 12.** The balance has a pan on one side and a set of beams on the other side. Each beam has a rider that slides on the beam.

When using a triple-beam balance, place an object on the pan. Slide the largest rider along its beam until the pointer drops below zero. Then move it back one notch. Repeat the process for each rider proceeding from the larger to smaller until the pointer swings an equal distance above and below the zero point. Sum the masses on each beam to find the mass of the object. Move all riders back to zero when finished.

Instead of putting materials directly on the balance, scientists often take a tare of a container. A tare is the mass of a container into which objects or substances are placed for measuring their masses. To mass objects or substances, find the mass of a clean container. Remove the container from the pan, and place the object or substances in the container. Find the mass of the container with the materials in it. Subtract the mass of the empty container from the mass of the filled container to find the mass of the materials you are using.

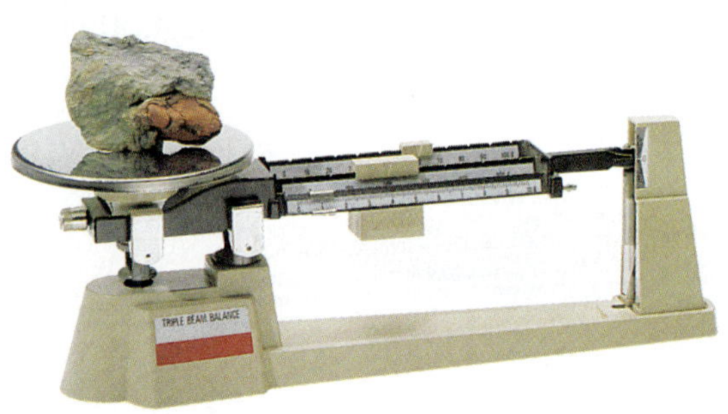

Figure 12 A triple-beam balance is used to determine the mass of an object.

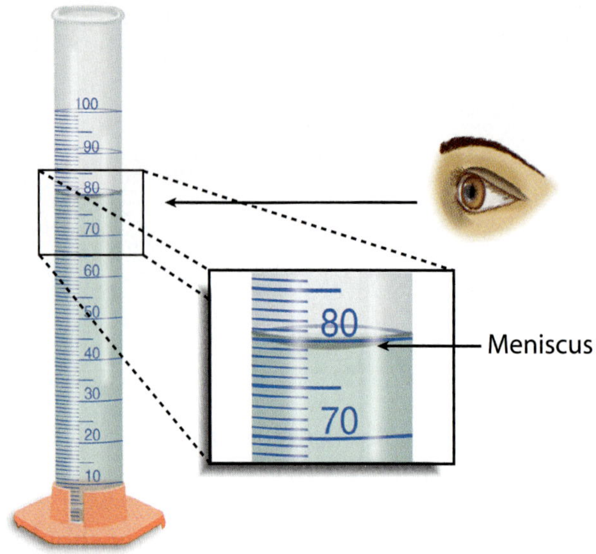

Figure 13 Graduated cylinders measure liquid volume.

Liquid Volume To measure liquids, the unit used is the liter. When a smaller unit is needed, scientists might use a milliliter. Because a milliliter takes up the volume of a cube measuring 1 cm on each side it also can be called a cubic centimeter ($cm^3 = cm \times cm \times cm$).

You can use beakers and graduated cylinders to measure liquid volume. A graduated cylinder, shown in **Figure 13,** is marked from bottom to top in milliliters. In lab, you might use a 10-mL graduated cylinder or a 100-mL graduated cylinder. When measuring liquids, notice that the liquid has a curved surface. Look at the surface at eye level, and measure the bottom of the curve. This is called the meniscus. The graduated cylinder in **Figure 13** contains 79.0 mL, or 79.0 cm^3, of a liquid.

Temperature Scientists often measure temperature using the Celsius scale. Pure water has a freezing point of 0°C and boiling point of 100°C. The unit of measurement is degrees Celsius. Two other scales often used are the Fahrenheit and Kelvin scales.

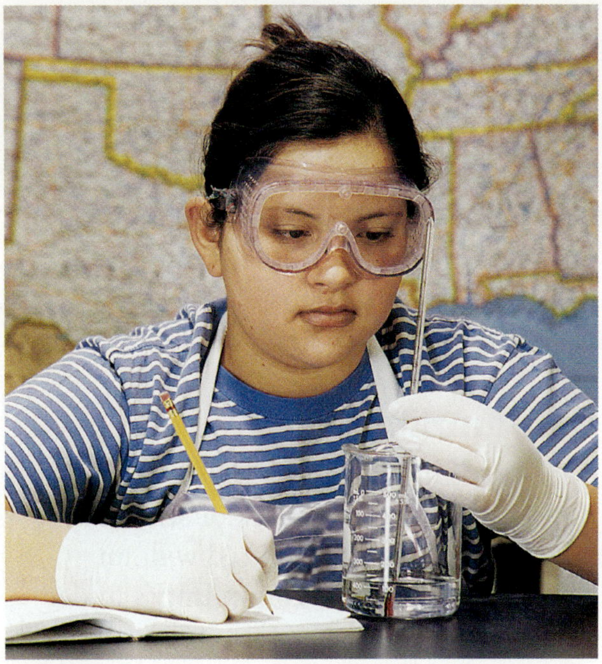

Figure 14 A thermometer measures the temperature of an object.

Scientists use a thermometer to measure temperature. Most thermometers in a laboratory are glass tubes with a bulb at the bottom end containing a liquid such as colored alcohol. The liquid rises or falls with a change in temperature. To read a glass thermometer like the thermometer in **Figure 14,** rotate it slowly until a red line appears. Read the temperature where the red line ends.

Form Operational Definitions An operational definition defines an object by how it functions, works, or behaves. For example, when you are playing hide and seek and a tree is home base, you have created an operational definition for a tree.

Objects can have more than one operational definition. For example, a ruler can be defined as a tool that measures the length of an object (how it is used). It can also be a tool with a series of marks used as a standard when measuring (how it works).

Analyze the Data

To determine the meaning of your observations and investigation results, you will need to look for patterns in the data. Then you must think critically to determine what the data mean. Scientists use several approaches when they analyze the data they have collected and recorded. Each approach is useful for identifying specific patterns.

Interpret Data The word *interpret* means "to explain the meaning of something." When analyzing data from an experiement, try to find out what the data show. Identify the control group and the test group to see whether or not changes in the independent variable have had an effect. Look for differences in the dependent variable between the control and test groups.

Classify Sorting objects or events into groups based on common features is called classifying. When classifying, first observe the objects or events to be classified. Then select one feature that is shared by some members in the group, but not by all. Place those members that share that feature in a subgroup. You can classify members into smaller and smaller subgroups based on characteristics. Remember that when you classify, you are grouping objects or events for a purpose. Keep your purpose in mind as you select the features to form groups and subgroups.

Compare and Contrast Observations can be analyzed by noting the similarities and differences between two more objects or events that you observe. When you look at objects or events to see how they are similar, you are comparing them. Contrasting is looking for differences in objects or events.

Recognize Cause and Effect A cause is a reason for an action or condition. The effect is that action or condition. When two events happen together, it is not necessarily true that one event caused the other. Scientists must design a controlled investigation to recognize the exact cause and effect.

Draw Conclusions

When scientists have analyzed the data they collected, they proceed to draw conclusions about the data. These conclusions are sometimes stated in words similar to the hypothesis that you formed earlier. They may confirm a hypothesis, or lead you to a new hypothesis.

Infer Scientists often make inferences based on their observations. An inference is an attempt to explain observations or to indicate a cause. An inference is not a fact, but a logical conclusion that needs further investigation. For example, you may infer that a fire has caused smoke. Until you investigate, however, you do not know for sure.

Apply When you draw a conclusion, you must apply those conclusions to determine whether the data supports the hypothesis. If your data do not support your hypothesis, it does not mean that the hypothesis is wrong. It means only that the result of the investigation did not support the hypothesis. Maybe the experiment needs to be redesigned, or some of the initial observations on which the hypothesis was based were incomplete or biased. Perhaps more observation or research is needed to refine your hypothesis. A successful investigation does not always come out the way you originally predicted.

Avoid Bias Sometimes a scientific investigation involves making judgments. When you make a judgment, you form an opinion. It is important to be honest and not to allow any expectations of results to bias your judgments. This is important throughout the entire investigation, from researching to collecting data to drawing conclusions.

Communicate

The communication of ideas is an important part of the work of scientists. A discovery that is not reported will not advance the scientific community's understanding or knowledge. Communication among scientists also is important as a way of improving their investigations.

Scientists communicate in many ways, from writing articles in journals and magazines that explain their investigations and experiments, to announcing important discoveries on television and radio. Scientists also share ideas with colleagues on the Internet or present them as lectures, like the student is doing in **Figure 15.**

Figure 15 A student communicates to his peers about his investigation.

Science Skill Handbook

SAFETY SYMBOLS	HAZARD	EXAMPLES	PRECAUTION	REMEDY
DISPOSAL	Special disposal procedures need to be followed.	certain chemicals, living organisms	Do not dispose of these materials in the sink or trash can.	Dispose of wastes as directed by your teacher.
BIOLOGICAL	Organisms or other biological materials that might be harmful to humans	bacteria, fungi, blood, unpreserved tissues, plant materials	Avoid skin contact with these materials. Wear mask or gloves.	Notify your teacher if you suspect contact with material. Wash hands thoroughly.
EXTREME TEMPERATURE	Objects that can burn skin by being too cold or too hot	boiling liquids, hot plates, dry ice, liquid nitrogen	Use proper protection when handling.	Go to your teacher for first aid.
SHARP OBJECT	Use of tools or glassware that can easily puncture or slice skin	razor blades, pins, scalpels, pointed tools, dissecting probes, broken glass	Practice common-sense behavior and follow guidelines for use of the tool.	Go to your teacher for first aid.
FUME	Possible danger to respiratory tract from fumes	ammonia, acetone, nail polish remover, heated sulfur, moth balls	Make sure there is good ventilation. Never smell fumes directly. Wear a mask.	Leave foul area and notify your teacher immediately.
ELECTRICAL	Possible danger from electrical shock or burn	improper grounding, liquid spills, short circuits, exposed wires	Double-check setup with teacher. Check condition of wires and apparatus.	Do not attempt to fix electrical problems. Notify your teacher immediately.
IRRITANT	Substances that can irritate the skin or mucous membranes of the respiratory tract	pollen, moth balls, steel wool, fiberglass, potassium permanganate	Wear dust mask and gloves. Practice extra care when handling these materials.	Go to your teacher for first aid.
CHEMICAL	Chemicals can react with and destroy tissue and other materials	bleaches such as hydrogen peroxide; acids such as sulfuric acid, hydrochloric acid; bases such as ammonia, sodium hydroxide	Wear goggles, gloves, and an apron.	Immediately flush the affected area with water and notify your teacher.
TOXIC	Substance may be poisonous if touched, inhaled, or swallowed.	mercury, many metal compounds, iodine, poinsettia plant parts	Follow your teacher's instructions.	Always wash hands thoroughly after use. Go to your teacher for first aid.
FLAMMABLE	Flammable chemicals may be ignited by open flame, spark, or exposed heat.	alcohol, kerosene, potassium permanganate	Avoid open flames and heat when using flammable chemicals.	Notify your teacher immediately. Use fire safety equipment if applicable.
OPEN FLAME	Open flame in use, may cause fire.	hair, clothing, paper, synthetic materials	Tie back hair and loose clothing. Follow teacher's instruction on lighting and extinguishing flames.	Notify your teacher immediately. Use fire safety equipment if applicable.

Eye Safety
Proper eye protection should be worn at all times by anyone performing or observing science activities.

Clothing Protection
This symbol appears when substances could stain or burn clothing.

Animal Safety
This symbol appears when safety of animals and students must be ensured.

Handwashing
After the lab, wash hands with soap and water before removing goggles.

Safety in the Science Laboratory

The science laboratory is a safe place to work if you follow standard safety procedures. Being responsible for your own safety helps to make the entire laboratory a safer place for everyone. When performing any lab, read and apply the caution statements and safety symbol listed at the beginning of the lab.

General Safety Rules

1. Obtain your teacher's permission to begin all investigations and use laboratory equipment.
2. Study the procedure. Ask your teacher any questions. Be sure you understand safety symbols shown on the page.
3. Notify your teacher about allergies or other health conditions which can affect your participation in a lab.
4. Learn and follow use and safety procedures for your equipment. If unsure, ask your teacher.
5. Never eat, drink, chew gum, apply cosmetics, or do any personal grooming in the lab. Never use lab glassware as food or drink containers. Keep your hands away from your face and mouth.
6. Know the location and proper use of the safety shower, eye wash, fire blanket, and fire alarm.

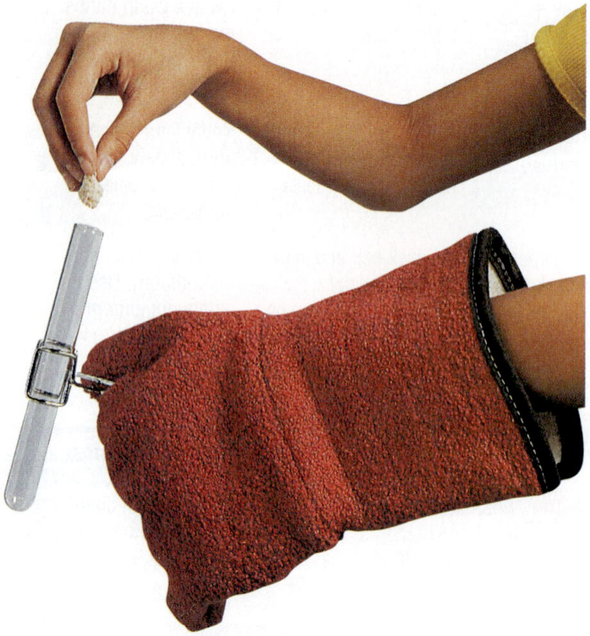

Prevent Accidents

1. Use the safety equipment provided to you. Goggles and a safety apron should be worn during investigations.
2. Do NOT use hair spray, mousse, or other flammable hair products. Tie back long hair and tie down loose clothing.
3. Do NOT wear sandals or other open-toed shoes in the lab.
4. Remove jewelry on hands and wrists. Loose jewelry, such as chains and long necklaces, should be removed to prevent them from getting caught in equipment.
5. Do not taste any substances or draw any material into a tube with your mouth.
6. Proper behavior is expected in the lab. Practical jokes and fooling around can lead to accidents and injury.
7. Keep your work area uncluttered.

Laboratory Work

1. Collect and carry all equipment and materials to your work area before beginning a lab.
2. Remain in your own work area unless given permission by your teacher to leave it.

3. Always slant test tubes away from yourself and others when heating them, adding substances to them, or rinsing them.
4. If instructed to smell a substance in a container, hold the container a short distance away and fan vapors towards your nose.
5. Do NOT substitute other chemicals/substances for those in the materials list unless instructed to do so by your teacher.
6. Do NOT take any materials or chemicals outside of the laboratory.
7. Stay out of storage areas unless instructed to be there and supervised by your teacher.

Laboratory Cleanup

1. Turn off all burners, water, and gas, and disconnect all electrical devices.
2. Clean all pieces of equipment and return all materials to their proper places.
3. Dispose of chemicals and other materials as directed by your teacher. Place broken glass and solid substances in the proper containers. Never discard materials in the sink.
4. Clean your work area.
5. Wash your hands with soap and water thoroughly BEFORE removing your goggles.

Emergencies

1. Report any fire, electrical shock, glassware breakage, spill, or injury, no matter how small, to your teacher immediately. Follow his or her instructions.
2. If your clothing should catch fire, STOP, DROP, and ROLL. If possible, smother it with the fire blanket or get under a safety shower. NEVER RUN.
3. If a fire should occur, turn off all gas and leave the room according to established procedures.
4. In most instances, your teacher will clean up spills. Do NOT attempt to clean up spills unless you are given permission and instructions to do so.
5. If chemicals come into contact with your eyes or skin, notify your teacher immediately. Use the eyewash or flush your skin or eyes with large quantities of water.
6. The fire extinguisher and first-aid kit should only be used by your teacher unless it is an extreme emergency and you have been given permission.
7. If someone is injured or becomes ill, only a professional medical provider or someone certified in first aid should perform first-aid procedures.

1 Water World

Time Required 20 minutes

Materials Have towels ready for any spilled water.

Teaching Strategies
- Before completing this lab, review basic metric volume measurements and their U.S. equivalents.
- 3 mL of water from the aquarium represents all of Earth's salt lakes.

Conclude and Apply
1. Earth's salt water
2. 0.75 percent (Excludes salt water and water locked in the ice caps.)

EXTRA Try at Home Labs

From Your Kitchen, Junk Drawer, or Yard

1 Water World

Real-World Question

Where is Earth's water?

Possible Materials
- 5-gallon bucket
- empty milk cartons (2)
- scissors
- 500-mL measuring cup
- drinking glasses (3)
- measuring spoons
- eyedropper or spoon

Procedure
1. Fill a 5-gallon (18.93 L) bucket with water.
2. Cut the milk cartons, to create 1-L open-topped containers.
3. Measure 659 mL of water from the bucket into a milk carton. This represents all the freshwater locked in the ice caps.
4. Pour 288 mL of water from the bucket into the other milk carton. This represents all of Earth's groundwater.
5. Take 0.5 teaspoon of water from the bucket and put it into a glass. This represents all the water in freshwater lakes.
6. Use the eyedropper or spoon to place 10 drops of water into a second glass. This represents all the water in Earth's atmosphere.
7. Place 2 drops of water into the third glass. This represents all Earth's rivers.

Conclude and Apply
1. Infer what the water left in the bucket represents.
2. Calculate the percentage of Earth's water readily available for drinking, farming, and other daily uses.

2 Compost Your Scraps

Real-World Question

How can you make a compost pile to reduce water pollution?

Possible Materials
- garden shovel
- garden rake
- garden hoe
- grass clippings
- gloves
- wheelbarrow
- food scraps
- fallen tree leaves

Procedure
1. Select a 2 m × 2 m plot of land in your garden or near the edge of your property and clear the space of grass and debris.
2. Remove about 10 cm of top soil from your space and pile it on one side.
3. Put grass clippings, scraps of food, and dead leaves into the space and cover them with just enough soil to cover the area.
4. Continue to pile organic materials on the compost pile each day covering each new layer with new soil.
5. Every couple weeks, use a shovel to turn over your pile and mix up the layers.

Conclude and Apply
1. List the organic materials you put in your compost pile.
2. Infer how a compost pile reduces water pollution.
3. Infer another environmental advantage to composting.

Adult supervision required for all labs.

These labs are available at bookh.msscience.com.

2 Compost Your Scraps

Time Required 90 minutes

Materials Check with the maintenance department at your school for the ideal location for a school compost pile.

Safety Precautions
- Students should thoroughly wash their hands after handling the soil and organic materials.
- Caution students to use the shovels with care.

Teaching Strategy Assign different students to care for the pile each week.

Conclude and Apply
1. Answers will vary.
2. Artificial fertilizers are a major source of water pollution. Using natural fertilizer created by composting reduces fertilizer run off into streams and rivers.
3. Compost piles reduce organic materials that take up landfill space.

Extra Try at Home Labs

3 Slow Water

Real-World Question
How does the water speed of groundwater compare to the water speed of a stream?

Possible Materials
- small stick
- meterstick
- tennis balls (2)
- stopwatch
- watch (with a second hand)
- calculator

Procedure
1. Measure a 10-m distance along the bank of a stream and mark off the distance with two tennis balls.
2. Drop a small stick into the center of the stream next to the first tennis ball.
3. Measure the time it takes the stick to move downstream between the two tennis balls.
4. Divide 10 m by the number of seconds it took the stick to move from the first tennis ball to the second.
5. Multiply this number by 3.6 to calculate the stream's water speed in kilometers per hour.

Conclude and Apply
1. What was the water speed of the stream?
2. The average speed of groundwater is 2 cm per day. Calculate the water speed of groundwater in kilometers per hour.
3. Infer why polluted groundwater takes much longer to clean itself than stream water.

4 That's Cold!

Real-World Question
How does salt affect the freezing temperature of water?

Possible Materials
- drinking glasses (2)
- ice cubes
- salt
- clock

Procedure
1. Place the same number of ice cubes into two different glasses. Make sure that the ice cubes have equal volume.
2. Sprinkle about 10 g of salt into one of the glasses.
3. Observe both glasses every 10 min until the ice is melted. Record the results.

Conclude and Apply
1. In which glass did the ice melt more quickly? Describe what you observed.
2. Infer which glass of water had the lower freezing point. How do you know?
3. Infer why ocean water gets colder than freshwater before freezing.

Adult supervision required for all labs.

Extra Try at Home Labs

3 Slow Water

Time Required 30 minutes

Materials Have students bring towels and extra shoes, socks, and pants.

Safety Precautions
- Students should never go near a stream after heavy rains or during flooding.
- An adult should accompany the students to the stream.

Teaching Strategies
- Students can measure and compare the stream's water speed in the rapids with the deep, slow moving water.
- Sample water speed derivation: 10 m/30 s × 3,600 s/1 h × 1 km/1,000 m = 1.2 km/h
- Groundwater speed calculation: 2 cm/day × 1 day/24 h × 1km/100,000 cm = 0.0000008 km/h

Conclude and Apply
1. Answers will vary. Fast-moving streams are between 1–4 km/h; slow moving streams are less than 1 km/h.
2. 0.0000008 km/h
3. Groundwater moves so slowly that long periods of time are needed to flush out contaminants.

4 That's Cold!

Time Required about two hours; varies with amount of ice used and room temperature

Materials It is important for students to use the same amount of ice in each glass. Four average ice cubes per glass works well.

Safety Precautions
- Caution students that the glasses can break.
- Students should not eat or drink any of the materials used in this lab.

Conclude and Apply
1. The ice melted more quickly in the glass that contained salt.
2. The water mixed with salt had the lower freezing point. The ice in this glass melted faster because there was a greater difference between its freezing point and room temperature.
3. Ocean water contains salt, so its freezing temperature is below 0°C, the freezing temperature of pure water.

5 Water Pressure

Time Required 30 minutes

Materials

- Plastic tubing can be purchased at a hardware store.
- A one-gallon plastic water jug works well.

Safety Precaution Caution students to use care when making nail holes.

Teaching Strategies

- Students are more familiar with the effects of air pressure than water pressure. Remind them about how their ears pop at high elevation. Pressure also decreases when higher in a water column.
- Water will rise up both tubes to a height that should be nearly equal to the height of the water in the jug.
- Water from the lower hole will rise to a higher column height because water pressure increases with depth.

Conclude and Apply

1. The water column will be higher for the lower hole because the water pressure is higher there. The water should rise from each hole to an elevation equal to top of the water in the jug.
2. Pressure increases with depth in the ocean. Submersibles must be built to withstand extremely high pressures.

Extra Try at Home Labs

Extra Try at Home Labs

5 Water Pressure

Real-World Question

How does water pressure change with depth?

Possible Materials

- plastic gallon milk jug
- 40-cm lengths of 1/4-in plastic tubing (2)
- nail
- duct tape
- metric ruler
- water

Procedure

1. Use a nail to make two holes in one side of a plastic jug. One hole should be near the bottom of the jug, and one should be closer to the top.
2. Insert a piece of plastic tubing into each hole and seal with duct tape.
3. Take your materials outside. While holding the ends of the tubing in the air, fill the jug with water.
4. Measure how high the water rises above the hole in each piece of tubing. Also measure the depth of water in the jug at each hole.

Conclude and Apply

1. How high was the water column in each piece of tubing? Was the water column higher for the top hole or the bottom hole? Explain.
2. How do you think pressure changes with depth in the ocean? How might this affect the way submersible vehicles are designed?

Adult supervision required for all labs.

These labs are available at bookh.msscience.com.

172 ◆ H STUDENT RESOURCES

Computer Skills

People who study science rely on computers, like the one in **Figure 16,** to record and store data and to analyze results from investigations. Whether you work in a laboratory or just need to write a lab report with tables, good computer skills are a necessity.

Using the computer comes with responsibility. Issues of ownership, security, and privacy can arise. Remember, if you did not author the information you are using, you must provide a source for your information. Also, anything on a computer can be accessed by others. Do not put anything on the computer that you would not want everyone to know. To add more security to your work, use a password.

Use a Word Processing Program

A computer program that allows you to type your information, change it as many times as you need to, and then print it out is called a word processing program. Word processing programs also can be used to make tables.

Figure 16 A computer will make reports neater and more professional looking.

Learn the Skill To start your word processing program, a blank document, sometimes called "Document 1," appears on the screen. To begin, start typing. To create a new document, click the *New* button on the standard tool bar. These tips will help you format the document.

- The program will automatically move to the next line; press *Enter* if you wish to start a new paragraph.
- Symbols, called non-printing characters, can be hidden by clicking the *Show/Hide* button on your toolbar.
- To insert text, move the cursor to the point where you want the insertion to go, click on the mouse once, and type the text.
- To move several lines of text, select the text and click the *Cut* button on your toolbar. Then position your cursor in the location that you want to move the cut text and click *Paste.* If you move to the wrong place, click *Undo.*
- The spell check feature does not catch words that are misspelled to look like other words, like "cold" instead of "gold." Always reread your document to catch all spelling mistakes.
- To learn about other word processing methods, read the user's manual or click on the *Help* button.
- You can integrate databases, graphics, and spreadsheets into documents by copying from another program and pasting it into your document, or by using desktop publishing (DTP). DTP software allows you to put text and graphics together to finish your document with a professional look. This software varies in how it is used and its capabilities.

Use a Database

A collection of facts stored in a computer and sorted into different fields is called a database. A database can be reorganized in any way that suits your needs.

Learn the Skill A computer program that allows you to create your own database is a database management system (DBMS). It allows you to add, delete, or change information. Take time to get to know the features of your database software.

- Determine what facts you would like to include and research to collect your information.
- Determine how you want to organize the information.
- Follow the instructions for your particular DBMS to set up fields. Then enter each item of data in the appropriate field.
- Follow the instructions to sort the information in order of importance.
- Evaluate the information in your database, and add, delete, or change as necessary.

Use the Internet

The Internet is a global network of computers where information is stored and shared. To use the Internet, like the students in **Figure 17,** you need a modem to connect your computer to a phone line and an Internet Service Provider account.

Learn the Skill To access internet sites and information, use a "Web browser," which lets you view and explore pages on the World Wide Web. Each page is its own site, and each site has its own address, called a URL. Once you have found a Web browser, follow these steps for a search (this also is how you search a database).

Figure 17 The Internet allows you to search a global network for a variety of information.

- Be as specific as possible. If you know you want to research "gold," don't type in "elements." Keep narrowing your search until you find what you want.
- Web sites that end in *.com* are commercial Web sites; *.org*, *.edu*, and *.gov* are non-profit, educational, or government Web sites.
- Electronic encyclopedias, almanacs, indexes, and catalogs will help locate and select relevant information.
- Develop a "home page" with relative ease. When developing a Web site, NEVER post pictures or disclose personal information such as location, names, or phone numbers. Your school or community usually can host your Web site. A basic understanding of HTML (hypertext mark-up language), the language of Web sites, is necessary. Software that creates HTML code is called authoring software, and can be downloaded free from many Web sites. This software allows text and pictures to be arranged as the software is writing the HTML code.

Use a Spreadsheet

A spreadsheet, shown in **Figure 18,** can perform mathematical functions with any data arranged in columns and rows. By entering a simple equation into a cell, the program can perform operations in specific cells, rows, or columns.

Learn the Skill Each column (vertical) is assigned a letter, and each row (horizontal) is assigned a number. Each point where a row and column intersect is called a cell, and is labeled according to where it is located—Column A, Row 1 (A1).

- Decide how to organize the data, and enter it in the correct row or column.
- Spreadsheets can use standard formulas or formulas can be customized to calculate cells.
- To make a change, click on a cell to make it activate, and enter the edited data or formula.
- Spreadsheets also can display your results in graphs. Choose the style of graph that best represents the data.

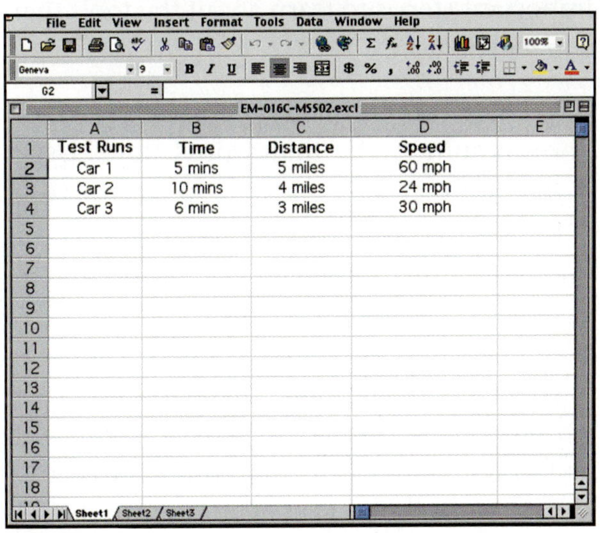

Figure 18 A spreadsheet allows you to perform mathematical operations on your data.

Use Graphics Software

Adding pictures, called graphics, to your documents is one way to make your documents more meaningful and exciting. This software adds, edits, and even constructs graphics. There is a variety of graphics software programs. The tools used for drawing can be a mouse, keyboard, or other specialized devices. Some graphics programs are simple. Others are complicated, called computer-aided design (CAD) software.

Learn the Skill It is important to have an understanding of the graphics software being used before starting. The better the software is understood, the better the results. The graphics can be placed in word-processing document.

- Clip art can be found on a variety of internet sites, and on CDs. These images can be copied and pasted into your document.
- When beginning, try editing existing drawings, then work up to creating drawings.
- The images are made of tiny rectangles of color called pixels. Each pixel can be altered.
- Digital photography is another way to add images. The photographs in the memory of a digital camera can be downloaded into a computer, then edited and added to the document.
- Graphics software also can allow animation. The software allows drawings to have the appearance of movement by connecting basic drawings automatically. This is called in-betweening, or tweening.
- Remember to save often.

Presentation Skills

Develop Multimedia Presentations

Most presentations are more dynamic if they include diagrams, photographs, videos, or sound recordings, like the one shown in **Figure 19.** A multimedia presentation involves using stereos, overhead projectors, televisions, computers, and more.

Learn the Skill Decide the main points of your presentation, and what types of media would best illustrate those points.

- Make sure you know how to use the equipment you are working with.
- Practice the presentation using the equipment several times.
- Enlist the help of a classmate to push play or turn lights out for you. Be sure to practice your presentation with him or her.
- If possible, set up all of the equipment ahead of time, and make sure everything is working properly.

Figure 19 These students are engaging the audience using a variety of tools.

Computer Presentations

There are many different interactive computer programs that you can use to enhance your presentation. Most computers have a compact disc (CD) drive that can play both CDs and digital video discs (DVDs). Also, there is hardware to connect a regular CD, DVD, or VCR. These tools will enhance your presentation.

Another method of using the computer to aid in your presentation is to develop a slide show using a computer program. This can allow movement of visuals at the presenter's pace, and can allow for visuals to build on one another.

Learn the Skill In order to create multimedia presentations on a computer, you need to have certain tools. These may include traditional graphic tools and drawing programs, animation programs, and authoring systems that tie everything together. Your computer will tell you which tools it supports. The most important step is to learn about the tools that you will be using.

- Often, color and strong images will convey a point better than words alone. Use the best methods available to convey your point.
- As with other presentations, practice many times.
- Practice your presentation with the tools you and any assistants will be using.
- Maintain eye contact with the audience. The purpose of using the computer is not to prompt the presenter, but to help the audience understand the points of the presentation.

Math Review

Use Fractions

A fraction compares a part to a whole. In the fraction $\frac{2}{3}$, the 2 represents the part and is the numerator. The 3 represents the whole and is the denominator.

Reduce Fractions To reduce a fraction, you must find the largest factor that is common to both the numerator and the denominator, the greatest common factor (GCF). Divide both numbers by the GCF. The fraction has then been reduced, or it is in its simplest form.

Example Twelve of the 20 chemicals in the science lab are in powder form. What fraction of the chemicals used in the lab are in powder form?

Step 1 Write the fraction.
$$\frac{\text{part}}{\text{whole}} = \frac{12}{20}$$

Step 2 To find the GCF of the numerator and denominator, list all of the factors of each number.
Factors of 12: 1, 2, 3, 4, 6, 12 (the numbers that divide evenly into 12)
Factors of 20: 1, 2, 4, 5, 10, 20 (the numbers that divide evenly into 20)

Step 3 List the common factors.
1, 2, 4.

Step 4 Choose the greatest factor in the list.
The GCF of 12 and 20 is 4.

Step 5 Divide the numerator and denominator by the GCF.
$$\frac{12 \div 4}{20 \div 4} = \frac{3}{5}$$

In the lab, $\frac{3}{5}$ of the chemicals are in powder form.

Practice Problem At an amusement park, 66 of 90 rides have a height restriction. What fraction of the rides, in its simplest form, has a height restriction?

Add and Subtract Fractions To add or subtract fractions with the same denominator, add or subtract the numerators and write the sum or difference over the denominator. After finding the sum or difference, find the simplest form for your fraction.

Example 1 In the forest outside your house, $\frac{1}{8}$ of the animals are rabbits, $\frac{3}{8}$ are squirrels, and the remainder are birds and insects. How many are mammals?

Step 1 Add the numerators.
$$\frac{1}{8} + \frac{3}{8} = \frac{(1+3)}{8} = \frac{4}{8}$$

Step 2 Find the GCF.
$$\frac{4}{8} \text{ (GCF, 4)}$$

Step 3 Divide the numerator and denominator by the GCF.
$$\frac{4}{4} = 1,\ \frac{8}{4} = 2$$

$\frac{1}{2}$ of the animals are mammals.

Example 2 If $\frac{7}{16}$ of the Earth is covered by freshwater, and $\frac{1}{16}$ of that is in glaciers, how much freshwater is not frozen?

Step 1 Subtract the numerators.
$$\frac{7}{16} - \frac{1}{16} = \frac{(7-1)}{16} = \frac{6}{16}$$

Step 2 Find the GCF.
$$\frac{6}{16} \text{ (GCF, 2)}$$

Step 3 Divide the numerator and denominator by the GCF.
$$\frac{6}{2} = 3,\ \frac{16}{2} = 8$$

$\frac{3}{8}$ of the freshwater is not frozen.

Practice Problem A bicycle rider is going 15 km/h for $\frac{4}{9}$ of his ride, 10 km/h for $\frac{2}{9}$ of his ride, and 8 km/h for the remainder of the ride. How much of his ride is he going over 8 km/h?

Math Skill Handbook

Reduce Fractions

$$\frac{66 \div 6}{90 \div 6} = \frac{11}{15}$$

Add and Subtract Fractions

$$\frac{4}{9} + \frac{2}{9} = \frac{6}{9}$$

$$\frac{6 \div 3}{9 \div 3} = \frac{2}{3}$$

Math Skill Handbook

MATH SKILL HANDBOOK H ◆ 177

Unlike Denominators

Problem 1

$1 \times 5 = 5, 8 \times 5 = 40$

$1 \times 4 = 4, 10 \times 4 = 40$

$\frac{5}{40} + \frac{4}{40} = \frac{9}{40}$

Problem 2

If $\frac{7}{10}$ are involuntary, the remainder are voluntary.

$\frac{10}{10} - \frac{7}{10} = \frac{3}{10}$

Unlike Denominators To add or subtract fractions with unlike denominators, first find the least common denominator (LCD). This is the smallest number that is a common multiple of both denominators. Rename each fraction with the LCD, and then add or subtract. Find the simplest form if necessary.

Example 1 A chemist makes a paste that is $\frac{1}{2}$ table salt (NaCl), $\frac{1}{3}$ sugar ($C_6H_{12}O_6$), and the rest water (H_2O). How much of the paste is a solid?

Step 1 Find the LCD of the fractions.

$\frac{1}{2} + \frac{1}{3}$ (LCD, 6)

Step 2 Rename each numerator and each denominator with the LCD.

$1 \times 3 = 3,\ 2 \times 3 = 6$

$1 \times 2 = 2,\ 3 \times 2 = 6$

Step 3 Add the numerators.

$\frac{3}{6} + \frac{2}{6} = \frac{(3 + 2)}{6} = \frac{5}{6}$

$\frac{5}{6}$ of the paste is a solid.

Example 2 The average precipitation in Grand Junction, CO, is $\frac{7}{10}$ inch in November, and $\frac{3}{5}$ inch in December. What is the total average precipitation?

Step 1 Find the LCD of the fractions.

$\frac{7}{10} + \frac{3}{5}$ (LCD, 10)

Step 2 Rename each numerator and each denominator with the LCD.

$7 \times 1 = 7,\ 10 \times 1 = 10$

$3 \times 2 = 6,\ 5 \times 2 = 10$

Step 3 Add the numerators.

$\frac{7}{10} + \frac{6}{10} = \frac{(7 + 6)}{10} = \frac{13}{10}$

$\frac{13}{10}$ inches total precipitation, or $1\frac{3}{10}$ inches.

Practice Problem On an electric bill, about $\frac{1}{8}$ of the energy is from solar energy and about $\frac{1}{10}$ is from wind power. How much of the total bill is from solar energy and wind power combined?

Example 3 In your body, $\frac{7}{10}$ of your muscle contractions are involuntary (cardiac and smooth muscle tissue). Smooth muscle makes $\frac{3}{15}$ of your muscle contractions. How many of your muscle contractions are made by cardiac muscle?

Step 1 Find the LCD of the fractions.

$\frac{7}{10} - \frac{3}{15}$ (LCD, 30)

Step 2 Rename each numerator and each denominator with the LCD.

$7 \times 3 = 21,\ 10 \times 3 = 30$

$3 \times 2 = 6,\ 15 \times 2 = 30$

Step 3 Subtract the numerators.

$\frac{21}{30} - \frac{6}{30} = \frac{(21 - 6)}{30} = \frac{15}{30}$

Step 4 Find the GCF.

$\frac{15}{30}$ (GCF, 15)

$\frac{1}{2}$

$\frac{1}{2}$ of all muscle contractions are cardiac muscle.

Example 4 Tony wants to make cookies that call for $\frac{3}{4}$ of a cup of flour, but he only has $\frac{1}{3}$ of a cup. How much more flour does he need?

Step 1 Find the LCD of the fractions.

$\frac{3}{4} - \frac{1}{3}$ (LCD, 12)

Step 2 Rename each numerator and each denominator with the LCD.

$3 \times 3 = 9,\ 4 \times 3 = 12$

$1 \times 4 = 4,\ 3 \times 4 = 12$

Step 3 Subtract the numerators.

$\frac{9}{12} - \frac{4}{12} = \frac{(9 - 4)}{12} = \frac{5}{12}$

$\frac{5}{12}$ of a cup of flour.

Practice Problem Using the information provided to you in Example 3 above, determine how many muscle contractions are voluntary (skeletal muscle).

Math Skill Handbook

178 ◆ H STUDENT RESOURCES

Multiply Fractions To multiply with fractions, multiply the numerators and multiply the denominators. Find the simplest form if necessary.

Example Multiply $\frac{3}{5}$ by $\frac{1}{3}$.

Step 1 Multiply the numerators and denominators.

$$\frac{3}{5} \times \frac{1}{3} = \frac{(3 \times 1)}{(5 \times 3)} = \frac{3}{15}$$

Step 2 Find the GCF.

$\frac{3}{15}$ (GCF, 3)

Step 3 Divide the numerator and denominator by the GCF.

$\frac{3}{3} = 1, \ \frac{15}{3} = 5$

$\frac{1}{5}$

$\frac{3}{5}$ multiplied by $\frac{1}{3}$ is $\frac{1}{5}$.

Practice Problem Multiply $\frac{3}{14}$ by $\frac{5}{16}$.

Find a Reciprocal Two numbers whose product is 1 are called multiplicative inverses, or reciprocals.

Example Find the reciprocal of $\frac{3}{8}$.

Step 1 Inverse the fraction by putting the denominator on top and the numerator on the bottom.

$\frac{8}{3}$

The reciprocal of $\frac{3}{8}$ is $\frac{8}{3}$.

Practice Problem Find the reciprocal of $\frac{4}{9}$.

Divide Fractions To divide one fraction by another fraction, multiply the dividend by the reciprocal of the divisor. Find the simplest form if necessary.

Example 1 Divide $\frac{1}{9}$ by $\frac{1}{3}$.

Step 1 Find the reciprocal of the divisor.

The reciprocal of $\frac{1}{3}$ is $\frac{3}{1}$.

Step 2 Multiply the dividend by the reciprocal of the divisor.

$$\frac{\frac{1}{9}}{\frac{1}{3}} = \frac{1}{9} \times \frac{3}{1} = \frac{(1 \times 3)}{(9 \times 1)} = \frac{3}{9}$$

Step 3 Find the GCF.

$\frac{3}{9}$ (GCF, 3)

Step 4 Divide the numerator and denominator by the GCF.

$\frac{3}{3} = 1, \ \frac{9}{3} = 3$

$\frac{1}{3}$

$\frac{1}{9}$ divided by $\frac{1}{3}$ is $\frac{1}{3}$.

Example 2 Divide $\frac{3}{5}$ by $\frac{1}{4}$.

Step 1 Find the reciprocal of the divisor.

The reciprocal of $\frac{1}{4}$ is $\frac{4}{1}$.

Step 2 Multiply the dividend by the reciprocal of the divisor.

$$\frac{\frac{3}{5}}{\frac{1}{4}} = \frac{3}{5} \times \frac{4}{1} = \frac{(3 \times 4)}{(5 \times 1)} = \frac{12}{5}$$

$\frac{3}{5}$ divided by $\frac{1}{4}$ is $\frac{12}{5}$ or $2\frac{2}{5}$.

Practice Problem Divide $\frac{3}{11}$ by $\frac{7}{10}$.

Math Skill Handbook

Multiply Fractions

$$\frac{3}{14} \times \frac{5}{16} = \frac{(3 \times 5)}{(14 \times 16)} = \frac{15}{224}$$

Find a Reciprocal

$\frac{9}{4}$

Divide Fractions

The reciprocal of $\frac{7}{10}$ is $\frac{10}{7}$.

$$\frac{3}{11} \times \frac{10}{7} = \frac{(3 \times 10)}{(11 \times 7)} = \frac{30}{77}$$

Math Skill Handbook

MATH SKILL HANDBOOK H ◆ 179

Use Ratios

$$\frac{100 \text{ cm}}{144 \text{ cm}} = \frac{100 \div 4}{144 \div 4} = \frac{25}{36}$$

25:36

Add or Subtract Decimals

$$\begin{array}{r} \scriptstyle 1 \\ 1.245 \\ +\ 3.842 \\ \hline 5.087 \end{array}$$

Use Ratios

When you compare two numbers by division, you are using a ratio. Ratios can be written 3 to 5, 3:5, or $\frac{3}{5}$. Ratios, like fractions, also can be written in simplest form.

Ratios can represent probabilities, also called odds. This is a ratio that compares the number of ways a certain outcome occurs to the number of outcomes. For example, if you flip a coin 100 times, what are the odds that it will come up heads? There are two possible outcomes, heads or tails, so the odds of coming up heads are 50:100. Another way to say this is that 50 out of 100 times the coin will come up heads. In its simplest form, the ratio is 1:2.

Example 1 A chemical solution contains 40 g of salt and 64 g of baking soda. What is the ratio of salt to baking soda as a fraction in simplest form?

Step 1 Write the ratio as a fraction.

$$\frac{\text{salt}}{\text{baking soda}} = \frac{40}{64}$$

Step 2 Express the fraction in simplest form.

The GCF of 40 and 64 is 8.

$$\frac{40}{64} = \frac{40 \div 8}{64 \div 8} = \frac{5}{8}$$

The ratio of salt to baking soda in the sample is 5:8.

Example 2 Sean rolls a 6-sided die 6 times. What are the odds that the side with a 3 will show?

Step 1 Write the ratio as a fraction.

$$\frac{\text{number of sides with a 3}}{\text{number of sides}} = \frac{1}{6}$$

Step 2 Multiply by the number of attempts.

$$\frac{1}{6} \times 6 \text{ attempts} = \frac{6}{6} \text{ attempts} = 1 \text{ attempt}$$

1 attempt out of 6 will show a 3.

Practice Problem Two metal rods measure 100 cm and 144 cm in length. What is the ratio of their lengths in simplest form?

Use Decimals

A fraction with a denominator that is a power of ten can be written as a decimal. For example, 0.27 means $\frac{27}{100}$. The decimal point separates the ones place from the tenths place.

Any fraction can be written as a decimal using division. For example, the fraction $\frac{5}{8}$ can be written as a decimal by dividing 5 by 8. Written as a decimal, it is 0.625.

Add or Subtract Decimals When adding and subtracting decimals, line up the decimal points before carrying out the operation.

Example 1 Find the sum of 47.68 and 7.80.

Step 1 Line up the decimal places when you write the numbers.

$$\begin{array}{r} 47.68 \\ +\ 7.80 \\ \hline \end{array}$$

Step 2 Add the decimals.

$$\begin{array}{r} 47.68 \\ +\ 7.80 \\ \hline 55.48 \end{array}$$

The sum of 47.68 and 7.80 is 55.48.

Example 2 Find the difference of 42.17 and 15.85.

Step 1 Line up the decimal places when you write the number.

$$\begin{array}{r} 42.17 \\ -15.85 \\ \hline \end{array}$$

Step 2 Subtract the decimals.

$$\begin{array}{r} 42.17 \\ -15.85 \\ \hline 26.32 \end{array}$$

The difference of 42.17 and 15.85 is 26.32.

Practice Problem Find the sum of 1.245 and 3.842.

Math Skill Handbook

180 ◆ H STUDENT RESOURCES

Multiply Decimals To multiply decimals, multiply the numbers like any other number, ignoring the decimal point. Count the decimal places in each factor. The product will have the same number of decimal places as the sum of the decimal places in the factors.

Example Multiply 2.4 by 5.9.

Step 1 Multiply the factors like two whole numbers.
$24 \times 59 = 1416$

Step 2 Find the sum of the number of decimal places in the factors. Each factor has one decimal place, for a sum of two decimal places.

Step 3 The product will have two decimal places.
14.16

The product of 2.4 and 5.9 is 14.16.

Practice Problem Multiply 4.6 by 2.2.

Divide Decimals When dividing decimals, change the divisor to a whole number. To do this, multiply both the divisor and the dividend by the same power of ten. Then place the decimal point in the quotient directly above the decimal point in the dividend. Then divide as you do with whole numbers.

Example Divide 8.84 by 3.4.

Step 1 Multiply both factors by 10.
$3.4 \times 10 = 34$, $8.84 \times 10 = 88.4$

Step 2 Divide 88.4 by 34.

$$\begin{array}{r} 2.6 \\ 34\overline{)88.4} \\ -68 \\ \hline 20\,4 \\ -20\,4 \\ \hline 0 \end{array}$$

8.84 divided by 3.4 is 2.6.

Practice Problem Divide 75.6 by 3.6.

Use Proportions

An equation that shows that two ratios are equivalent is a proportion. The ratios $\frac{2}{4}$ and $\frac{5}{10}$ are equivalent, so they can be written as $\frac{2}{4} = \frac{5}{10}$. This equation is a proportion.

When two ratios form a proportion, the cross products are equal. To find the cross products in the proportion $\frac{2}{4} = \frac{5}{10}$, multiply the 2 and the 10, and the 4 and the 5. Therefore $2 \times 10 = 4 \times 5$, or $20 = 20$.

Because you know that both proportions are equal, you can use cross products to find a missing term in a proportion. This is known as solving the proportion.

Example The heights of a tree and a pole are proportional to the lengths of their shadows. The tree casts a shadow of 24 m when a 6-m pole casts a shadow of 4 m. What is the height of the tree?

Step 1 Write a proportion.
$$\frac{\text{height of tree}}{\text{height of pole}} = \frac{\text{length of tree's shadow}}{\text{length of pole's shadow}}$$

Step 2 Substitute the known values into the proportion. Let h represent the unknown value, the height of the tree.
$$\frac{h}{6} = \frac{24}{4}$$

Step 3 Find the cross products.
$h \times 4 = 6 \times 24$

Step 4 Simplify the equation.
$4h = 144$

Step 5 Divide each side by 4.
$$\frac{4h}{4} = \frac{144}{4}$$
$h = 36$

The height of the tree is 36 m.

Practice Problem The ratios of the weights of two objects on the Moon and on Earth are in proportion. A rock weighing 3 N on the Moon weighs 18 N on Earth. How much would a rock that weighs 5 N on the Moon weigh on Earth?

Multiply Decimals

Multiply 4.6 and 2.2 by 10.

$46 \times 22 = 1012$

Each factor had one decimal place.

10.12

Divide Decimals

Multiply both factors by 10.

Divide 756 by 36.

$$\begin{array}{r} 21 \\ 36\overline{)756} \\ 72 \\ \hline 36 \\ 36 \\ \hline 0 \end{array}$$

Use Proportions

$$\frac{3}{18} = \frac{5}{w}$$

$w \times 3 = 5 \times 18$

$$\frac{3w}{3} = \frac{90}{3}$$

$w = 30$

Math Skill Handbook

MATH SKILL HANDBOOK H ◆ 181

Use Percentages

$$\frac{73}{365} = \frac{x}{100}$$

$$\frac{7300}{365} = \frac{365x}{365}$$

$$20\% = x$$

Solve One-Step Equations

$$h = gd$$

$$\frac{17.4}{12.3} = \frac{12.3d}{12.3}$$

$$1.41 = d$$

Use Percentages

The word *percent* means "out of one hundred." It is a ratio that compares a number to 100. Suppose you read that 77 percent of the Earth's surface is covered by water. That is the same as reading that the fraction of the Earth's surface covered by water is $\frac{77}{100}$. To express a fraction as a percent, first find the equivalent decimal for the fraction. Then, multiply the decimal by 100 and add the percent symbol.

Example Express $\frac{13}{20}$ as a percent.

Step 1 Find the equivalent decimal for the fraction.

$$\begin{array}{r} 0.65 \\ 20\overline{)13.00} \\ \underline{12\,0} \\ 1\,00 \\ \underline{1\,00} \\ 0 \end{array}$$

Step 2 Rewrite the fraction $\frac{13}{20}$ as 0.65.

Step 3 Multiply 0.65 by 100 and add the % sign.

$$0.65 \times 100 = 65 = 65\%$$

So, $\frac{13}{20} = 65\%$.

This also can be solved as a proportion.

Example Express $\frac{13}{20}$ as a percent.

Step 1 Write a proportion.

$$\frac{13}{20} = \frac{x}{100}$$

Step 2 Find the cross products.

$$1300 = 20x$$

Step 3 Divide each side by 20.

$$\frac{1300}{20} = \frac{20x}{20}$$

$$65\% = x$$

Practice Problem In one year, 73 of 365 days were rainy in one city. What percent of the days in that city were rainy?

Solve One-Step Equations

A statement that two things are equal is an equation. For example, $A = B$ is an equation that states that A is equal to B.

An equation is solved when a variable is replaced with a value that makes both sides of the equation equal. To make both sides equal the inverse operation is used. Addition and subtraction are inverses, and multiplication and division are inverses.

Example 1 Solve the equation $x - 10 = 35$.

Step 1 Find the solution by adding 10 to each side of the equation.

$$x - 10 = 35$$

$$x - 10 + 10 = 35 + 10$$

$$x = 45$$

Step 2 Check the solution.

$$x - 10 = 35$$

$$45 - 10 = 35$$

$$35 = 35$$

Both sides of the equation are equal, so $x = 45$.

Example 2 In the formula $a = bc$, find the value of c if $a = 20$ and $b = 2$.

Step 1 Rearrange the formula so the unknown value is by itself on one side of the equation by dividing both sides by b.

$$a = bc$$

$$\frac{a}{b} = \frac{bc}{b}$$

$$\frac{a}{b} = c$$

Step 2 Replace the variables a and b with the values that are given.

$$\frac{a}{b} = c$$

$$\frac{20}{2} = c$$

$$10 = c$$

Step 3 Check the solution.

$$a = bc$$

$$20 = 2 \times 10$$

$$20 = 20$$

Both sides of the equation are equal, so $c = 10$ is the solution when $a = 20$ and $b = 2$.

Practice Problem In the formula $h = gd$, find the value of d if $g = 12.3$ and $h = 17.4$.

Use Statistics

The branch of mathematics that deals with collecting, analyzing, and presenting data is statistics. In statistics, there are three common ways to summarize data with a single number—the mean, the median, and the mode.

The **mean** of a set of data is the arithmetic average. It is found by adding the numbers in the data set and dividing by the number of items in the set.

The **median** is the middle number in a set of data when the data are arranged in numerical order. If there were an even number of data points, the median would be the mean of the two middle numbers.

The **mode** of a set of data is the number or item that appears most often.

Another number that often is used to describe a set of data is the range. The **range** is the difference between the largest number and the smallest number in a set of data.

A **frequency table** shows how many times each piece of data occurs, usually in a survey. **Table 2** below shows the results of a student survey on favorite color.

Table 2 Student Color Choice

Color	Tally	Frequency
red	||||	4
blue	~~||||~~	5
black	||	2
green	|||	3
purple	~~||||~~ ||	7
yellow	~~||||~~ |	6

Based on the frequency table data, which color is the favorite?

Example The speeds (in m/s) for a race car during five different time trials are 39, 37, 44, 36, and 44.

To find the mean:

Step 1 Find the sum of the numbers.
$39 + 37 + 44 + 36 + 44 = 200$

Step 2 Divide the sum by the number of items, which is 5.
$200 \div 5 = 40$

The mean is 40 m/s.

To find the median:

Step 1 Arrange the measures from least to greatest.
36, 37, 39, 44, 44

Step 2 Determine the middle measure.
36, 37, <u>39</u>, 44, 44

The median is 39 m/s.

To find the mode:

Step 1 Group the numbers that are the same together.
44, 44, 36, 37, 39

Step 2 Determine the number that occurs most in the set.
<u>44, 44</u>, 36, 37, 39

The mode is 44 m/s.

To find the range:

Step 1 Arrange the measures from largest to smallest.
44, 44, 39, 37, 36

Step 2 Determine the largest and smallest measures in the set.
<u>44</u>, 44, 39, 37, <u>36</u>

Step 3 Find the difference between the largest and smallest measures.
$44 - 36 = 8$

The range is 8 m/s.

Practice Problem Find the mean, median, mode, and range for the data set 8, 4, 12, 8, 11, 14, 16.

Use Statistics

mean
$8 + 4 + 12 + 8 + 11 + 14 + 16 = 73$
$73 \div 7 = 10.4$

median
4, 8, 8, <u>11</u>, 12, 14, 16

mode
4, <u>8</u>, <u>8</u>, 11, 12, 14, 16

range
<u>4</u>, 8, 8, 11, 12, 14, <u>16</u>
$16 - 4 = 12$

Math Skill Handbook

MATH SKILL HANDBOOK H ◆ 183

Perimeter

Problem 1

$P = 2(18\text{ m} + 7\text{ m})$

$P = 2(25\text{ m})$

$P = 50\text{ m}$

Problem 2

$P = 1.6\text{ cm} + 2.4\text{ cm} + 2.4\text{ cm}$

$P = 6.4\text{ cm}$

Area of a Rectangle

$A = (4\text{ m} \times 4\text{ m})$

$A = 16\text{ m}^2$

Area of a Triangle

$A = \frac{1}{2}(27\text{ cm} \times 17\text{ cm})$

$A = \frac{1}{2}(459\text{ cm}^2)$

$A = 229.5\text{ cm}^2$

Math Skill Handbook

Use Geometry

The branch of mathematics that deals with the measurement, properties, and relationships of points, lines, angles, surfaces, and solids is called geometry.

Perimeter The **perimeter** (P) is the distance around a geometric figure. To find the perimeter of a rectangle, add the length and width and multiply that sum by two, or $2(l + w)$. To find perimeters of irregular figures, add the length of the sides.

Example 1 Find the perimeter of a rectangle that is 3 m long and 5 m wide.

Step 1 You know that the perimeter is 2 times the sum of the width and length.

$P = 2(3\text{ m} + 5\text{ m})$

Step 2 Find the sum of the width and length.

$P = 2(8\text{ m})$

Step 3 Multiply by 2.

$P = 16\text{ m}$

The perimeter is 16 m.

Example 2 Find the perimeter of a shape with sides measuring 2 cm, 5 cm, 6 cm, 3 cm.

Step 1 You know that the perimeter is the sum of all the sides.

$P = 2 + 5 + 6 + 3$

Step 2 Find the sum of the sides.

$P = 2 + 5 + 6 + 3$

$P = 16$

The perimeter is 16 cm.

Practice Problem Find the perimeter of a rectangle with a length of 18 m and a width of 7 m.

Practice Problem Find the perimeter of a triangle measuring 1.6 cm by 2.4 cm by 2.4 cm.

Area of a Rectangle The **area** (A) is the number of square units needed to cover a surface. To find the area of a rectangle, multiply the length times the width, or $l \times w$. When finding area, the units also are multiplied. Area is given in square units.

Example Find the area of a rectangle with a length of 1 cm and a width of 10 cm.

Step 1 You know that the area is the length multiplied by the width.

$A = (1\text{ cm} \times 10\text{ cm})$

Step 2 Multiply the length by the width. Also multiply the units.

$A = 10\text{ cm}^2$

The area is 10 cm^2.

Practice Problem Find the area of a square whose sides measure 4 m.

Area of a Triangle To find the area of a triangle, use the formula:

$$A = \frac{1}{2}(\text{base} \times \text{height})$$

The base of a triangle can be any of its sides. The height is the perpendicular distance from a base to the opposite endpoint, or vertex.

Example Find the area of a triangle with a base of 18 m and a height of 7 m.

Step 1 You know that the area is $\frac{1}{2}$ the base times the height.

$A = \frac{1}{2}(18\text{ m} \times 7\text{ m})$

Step 2 Multiply $\frac{1}{2}$ by the product of 18×7. Multiply the units.

$A = \frac{1}{2}(126\text{ m}^2)$

$A = 63\text{ m}^2$

The area is 63 m^2.

Practice Problem Find the area of a triangle with a base of 27 cm and a height of 17 cm.

184 ◆ H STUDENT RESOURCES

Circumference of a Circle The **diameter** (d) of a circle is the distance across the circle through its center, and the **radius** (r) is the distance from the center to any point on the circle. The radius is half of the diameter. The distance around the circle is called the **circumference** (C). The formula for finding the circumference is:

$C = 2\pi r$ *or* $C = \pi d$

The circumference divided by the diameter is always equal to 3.1415926... This nonterminating and nonrepeating number is represented by the Greek letter π (pi). An approximation often used for π is 3.14.

Example 1 Find the circumference of a circle with a radius of 3 m.

Step 1 You know the formula for the circumference is 2 times the radius times π.
$C = 2\pi(3)$

Step 2 Multiply 2 times the radius.
$C = 6\pi$

Step 3 Multiply by π.
$C = 19$ m

The circumference is 19 m.

Example 2 Find the circumference of a circle with a diameter of 24.0 cm.

Step 1 You know the formula for the circumference is the diameter times π.
$C = \pi(24.0)$

Step 2 Multiply the diameter by π.
$C = 75.4$ cm

The circumference is 75.4 cm.

Practice Problem Find the circumference of a circle with a radius of 19 cm.

Area of a Circle The formula for the area of a circle is:
$A = \pi r^2$

Example 1 Find the area of a circle with a radius of 4.0 cm.

Step 1 $A = \pi(4.0)^2$

Step 2 Find the square of the radius.
$A = 16\pi$

Step 3 Multiply the square of the radius by π.
$A = 50\ cm^2$

The area of the circle is 50 cm^2.

Example 2 Find the area of a circle with a radius of 225 m.

Step 1 $A = \pi(225)^2$

Step 2 Find the square of the radius.
$A = 50625\pi$

Step 3 Multiply the square of the radius by π.
$A = 158962.5$

The area of the circle is 158,962 m^2.

Example 3 Find the area of a circle whose diameter is 20.0 mm.

Step 1 You know the formula for the area of a circle is the square of the radius times π, and that the radius is half of the diameter.
$A = \pi\left(\frac{20.0}{2}\right)^2$

Step 2 Find the radius.
$A = \pi(10.0)^2$

Step 3 Find the square of the radius.
$A = 100\pi$

Step 4 Multiply the square of the radius by π.
$A = 314\ mm^2$

The area is 314 mm^2.

Practice Problem Find the area of a circle with a radius of 16 m.

Math Skill Handbook

Circumference of a Circle

$C = 2\pi r$
$C = 2\pi(19)$
$C = 38\pi$
$C = 119.3$

Area of a Circle

$A = \pi r^2$
$A = \pi(16\ m)^2$
$A = \pi\ 256\ m^2$
$A = 803.8\ m^2$

Math Skill Handbook

MATH SKILL HANDBOOK H ◆ 185

Volume

Problem 1

$V = 8\text{ m} \times 4\text{ m} \times 4\text{ m}$

$V = 128\text{ m}^3$

Problem 2

$V = \pi r^2 \times \text{height}$

$V = \left[\pi\left(\frac{1}{2} \times 7\right)^2\right] \times 16$

$V = [\pi(3.5)^2] \times 16$

$V = [\pi(12.25)] \times 16$

$V = 38.46 \times 16$

$V = 615.36$

Math Skill Handbook

Volume The measure of space occupied by a solid is the **volume** (V). To find the volume of a rectangular solid multiply the length times width times height, or $V = l \times w \times h$. It is measured in cubic units, such as cubic centimeters (cm^3).

Example Find the volume of a rectangular solid with a length of 2.0 m, a width of 4.0 m, and a height of 3.0 m.

Step 1 You know the formula for volume is the length times the width times the height.

$V = 2.0\text{ m} \times 4.0\text{ m} \times 3.0\text{ m}$

Step 2 Multiply the length times the width times the height.

$V = 24\text{ m}^3$

The volume is 24 m^3.

Practice Problem Find the volume of a rectangular solid that 8 m long, 4 m wide, and 4 m high.

To find the volume of other solids, multiply the area of the base times the height.

Example 1 Find the volume of a solid that has a triangular base with a length of 8.0 m and a height of 7.0 m. The height of the entire solid is 15.0 m.

Step 1 You know that the base is a triangle, and the area of a triangle is $\frac{1}{2}$ the base times the height, and the volume is the area of the base times the height.

$V = \left[\frac{1}{2}(b \times h)\right] \times 15$

Step 2 Find the area of the base.

$V = \left[\frac{1}{2}(8 \times 7)\right] \times 15$

$V = \left(\frac{1}{2} \times 56\right) \times 15$

Step 3 Multiply the area of the base by the height of the solid.

$V = 28 \times 15$

$V = 420\text{ m}^3$

The volume is 420 m^3.

Example 2 Find the volume of a cylinder that has a base with a radius of 12.0 cm, and a height of 21.0 cm.

Step 1 You know that the base is a circle, and the area of a circle is the square of the radius times π, and the volume is the area of the base times the height.

$V = (\pi r^2) \times 21$

$V = (\pi 12^2) \times 21$

Step 2 Find the area of the base.

$V = 144\pi \times 21$

$V = 452 \times 21$

Step 3 Multiply the area of the base by the height of the solid.

$V = 9490\text{ cm}^3$

The volume is 9490 cm^3.

Example 3 Find the volume of a cylinder that has a diameter of 15 mm and a height of 4.8 mm.

Step 1 You know that the base is a circle with an area equal to the square of the radius times π. The radius is one-half the diameter. The volume is the area of the base times the height.

$V = (\pi r^2) \times 4.8$

$V = \left[\pi\left(\frac{1}{2} \times 15\right)^2\right] \times 4.8$

$V = (\pi 7.5^2) \times 4.8$

Step 2 Find the area of the base.

$V = 56.25\pi \times 4.8$

$V = 176.63 \times 4.8$

Step 3 Multiply the area of the base by the height of the solid.

$V = 847.8$

The volume is 847.8 mm^3.

Practice Problem Find the volume of a cylinder with a diameter of 7 cm in the base and a height of 16 cm.

Math Skill Handbook

Science Applications

Measure in SI

The metric system of measurement was developed in 1795. A modern form of the metric system, called the International System (SI), was adopted in 1960 and provides the standard measurements that all scientists around the world can understand.

The SI system is convenient because unit sizes vary by powers of 10. Prefixes are used to name units. Look at **Table 3** for some common SI prefixes and their meanings.

Table 3 Some SI Prefixes

Prefix	Symbol	Meaning	
kilo-	k	1,000	thousand
hecto-	h	100	hundred
deka-	da	10	ten
deci-	d	0.1	tenth
centi-	c	0.01	hundredth
milli-	m	0.001	thousandth

Example How many grams equal one kilogram?

Step 1 Find the prefix *kilo* in **Table 3.**

Step 2 Using **Table 3,** determine the meaning of *kilo.* According to the table, it means 1,000. When the prefix *kilo* is added to a unit, it means that there are 1,000 of the units in a "*kilo*unit."

Step 3 Apply the prefix to the units in the question. The units in the question are grams. There are 1,000 grams in a kilogram.

Practice Problem Is a milligram larger or smaller than a gram? How many of the smaller units equal one larger unit? What fraction of the larger unit does one smaller unit represent?

Dimensional Analysis

Convert SI Units In science, quantities such as length, mass, and time sometimes are measured using different units. A process called dimensional analysis can be used to change one unit of measure to another. This process involves multiplying your starting quantity and units by one or more conversion factors. A conversion factor is a ratio equal to one and can be made from any two equal quantities with different units. If 1,000 mL equal 1 L then two ratios can be made.

$$\frac{1{,}000 \text{ mL}}{1 \text{ L}} = \frac{1 \text{ L}}{1{,}000 \text{ mL}} = 1$$

One can covert between units in the SI system by using the equivalents in **Table 3** to make conversion factors.

Example 1 How many cm are in 4 m?

Step 1 Write conversion factors for the units given. From **Table 3,** you know that 100 cm = 1 m. The conversion factors are

$$\frac{100 \text{ cm}}{1 \text{ m}} \text{ and } \frac{1 \text{ m}}{100 \text{ cm}}$$

Step 2 Decide which conversion factor to use. Select the factor that has the units you are converting from (m) in the denominator and the units you are converting to (cm) in the numerator.

$$\frac{100 \text{ cm}}{1 \text{ m}}$$

Step 3 Multiply the starting quantity and units by the conversion factor. Cancel the starting units with the units in the denominator. There are 400 cm in 4 m.

$$4 \cancel{\text{m}} \times \frac{100 \text{ cm}}{1 \cancel{\text{m}}} = 400 \text{ cm}$$

Practice Problem How many milligrams are in one kilogram? (Hint: You will need to use two conversion factors from **Table 3.**)

Measure in SI

smaller; 1000; one thousandth

Dimensional Analysis

$$x \text{ mg} = 1 \cancel{\text{kg}} \times \frac{1000 \cancel{\text{g}}}{1 \cancel{\text{kg}}} \times \frac{1000 \text{ mg}}{1 \cancel{\text{g}}} =$$

1,000,000 mg

1,000,000 mg = 1 kg

Math Skill Handbook

Convert Between Unit Systems

$$\frac{(1\ \text{in})^3}{(2.54\ \text{cm})^3} = \frac{1\ \text{in} \times 1\ \text{in} \times 1\ \text{in}}{2.54\ \text{cm} \times 2.54\ \text{cm} \times 2.54\ \text{cm}} = \frac{1\ \text{in}^3}{16.39\ \text{cm}^3}$$

Table 4 Unit System Equivalents

Type of Measurement	Equivalent
Length	1 in = 2.54 cm 1 yd = 0.91 m 1 mi = 1.61 km
Mass and Weight*	1 oz = 28.35 g 1 lb = 0.45 kg 1 ton (short) = 0.91 tonnes (metric tons) 1 lb = 4.45 N
Volume	1 in^3 = 16.39 cm^3 1 qt = 0.95 L 1 gal = 3.78 L
Area	1 in^2 = 6.45 cm^2 1 yd^2 = 0.83 m^2 1 mi^2 = 2.59 km^2 1 acre = 0.40 hectares
Temperature	$°C = \frac{(°F - 32)}{1.8}$ K = °C + 273

*Weight is measured in standard Earth gravity.

Convert Between Unit Systems **Table 4** gives a list of equivalents that can be used to convert between English and SI units.

Example If a meterstick has a length of 100 cm, how long is the meterstick in inches?

Step 1 Write the conversion factors for the units given. From **Table 4,** 1 in = 2.54 cm.

$$\frac{1\ \text{in}}{2.54\ \text{cm}} \quad \textit{and} \quad \frac{2.54\ \text{cm}}{1\ \text{in}}$$

Step 2 Determine which conversion factor to use. You are converting from cm to in. Use the conversion factor with cm on the bottom.

$$\frac{1\ \text{in}}{2.54\ \text{cm}}$$

Step 3 Multiply the starting quantity and units by the conversion factor. Cancel the starting units with the units in the denominator. Round your answer based on the number of significant figures in the conversion factor.

$$100\ \cancel{\text{cm}} \times \frac{1\ \text{in}}{2.54\ \cancel{\text{cm}}} = 39.37\ \text{in}$$

The meterstick is 39.4 in long.

Practice Problem A book has a mass of 5 lbs. What is the mass of the book in kg?

Practice Problem Use the equivalent for in and cm (1 in = 2.54 cm) to show how 1 in^3 = 16.39 cm^3.

Math Skill Handbook

Math Skill Handbook

Precision and Significant Digits

When you make a measurement, the value you record depends on the precision of the measuring instrument. This precision is represented by the number of significant digits recorded in the measurement. When counting the number of significant digits, all digits are counted except zeros at the end of a number with no decimal point such as 2,050, and zeros at the beginning of a decimal such as 0.03020. When adding or subtracting numbers with different precision, round the answer to the smallest number of decimal places of any number in the sum or difference. When multiplying or dividing, the answer is rounded to the smallest number of significant digits of any number being multiplied or divided.

Example The lengths 5.28 and 5.2 are measured in meters. Find the sum of these lengths and record your answer using the correct number of significant digits.

Step 1 Find the sum.

5.28 m	2 digits after the decimal
+ 5.2 m	1 digit after the decimal
10.48 m	

Step 2 Round to one digit after the decimal because the least number of digits after the decimal of the numbers being added is 1.

The sum is 10.5 m.

Practice Problem How many significant digits are in the measurement 7,071,301 m? How many significant digits are in the measurement 0.003010 g?

Practice Problem Multiply 5.28 and 5.2 using the rule for multiplying and dividing. Record the answer using the correct number of significant digits.

Scientific Notation

Many times numbers used in science are very small or very large. Because these numbers are difficult to work with scientists use scientific notation. To write numbers in scientific notation, move the decimal point until only one non-zero digit remains on the left. Then count the number of places you moved the decimal point and use that number as a power of ten. For example, the average distance from the Sun to Mars is 227,800,000,000 m. In scientific notation, this distance is 2.278×10^{11} m. Because you moved the decimal point to the left, the number is a positive power of ten.

The mass of an electron is about 0.000 000 000 000 000 000 000 000 000 000 911 kg. Expressed in scientific notation, this mass is 9.11×10^{-31} kg. Because the decimal point was moved to the right, the number is a negative power of ten.

Example Earth is 149,600,000 km from the Sun. Express this in scientific notation.

Step 1 Move the decimal point until one non-zero digit remains on the left.
1.496 000 00

Step 2 Count the number of decimal places you have moved. In this case, eight.

Step 3 Show that number as a power of ten, 10^8.

The Earth is 1.496×10^8 km from the Sun.

Practice Problem How many significant digits are in 149,600,000 km? How many significant digits are in 1.496×10^8 km?

Practice Problem Parts used in a high performance car must be measured to 7×10^{-6} m. Express this number as a decimal.

Practice Problem A CD is spinning at 539 revolutions per minute. Express this number in scientific notation.

Precision and Significant Digits

Problem 1
7; 4

Problem 2
$5.28 \times 5.2 = 27.456$

5.28 has 3 significant digits.

5.2 has 2 significant digits.

When multiplying and dividing, the answer is rounded to the smallest number of significant digits of the numbers being multiplied or divided—in this case, 2.

27.456 is rounded to 27.

Scientific Notation

Problem 1
4; 4

Problem 2
0.000007

Problem 3
5.39×10^2

Line Graph

x	y
3	52
6	72
9	83
12	86

Puppy Growth

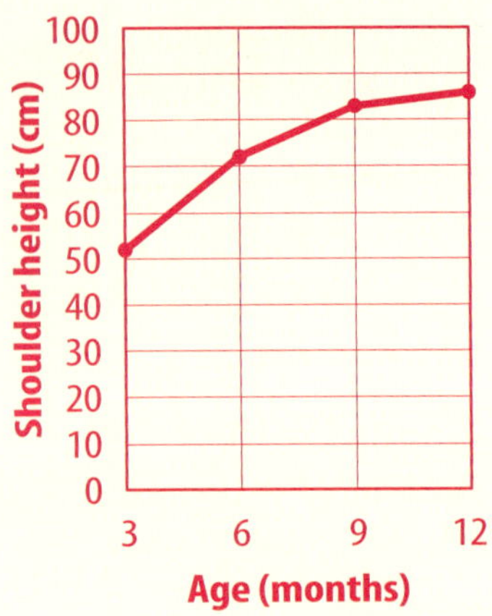

Math Skill Handbook

Make and Use Graphs

Data in tables can be displayed in a graph—a visual representation of data. Common graph types include line graphs, bar graphs, and circle graphs.

Line Graph A line graph shows a relationship between two variables that change continuously. The independent variable is changed and is plotted on the *x*-axis. The dependent variable is observed, and is plotted on the *y*-axis.

Example Draw a line graph of the data below from a cyclist in a long-distance race.

Table 5 Bicycle Race Data

Time (h)	Distance (km)
0	0
1	8
2	16
3	24
4	32
5	40

Step 1 Determine the *x*-axis and *y*-axis variables. Time varies independently of distance and is plotted on the *x*-axis. Distance is dependent on time and is plotted on the *y*-axis.

Step 2 Determine the scale of each axis. The *x*-axis data ranges from 0 to 5. The *y*-axis data ranges from 0 to 40.

Step 3 Using graph paper, draw and label the axes. Include units in the labels.

Step 4 Draw a point at the intersection of the time value on the *x*-axis and corresponding distance value on the *y*-axis. Connect the points and label the graph with a title, as shown in **Figure 20.**

Distance v. Time

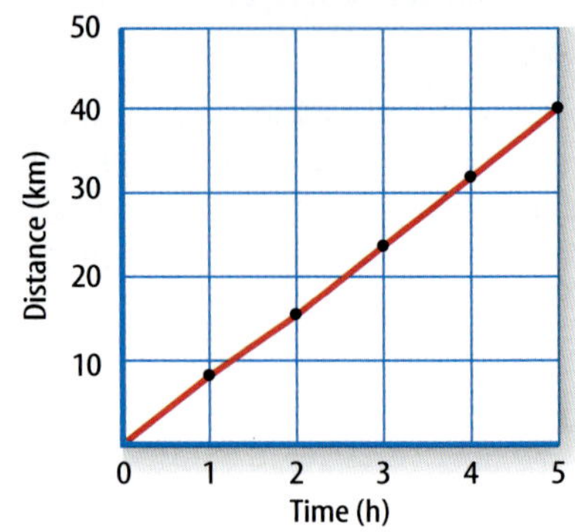

Figure 20 This line graph shows the relationship between distance and time during a bicycle ride.

Practice Problem A puppy's shoulder height is measured during the first year of her life. The following measurements were collected: (3 mo, 52 cm), (6 mo, 72 cm), (9 mo, 83 cm), (12 mo, 86). Graph this data.

Find a Slope The slope of a straight line is the ratio of the vertical change, rise, to the horizontal change, run.

$$\text{Slope} = \frac{\text{vertical change (rise)}}{\text{horizontal change (run)}} = \frac{\text{change in } y}{\text{change in } x}$$

Example Find the slope of the graph in **Figure 20.**

Step 1 You know that the slope is the change in *y* divided by the change in *x*.

$$\text{Slope} = \frac{\text{change in } y}{\text{change in } x}$$

Step 2 Determine the data points you will be using. For a straight line, choose the two sets of points that are the farthest apart.

$$\text{Slope} = \frac{(40-0)\text{ km}}{(5-0)\text{ hr}}$$

Step 3 Find the change in *y* and *x*.

$$\text{Slope} = \frac{40\text{ km}}{5\text{h}}$$

Step 4 Divide the change in *y* by the change in *x*.

$$\text{Slope} = \frac{8\text{ km}}{\text{h}}$$

The slope of the graph is 8 km/h.

190 ◆ H STUDENT RESOURCES

Bar Graph To compare data that does not change continuously you might choose a bar graph. A bar graph uses bars to show the relationships between variables. The *x*-axis variable is divided into parts. The parts can be numbers such as years, or a category such as a type of animal. The *y*-axis is a number and increases continuously along the axis.

Example A recycling center collects 4.0 kg of aluminum on Monday, 1.0 kg on Wednesday, and 2.0 kg on Friday. Create a bar graph of this data.

Step 1 Select the *x*-axis and *y*-axis variables. The measured numbers (the masses of aluminum) should be placed on the *y*-axis. The variable divided into parts (collection days) is placed on the *x*-axis.

Step 2 Create a graph grid like you would for a line graph. Include labels and units.

Step 3 For each measured number, draw a vertical bar above the *x*-axis value up to the *y*-axis value. For the first data point, draw a vertical bar above Monday up to 4.0 kg.

Aluminum Collected During Week

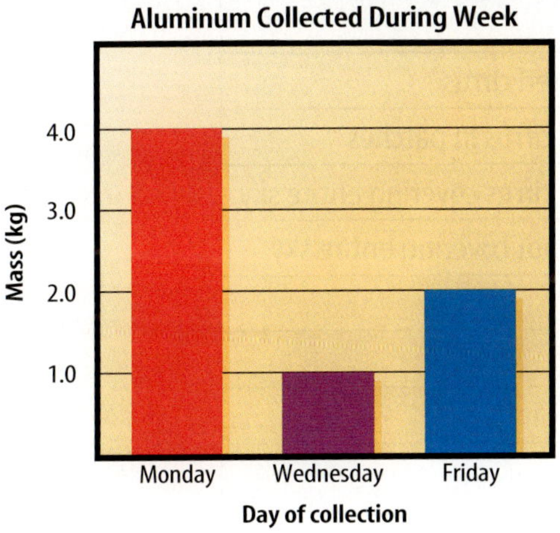

Practice Problem Draw a bar graph of the gases in air: 78% nitrogen, 21% oxygen, 1% other gases.

Circle Graph To display data as parts of a whole, you might use a circle graph. A circle graph is a circle divided into sections that represent the relative size of each piece of data. The entire circle represents 100%, half represents 50%, and so on.

Example Air is made up of 78% nitrogen, 21% oxygen, and 1% other gases. Display the composition of air in a circle graph.

Step 1 Multiply each percent by 360° and divide by 100 to find the angle of each section in the circle.

$$78\% \times \frac{360°}{100} = 280.8°$$

$$21\% \times \frac{360°}{100} = 75.6°$$

$$1\% \times \frac{360°}{100} = 3.6°$$

Step 2 Use a compass to draw a circle and to mark the center of the circle. Draw a straight line from the center to the edge of the circle.

Step 3 Use a protractor and the angles you calculated to divide the circle into parts. Place the center of the protractor over the center of the circle and line the base of the protractor over the straight line.

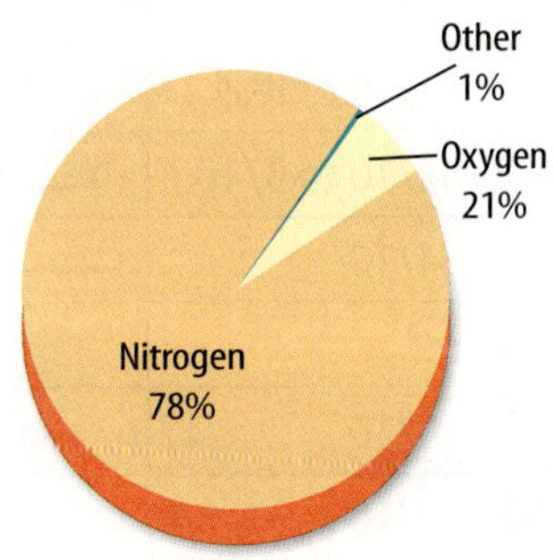

Practice Problem Draw a circle graph to represent the amount of aluminum collected during the week shown in the bar graph to the left.

Bar Graph

Composition of Air

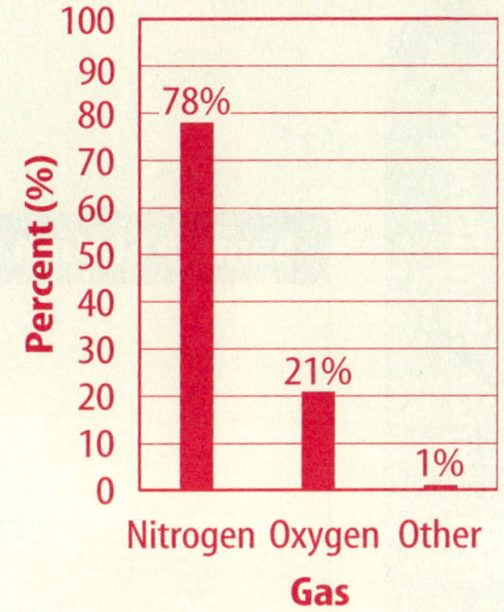

Circle Graph

The total amount of aluminum collected is:

$4.0\ \text{kg} + 1.0\ \text{kg} + 2.0\ \text{kg} = 7.0\ \text{kg}$

$\frac{4.0\ \text{kg}}{7.0\ \text{kg}} = \frac{x}{360°};\ x = 206°$

$\frac{1.0\ \text{kg}}{7.0\ \text{kg}} = \frac{x}{360°};\ x = 51°$

$\frac{2.0\ \text{kg}}{7.0\ \text{kg}} = \frac{x}{360°};\ x = 103°$

Math Skill Handbook

MATH SKILL HANDBOOK H ◆ 191

Weather Map Symbols

Sample Station Model

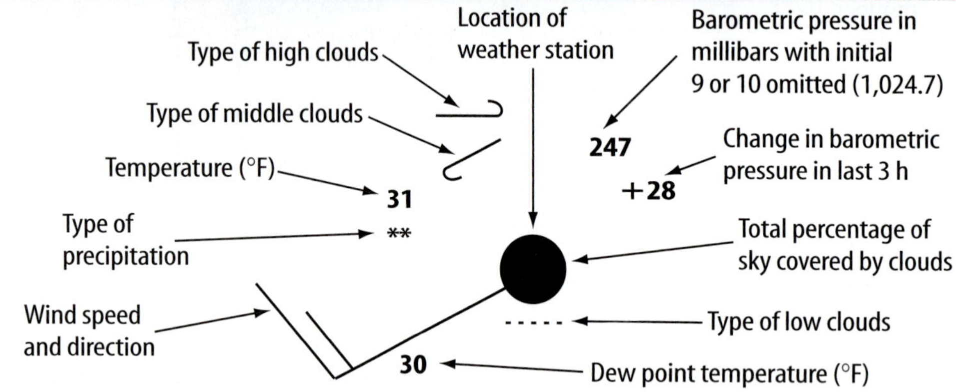

Sample Plotted Report at Each Station

Precipitation	Wind Speed and Direction	Sky Coverage	Some Types of High Clouds
≡ Fog	○ 0 calm	○ No cover	Scattered cirrus
★ Snow	1–2 knots	1/10 or less	Dense cirrus in patches
● Rain	3–7 knots	2/10 to 3/10	Veil of cirrus covering entire sky
Thunderstorm	8–12 knots	4/10	Cirrus not covering entire sky
, Drizzle	13–17 knots	–	
∇ Showers	18–22 knots	6/10	
	23–27 knots	7/10	
	48–52 knots	Overcast with openings	
	1 knot = 1.852 km/h	● Completely overcast	

Some Types of Middle Clouds	Some Types of Low Clouds	Fronts and Pressure Systems	
Thin altostratus layer	Cumulus of fair weather	(H) or High (L) or Low	Center of high- or low-pressure system
Thick altostratus layer	Stratocumulus		Cold front
Thin altostratus in patches	----- Fractocumulus of bad weather		Warm front
Thin altostratus in bands	— Stratus of fair weather		Occluded front
			Stationary front

Rocks

Rocks		
Rock Type	**Rock Name**	**Characteristics**
Igneous (intrusive)	Granite	Large mineral grains of quartz, feldspar, hornblende, and mica. Usually light in color.
	Diorite	Large mineral grains of feldspar, hornblende, and mica. Less quartz than granite. Intermediate in color.
	Gabbro	Large mineral grains of feldspar, augite, and olivine. No quartz. Dark in color.
Igneous (extrusive)	Rhyolite	Small mineral grains of quartz, feldspar, hornblende, and mica, or no visible grains. Light in color.
	Andesite	Small mineral grains of feldspar, hornblende, and mica or no visible grains. Intermediate in color.
	Basalt	Small mineral grains of feldspar, augite, and possibly olivine or no visible grains. No quartz. Dark in color.
	Obsidian	Glassy texture. No visible grains. Volcanic glass. Fracture looks like broken glass.
	Pumice	Frothy texture. Floats in water. Usually light in color.
Sedimentary (detrital)	Conglomerate	Coarse grained. Gravel or pebble-size grains.
	Sandstone	Sand-sized grains 1/16 to 2 mm.
	Siltstone	Grains are smaller than sand but larger than clay.
	Shale	Smallest grains. Often dark in color. Usually platy.
Sedimentary (chemical or organic)	Limestone	Major mineral is calcite. Usually forms in oceans and lakes. Often contains fossils.
	Coal	Forms in swampy areas. Compacted layers of organic material, mainly plant remains.
Sedimentary (chemical)	Rock Salt	Commonly forms by the evaporation of seawater.
Metamorphic (foliated)	Gneiss	Banding due to alternate layers of different minerals, of different colors. Parent rock often is granite.
	Schist	Parallel arrangement of sheetlike minerals, mainly micas. Forms from different parent rocks.
	Phyllite	Shiny or silky appearance. May look wrinkled. Common parent rocks are shale and slate.
	Slate	Harder, denser, and shinier than shale. Common parent rock is shale.
Metamorphic (nonfoliated)	Marble	Calcite or dolomite. Common parent rock is limestone.
	Soapstone	Mainly of talc. Soft with greasy feel.
	Quartzite	Hard with interlocking quartz crystals. Common parent rock is sandstone.

Minerals

Minerals					
Mineral (formula)	**Color**	**Streak**	**Hardness**	**Breakage Pattern**	**Uses and Other Properties**
Graphite (C)	black to gray	black to gray	1–1.5	basal cleavage (scales)	pencil lead, lubricants for locks, rods to control some small nuclear reactions, battery poles
Galena (PbS)	gray	gray to black	2.5	cubic cleavage perfect	source of lead, used for pipes, shields for X rays, fishing equipment sinkers
Hematite (Fe_2O_3)	black or reddish-brown	reddish-brown	5.5–6.5	irregular fracture	source of iron; converted to pig iron, made into steel
Magnetite (Fe_3O_4)	black	black	6	conchoidal fracture	source of iron, attracts a magnet
Pyrite (FeS_2)	light, brassy, yellow	greenish-black	6–6.5	uneven fracture	fool's gold
Talc ($Mg_3 Si_4O_{10} (OH)_2$)	white, greenish	white	1	cleavage in one direction	used for talcum powder, sculptures, paper, and tabletops
Gypsum ($CaSO_4 \cdot 2H_2O$)	colorless, gray, white, brown	white	2	basal cleavage	used in plaster of paris and dry wall for building construction
Sphalerite (ZnS)	brown, reddish-brown, greenish	light to dark brown	3.5–4	cleavage in six directions	main ore of zinc; used in paints, dyes, and medicine
Muscovite ($KAl_3Si_3 O_{10}(OH)_2$)	white, light gray, yellow, rose, green	colorless	2–2.5	basal cleavage	occurs in large, flexible plates; used as an insulator in electrical equipment, lubricant
Biotite ($K(Mg,Fe)_3 (AlSi_3O_{10}) (OH)_2$)	black to dark brown	colorless	2.5–3	basal cleavage	occurs in large, flexible plates
Halite (NaCl)	colorless, red, white, blue	colorless	2.5	cubic cleavage	salt; soluble in water; a preservative

Minerals

Minerals					
Mineral (formula)	**Color**	**Streak**	**Hardness**	**Breakage Pattern**	**Uses and Other Properties**
Calcite ($CaCO_3$)	colorless, white, pale blue	colorless, white	3	cleavage in three directions	fizzes when HCl is added; used in cements and other building materials
Dolomite ($CaMg\ (CO_3)_2$)	colorless, white, pink, green, gray, black	white	3.5–4	cleavage in three directions	concrete and cement; used as an ornamental building stone
Fluorite (CaF_2)	colorless, white, blue, green, red, yellow, purple	colorless	4	cleavage in four directions	used in the manufacture of optical equipment; glows under ultraviolet light
Hornblende ($(CaNa)_{2\text{-}3}(Mg,Al,Fe)_5\text{-}(Al,Si)_2Si_6O_{22}(OH)_2$)	green to black	gray to white	5–6	cleavage in two directions	will transmit light on thin edges; 6-sided cross section
Feldspar ($(KAlSi_3O_8)$ $(NaAlSi_3O_8)$, $(CaAl_2Si_2O_8)$)	colorless, white to gray, green	colorless	6	two cleavage planes meet at 90° angle	used in the manufacture of ceramics
Augite ($((Ca,Na)(Mg,Fe,Al)(Al,Si)_2O_6)$)	black	colorless	6	cleavage in two directions	square or 8-sided cross section
Olivine ($((Mg,Fe)_2SiO_4)$)	olive, green	none	6.5–7	conchoidal fracture	gemstones, refractory sand
Quartz (SiO_2)	colorless, various colors	none	7	conchoidal fracture	used in glass manufacture, electronic equipment, radios, computers, watches, gemstones

Reference Handbooks

PERIODIC TABLE OF THE ELEMENTS

Columns of elements are called groups. Elements in the same group have similar chemical properties.

Element — Hydrogen
Atomic number — 1
Symbol — H
Atomic mass — 1.008
State of matter

Gas
Liquid
Solid
Synthetic

The first three symbols tell you the state of matter of the element at room temperature. The fourth symbol identifies elements that are not present in significant amounts on Earth. Useful amounts are made synthetically.

	1	2	3	4	5	6	7	8	9
1	Hydrogen 1 H 1.008								
2	Lithium 3 Li 6.941	Beryllium 4 Be 9.012							
3	Sodium 11 Na 22.990	Magnesium 12 Mg 24.305							
4	Potassium 19 K 39.098	Calcium 20 Ca 40.078	Scandium 21 Sc 44.956	Titanium 22 Ti 47.867	Vanadium 23 V 50.942	Chromium 24 Cr 51.996	Manganese 25 Mn 54.938	Iron 26 Fe 55.845	Cobalt 27 Co 58.933
5	Rubidium 37 Rb 85.468	Strontium 38 Sr 87.62	Yttrium 39 Y 88.906	Zirconium 40 Zr 91.224	Niobium 41 Nb 92.906	Molybdenum 42 Mo 95.94	Technetium 43 Tc (98)	Ruthenium 44 Ru 101.07	Rhodium 45 Rh 102.906
6	Cesium 55 Cs 132.905	Barium 56 Ba 137.327	Lanthanum 57 La 138.906	Hafnium 72 Hf 178.49	Tantalum 73 Ta 180.948	Tungsten 74 W 183.84	Rhenium 75 Re 186.207	Osmium 76 Os 190.23	Iridium 77 Ir 192.217
7	Francium 87 Fr (223)	Radium 88 Ra (226)	Actinium 89 Ac (227)	Rutherfordium 104 Rf (261)	Dubnium 105 Db (262)	Seaborgium 106 Sg (266)	Bohrium 107 Bh (264)	Hassium 108 Hs (277)	Meitnerium 109 Mt (268)

The number in parentheses is the mass number of the longest-lived isotope for that element.

Rows of elements are called periods. Atomic number increases across a period.

The arrow shows where these elements would fit into the periodic table. They are moved to the bottom of the table to save space.

Lanthanide series	Cerium 58 Ce 140.116	Praseodymium 59 Pr 140.908	Neodymium 60 Nd 144.24	Promethium 61 Pm (145)	Samarium 62 Sm 150.36
Actinide series	Thorium 90 Th 232.038	Protactinium 91 Pa 231.036	Uranium 92 U 238.029	Neptunium 93 Np (237)	Plutonium 94 Pu (244)

Metal

Metalloid

Nonmetal

The color of an element's block tells you if the element is a metal, nonmetal, or metalloid.

Science Online

Visit bookh.msscience.com for updates to the periodic table.

10	11	12	13	14	15	16	17	18
								Helium 2 He 4.003
			Boron 5 B 10.811	Carbon 6 C 12.011	Nitrogen 7 N 14.007	Oxygen 8 O 15.999	Fluorine 9 F 18.998	Neon 10 Ne 20.180
			Aluminum 13 Al 26.982	Silicon 14 Si 28.086	Phosphorus 15 P 30.974	Sulfur 16 S 32.065	Chlorine 17 Cl 35.453	Argon 18 Ar 39.948
Nickel 28 Ni 58.693	Copper 29 Cu 63.546	Zinc 30 Zn 65.409	Gallium 31 Ga 69.723	Germanium 32 Ge 72.64	Arsenic 33 As 74.922	Selenium 34 Se 78.96	Bromine 35 Br 79.904	Krypton 36 Kr 83.798
Palladium 46 Pd 106.42	Silver 47 Ag 107.868	Cadmium 48 Cd 112.411	Indium 49 In 114.818	Tin 50 Sn 118.710	Antimony 51 Sb 121.760	Tellurium 52 Te 127.60	Iodine 53 I 126.904	Xenon 54 Xe 131.293
Platinum 78 Pt 195.078	Gold 79 Au 196.967	Mercury 80 Hg 200.59	Thallium 81 Tl 204.383	Lead 82 Pb 207.2	Bismuth 83 Bi 208.980	Polonium 84 Po (209)	Astatine 85 At (210)	Radon 86 Rn (222)
Darmstadtium 110 Ds (281)	Unununium * 111 Uuu (272)	Ununbium * 112 Uub (285)		Ununquadium * 114 Uuq (289)		** 116		** 118

* The names and symbols for elements 111–114 are temporary. Final names will be selected when the elements' discoveries are verified.

** Elements 116 and 118 were thought to have been created. The claim was retracted because the experimental results could not be repeated.

Europium 63 Eu 151.964	Gadolinium 64 Gd 157.25	Terbium 65 Tb 158.925	Dysprosium 66 Dy 162.500	Holmium 67 Ho 164.930	Erbium 68 Er 167.259	Thulium 69 Tm 168.934	Ytterbium 70 Yb 173.04	Lutetium 71 Lu 174.967
Americium 95 Am (243)	Curium 96 Cm (247)	Berkelium 97 Bk (247)	Californium 98 Cf (251)	Einsteinium 99 Es (252)	Fermium 100 Fm (257)	Mendelevium 101 Md (258)	Nobelium 102 No (259)	Lawrencium 103 Lr (262)

Topographic Map Symbols

Topographic Map Symbols			
	Primary highway, hard surface		Index contour
	Secondary highway, hard surface		Supplementary contour
	Light-duty road, hard or improved surface		Intermediate contour
	Unimproved road		Depression contours
	Railroad: single track		
	Railroad: multiple track		Boundaries: national
	Railroads in juxtaposition		State
			County, parish, municipal
	Buildings		Civil township, precinct, town, barrio
cem	Schools, church, and cemetery		Incorporated city, village, town, hamlet
	Buildings (barn, warehouse, etc.)		Reservation, national or state
	Wells other than water (labeled as to type)		Small park, cemetery, airport, etc.
	Tanks: oil, water, etc. (labeled only if water)		Land grant
	Located or landmark object; windmill		Township or range line, U.S. land survey
	Open pit, mine, or quarry; prospect		
			Township or range line, approximate location
	Marsh (swamp)		
	Wooded marsh		Perennial streams
	Woods or brushwood		Elevated aqueduct
	Vineyard		Water well and spring
	Land subject to controlled inundation		Small rapids
	Submerged marsh		Large rapids
	Mangrove		Intermittent lake
	Orchard		Intermittent stream
	Scrub		Aqueduct tunnel
	Urban area		Glacier
			Small falls
x7369	Spot elevation		Large falls
670	Water elevation		Dry lake bed

Glossary/Glosario

Cómo usar el glosario en español:
1. Busca el término en inglés que desees encontrar.
2. El término en español, junto con la definición, se encuentran en la columna de la derecha.

Pronunciation Key

Use the following key to help you sound out words in the glossary.

a	back (BAK)	**ew**	**food** (FEWD)
ay	day (DAY)	**yoo**	**pure** (PYOOR)
ah	father (FAH thur)	**yew**	few (FYEW)
ow	flower (FLOW ur)	**uh**	comma (CAH muh)
ar	car (CAR)	**u** (+ con)	rub (RUB)
e	less (LES)	**sh**	**sh**elf (SHELF)
ee	leaf (LEEF)	**ch**	nature (NAY chur)
ih	trip (TRIHP)	**g**	gift (GIHFT)
i (i + con + e)	idea (i DEE uh)	**j**	gem (JEM)
oh	go (GOH)	**ing**	sing (SING)
aw	soft (SAWFT)	**zh**	vision (VIH zhun)
or	**or**bit (OR buht)	**k**	cake (KAYK)
oy	**coin** (COYN)	**s**	seed, cent (SEED, SENT)
oo	**foot** (FOOT)	**z**	zone, raise (ZOHN, RAYZ)

English — **A** — **Español**

abyssal (uh BIH sul) plain: flat seafloor area from 4,000 m to 6,000 m below the ocean surface, formed by the deposition of sediments. (p. 129)

planicie abisal: área plana del suelo marino entre 4000 y 6000 metros por debajo de la superficie del océano, formada por deposición de sedimentos. (p. 129)

aquifer: layer of rock or sediment that allows groundwater to flow easily through its connecting openings. (pp. 23, 70)

acuífero: capa de rocas o sedimentos que permite al agua subterránea fluir con facilidad a través de sus aperturas de conexión. (pp. 23, 70)

artesian well: well drilled into a pressurized aquifer that supplies fresh water, usually without pumping. (p. 73)

pozo artesano: pozo perforado en un acuífero a presión, el cual suministra agua dulce usualmente sin necesidad de bombeo. (p. 73)

B

basin: low area on Earth in which an ocean formed when the area filled with water from torrential rains. (p. 101)

depresión: área baja de la Tierra en la que se forma un océano cuando el área es llenada con agua proveniente de lluvias torrenciales. (p. 101)

benthos: marine plants and animals that live on or in the ocean floor. (p. 139)

bentos: plantas y animales marinos que subsisten o viven en el suelo del océano. (p. 139)

bioremediation: process that uses living organisms to remove pollutants. (p. 83)

bioremediación: proceso que utiliza organismos vivos para eliminar contaminantes. (p. 83)

braided stream: stream with many interlacing channels that are separated by bars and islands. (p. 38)

río de cauces interconectados: corriente con numerosos canales entrecruzados separados por barras e islas. (p. 38)

breaker: collapsing ocean wave that forms in shallow water and breaks onto the shore. (p. 111)

rompiente: ola oceánica colapsante que se forma en aguas poco profundas y rompe en la orilla. (p. 111)

C

cave: underground chamber that opens to the surface and is formed when slightly acidic groundwater dissolves compounds in rock. (p. 85)

chemosynthesis (kee moh SIHN thuh sihs): food-making process using sulfur or nitrogen compounds, rather than light energy from the Sun, that is used by bacteria living near hydrothermal vents. (p. 137)

cohesion: attraction between water molecules that allows water to form drops and keeps it liquid at room temperature. (p. 12)

continental shelf: gradually sloping end of a continent that extends beneath the ocean. (p. 128)

continental slope: ocean basin feature that dips steeply down from the continental shelf. (p. 129)

Coriolis effect: causes moving air and water to turn left in the southern hemisphere and turn right in the northern hemisphere due to Earth's rotation. (p. 105)

crest: highest point of a wave. (p. 110)

cueva: cámara subterránea que se abre en la superficie y se forma cuando aguas subterráneas ligeramente ácidas disuelven compuestos de las rocas. (p. 85)

quimiosíntesis: proceso de producción de alimentos que utiliza compuestos de azufre o nitrógeno en lugar de energía solar; este proceso es utilizado por las bacterias que viven cerca de los conductos hidrotérmicos. (p. 137)

cohesión: atracción entre moléculas de agua que permite al agua formar gotas y mantenerse en estado líquido a temperatura ambiente. (p. 12)

plataforma continental: extremo gradualmente inclinado de un continente que se extiende por debajo del océano. (p. 128)

talud continental: depresión oceánica característica que se sumerge abruptamente desde la plataforma continental. (p. 129)

efecto de Coriolis: causa el movimiento del aire y el agua hacia la izquierda en el hemisferio sur y hacia la derecha en el hemisferio norte, debido a la rotación de la Tierra. (p. 105)

cresta: el punto más alto de una ola. (p. 110)

D

density: amount of mass in a unit of volume. (p. 11)

density current: circulation pattern in the ocean that forms when a mass of more dense seawater sinks beneath less dense seawater. (p. 107)

drainage basin: specific area that is drained by a stream. (p. 38)

dripstone: deposits of calcium carbonate, such as stalagmites and stalactites, that are left behind when groundwater drips and evaporates inside caves. (p. 87)

densidad: cantidad de masa en una unidad de volumen. (p. 11)

corriente de densidad: patrón de circulación en el océano que se forma cuando una masa de agua marina más densa se hunde por debajo del agua marina menos densa. (p. 107)

cuenca fluvial: área específica drenada por una corriente. (p. 38)

concreciones calcáreas: depósitos de carbonato de calcio, tales como las estalagmitas y estalactitas, producto del goteo y evaporación de las aguas subterráneas dentro de las cuevas. (p. 87)

E

estuary: area where a river meets the ocean that contains a mixture of freshwater and ocean water and provides an important habitat to many marine organisms. (p. 142)

estuario: área donde un río desemboca en el océano, contiene una mezcla de agua dulce y agua salada y proporciona un hábitat importante para muchos organismos marinos. (p. 142)

eutrophication (yew troh fuh KAY shun): natural process that ultimately turns a lake into dry land over time through an increase in sediment, nutrients, and organisms. (p. 47)

eutrofización: proceso natural que con el tiempo conduce a que un lago se convierta en terreno seco mediante el incremento de sedimentos, nutrientes y organismos. (p. 47)

F

floodplain: flat, level area of a stream valley covered by water during a flood. (p. 42)

planicie aluvial/de inundaciones: área plana y llana del valle de una corriente cubierta por agua durante una inundación. (p. 42)

G

geyser: hot spring that erupts periodically and shoots water and steam into the air. (p. 74)

groundwater: water that soaks into the ground and collects in pores and empty spaces and is an important source of drinking water. (pp. 23, 68)

géiser: manantial caliente que hace erupción periódicamente y dispara agua y vapor al aire. (p. 74)

agua subterránea: agua que se difunde en el suelo y se acumula en poros y espacios vacios seindo una fuente importante de agua potable. (pp. 23, 68)

I

irrigation: process of piping water from somewhere else to grow crops. (p. 17)

irrigación: proceso de conducir agua mediante tuberías de un punto a otro para cultivar. (p. 17)

L

load: Earth material that is carried by a stream, either suspended in the water, dissolved in the water, or rolling, bouncing, or sliding along the stream bed. (p. 39)

carga: material terrestre arrastrado por una corriente, ya sea suspendido o disuelto en el agua, rodando, rebotando o deslizándose junto con la corriente. (p. 39)

M

meandering stream: twisting stream with many curves that erodes on the outside of meanders and deposits sediment on the inside of meanders. (p. 37)

mid-ocean ridge: area where new ocean floor is formed when lava erupts through cracks in Earth's crust. (p. 130)

corriente meandriforme: corriente que sinuosa con numerosas curvas que erosiona la parte exterior de los meandros y deposita sedimentos en su interior. (p. 37)

surco en mitad del océano: área donde se forma el nuevo suelo oceánico cuando la lava brota a través de grietas en la corteza terrestre. (p. 130)

N

nekton: marine organisms that actively swim in the ocean. (p. 138)

nonpoint source pollution: pollution that enters water from a large area and cannot be traced to a single location. (p. 55)

nutrients: compounds, such as nitrates, that are released into lake water and used by plants, algae, and some plankton for growth. (p. 47)

necton: organismos marinos que nadan activamente en el océano. (p. 138)

contaminación sin fuente puntual: contaminación del agua proveniente de un área extensa y cuyo origen no se puede señalar en un solo lugar. (p. 55)

nutrientes: compuestos, como los nitratos, depositados en aguas lacustres y utilizados por plantas, algas y parte del plankton para su crecimiento. (p. 47)

P

permeable: describes rock that allows groundwater to flow through it because it contains many well-connected pores or cracks. (p. 69)

photosynthesis: food-making process using light energy from the Sun, carbon dioxide, and water. (p. 135)

plankton: marine organisms that drift in ocean currents. (p. 138)

point bar: pile of sand and gravel deposited by a meandering stream on the inside of a meander. (p. 37)

point source pollution: pollution that enters water from a specific location and can be controlled or treated before it enters a body of water. (p. 54)

polar molecule: molecule with a slightly positive end and a slightly negative end as a result of electrons being shared unequally. (p. 12)

pollution: introduction of wastes to an environment, such as sewage and chemicals, that can damage organisms. (pp. 76, 143)

porosity: volume of pore space divided by the volume of a rock or soil sample. (p. 69)

permeable: describe aquellas rocas que permiten el flujo de las aguas subterráneas debido a que contienen numerosos poros o grietas comunicados entre sí. (p. 69)

fotosíntesis: proceso de producción de alimentos usando la energía luminosa del sol, dióxido de carbono y agua. (p. 135)

plancton: organismos marinos que se desplazan a la deriva en las corrientes oceánicas. (p. 138)

punto de barra: acumulación de arena y gravilla depositada por una corriente meandriforme en la parte interna de un meandro. (p. 37)

contaminación de fuente puntual: contaminación del agua proveniente de un lugar específico y que puede ser controlada o tratada antes de que entre a una masa de agua. (p. 54)

molécula polar: molécula con un extremo ligeramente positivo y otro ligeramente negativo como resultado de compartir electrones de manera desigual. (p. 12)

contaminación: introducción de desechos al medio ambiente, como aguas residuales y químicos, que pueden causar daño a los organismos. (pp. 76, 143)

porosidad: volumen del espacio que ocupan los poros dividido entre el volumen total de una muestra de rocas o suelo. (p. 69)

R

reef: rigid, wave-resistant, ocean margin habitat built by corals from skeletal materials and calcium. (p. 142)

arrecife: hábitat de los márgenes oceánicos, rígido y resistente a las olas; es generado por los corales a partir de materiales esqueléticos y calcio. (p. 142)

runoff: rainwater that does not sink into the ground, but instead runs across Earth's surface until it flows into a stream. (p. 36)

escorrentía: agua de lluvia que no es absorbida por el suelo sino que recorre la superficie terrestre hasta incorporarse a una corriente. (p. 36)

S

salinity (say LIHN ut ee): a measure of the amount of salts dissolved in seawater. (p. 102)

sanitary landfill: landfill lined with plastic or concrete, or located in clay-rich soil; reduces the chance of hazardous wastes leaking into the surrounding soil and groundwater. (p. 78)

sinkhole: depression formed when a cave's roof is no longer able to support the land above it and the land collapses into the cave. (p. 88)

soil water: groundwater that is trapped within openings in the soil and keeps plants and crops alive. (p. 23)

specific heat: amount of energy needed to raise the temperature of 1 kg of a substance by 1°C. (p. 14)

stream discharge: volume of water that flows through a stream in a certain amount of time. (p. 41)

subsidence: occurs when water no longer fills the pores in an aquifer and the land above the aquifer sinks. (p. 84)

surface current: wind-powered ocean current that moves water horizontally, parallel to Earth's surface, and moves only the upper few hundred meters of seawater. (p. 104)

surface water: all the freshwater at Earth's surface, including streams, rivers, lakes, and reservoirs. (p. 24)

salinidad: medida de la cantidad de sales disueltas en el agua marina. (p. 102)

relleno sanitario: relleno limitado con plástico o concreto o ubicado en suelos ricos en arcillas; reduce la posibilidad de que los residuos peligrosos se filtren hacia los suelos vecinos o aguas subterráneas. (p. 78)

hundimiento: depresión formada cuando el techo de una cueva ya no es capaz de sostener la tierra de la parte superior y ésta se colapsa hacia el interior de la cueva. (p. 88)

agua del suelo: agua subterránea atrapadas en las aberturas del suelo y que mantiene vivas a las plantas y cultivos. (p. 23)

calor específico: cantidad de energía necesaria para subir un grado centígrado la temperatura de un kilogramo de una sustancia. (p. 14)

descarga fluvial: volumen de agua que fluye a través de una corriente durante cierto periodo de tiempo. (p. 41)

subsidencia: ocurre cuando el agua deja de llenar los poros de un acuífero y la tierra encima del acuífero se hunde. (p. 84)

corriente de superficie: corriente oceánica empujada por el viento que mueve el agua horizontalmente, paralela a la superficie de la Tierra, y mueve sólo unos cientos de metros de la parte superior del agua marina. (p. 104)

agua de la superficie: toda el agua dulce de la superficie terrestre, incluyendo arroyos, ríos, lagunas y embalses. (p. 24)

T

tidal range: the difference between the level of the ocean at high tide and the level at low tide. (p. 114)

tide: daily rise and fall in sea level caused, for the most part, by the interaction of gravity in the Earth-Moon system. (p. 113)

trench: long, narrow, steep-sided depression in the seafloor formed where one crustal plate sinks beneath another. (p. 131)

rango de la marea: la diferencia entre el nivel del océano en marea alta y marea baja. (p. 114)

marea: elevación y disminución diaria del nivel del mar causada, en su mayor parte, por la interacción de la gravedad en el sistema Tierra-Luna. (p. 113)

fosa: depresión estrecha, alargada y de bordes pronunciados en el suelo marino; se forma cuando una placa de la corteza se hunde por debajo de otra. (p. 131)

Glossary/Glosario

trough (TRAWF): lowest point of a wave. (p. 110)

turnover: the mixing of surface water with bottom water, circulating oxygen and nutrients throughout a lake. (p. 49)

seno: el punto más bajo de una ola. (p. 110)

volcamiento: mezcla del agua de la superficie con el agua del fondo, haciendo circular oxígeno y nutrientes en un lago. (p. 49)

U

upwelling: vertical circulation in the ocean that brings deep, cold water to the ocean surface. (p. 107)

solevantamiento: circulación vertical en el océano que trae el agua fría de las profundidades a la superficie del océano. (p. 107)

W

water conservation: the careful use and protection of water. (p. 20)

water table: top surface of the zone of saturation; also the surface of lakes and rivers. (p. 70)

wave: rhythmic movement that carries energy through matter or space; can be described by its crest, trough, wavelength, and wave height. (p. 110)

wetland: area of land covered by water during part of the year and recognizable by the types of soils and plants that grow there. (p. 51)

conservación del agua: uso y protección cuidadosa del agua. (p. 20)

nivel freático/agua de superficie: superficie superior de la zona de saturación; también define la superficie de los lagos y ríos. (p. 70)

ola: movimiento rítmico que lleva energía a través de la materia o del espacio; puede describirse por su cresta, valle, longitud de la ola y altura de la ola. (p. 110)

humedal: área de terreno cubierta por agua durante una parte del año y reconocible por los tipos de suelos y las plantas que allí crecen. (p. 51)

Z

zone of saturation: in an aquifer, the zone where the pores are full of water. (p. 70)

zona de saturación: en un acuífero, la zona donde los poros están llenos de agua. (p. 70)

Italic numbers = illustration/photo **Bold numbers = vocabulary term**
lab = indicates a page on which the entry is used in a lab
act = indicates a page on which the entry is used in an activity

Credits

Magnification Key: Magnifications listed are the magnifications at which images were originally photographed.
LM–Light Microscope
SEM–Scanning Electron Microscope
TEM–Transmission Electron Microscope

Acknowledgments: Glencoe would like to acknowledge the artists and agencies who participated in illustrating this program: Absolute Science Illustration; Andrew Evansen; Argosy; Articulate Graphics; Craig Attebery represented by Frank & Jeff Lavaty; CHK America; John Edwards and Associates; Gagliano Graphics; Pedro Julio Gonzalez represented by Melissa Turk & The Artist Network; Robert Hynes represented by Mendola Ltd.; Morgan Cain & Associates; JTH Illustration; Laurie O'Keefe; Matthew Pippin represented by Beranbaum Artist's Representative; Precision Graphics; Publisher's Art; Rolin Graphics, Inc.; Wendy Smith represented by Melissa Turk & The Artist Network; Kevin Torline represented by Berendsen and Associates, Inc.; WILDlife ART; Phil Wilson represented by Cliff Knecht Artist Representative; Zoo Botanica.

Photo Credits

Cover Craig Tuttle/Getty Images; **i ii** Craig Tuttle/Getty Images; **iv** (bkgd)John Evans, (inset)Craig Tuttle/Getty Images; **v** (t)PhotoDisc, (b)John Evans; **vi** (l)John Evans, (r)Geoff Butler; **vii** (l)John Evans, (r)PhotoDisc; **viii** PhotoDisc; **ix** Aaron Haupt Photography; **x** Robin Karpan; **xi** Jeff Rotman/Peter Arnold, Inc.; **xii** Peter Skinner/Photo Researchers; **1** Chris Lisle/CORBIS; **2** TSADO/NOAA/Tom Stack & Assoc.; **4** (t)TSADO/NOAA/Tom Stack & Assoc., (b)Ralph White/CORBIS; **5** (t)file photo, (b)NOAA; **6–7** Grafton Marshall Smith/CORBIS; **7** Matt Meadows; **8** NASA; **9** (l)Matt Meadows, (r)Doug Martin; **10** Aaron Haupt; **11** Dominic Oldershaw; **12** Nick Daly/Stone/Getty Images; **13** (t)David Muench/CORBIS, (b)NASA; **15 16** Matt Meadows; **17** James L. Amos/Peter Arnold, Inc; **18** Brian Yarvin/Peter Arnold, Inc.; **19** (tl)Jack Fields/CORBIS, (tr)Vince Streano/CORBIS, (bl)CORBIS, (br)Susan E. Benson/Stock Connection/PictureQuest; **20** (l)Jerry Howard/Stock Boston, (r)Ted Streshinsky/Photo 20-20 PictureQuest; **24** SuperStock; **26** Steve Callahan/Visuals Unlimited; **27** Dominic Oldershaw; **28** Jose Fuste Raga/The Stock Market/CORBIS; **29** (tl)Martin Bydalek/Stone/Getty Images, (tr)Zig Leszczynski/Earth Scenes, (br)Roland Seitre/Peter Arnold, Inc.; **31** Dominic Oldershaw; **32** Ted Streshinsky /Photo 20-20/PictureQuest; **33** Maslowski/Visuals Unlimited; **34–35** Stuart & Cynthia Pernick/SuperStock; **38** James L. Amos/CORBIS; **40** (t)Eastcott/Momatiuk/Earth Scenes, (b)Robin Karpan; **41** (l)Calvin Larsen/Photo Researchers, (r)Martin G. Miller/Visuals Unlimited; **43** Raymone Gehman/CORBIS; **44** (l)NASA, (r)Peter Arnold/Peter Arnold, Inc.; **45** (l)Victor Englebert/Photo Researchers, (r)Lowell Georgia/Science Source/Photo Researchers; **48** (t)David R. Frazier/Photo Researchers, (cl)Jerome Wexler/Visuals Unlimited, (cr)Bazaszewski/Visuals Unlimited, (b)Mark Wright/Photo Researchers; **50** Richard Hutchings; **51** Jeremy Woodhouse/DRK Photo; **52** (t)Tom Bean/DRK Photo, (b)Robert & Linda Mitchell; **53** Courtesy the City of San Diego CA; **54** Simon Fraser/Science Photo Library/Photo Researchers; **55** Tom & Pat Lesson/DRK Photo; **57** KS Studios; **58** Aaron Haupt; **61** (l)C.C. Lockwood/Earth Scenes, (r)Marc Epstein/DRK Photo; **62** Martin G. Miller/Visuals Unlimited; **64** Claus Meyer/Minden Pictures; **66–67** Jack Dykinga/Stone/Getty Images; **68** A.J. Copley/Visuals Unlimited; **72** (t)Willard Luce/Earth Scenes, (b)Fred Habegger from Grant Heilman; **74** C. Alan Chapman/Visuals Unlimited; **75** Mark Burnett; **76** Salt Institute; **80** Lowell Georgia/CORBIS; **83** Fritz Prenzel/Earth Scenes; **85** James Jasek; **87** (l)Runk/Schoenberger from Grant Heilman, (tr)M.L. Sinibaldi/The Stock Market/CORBIS, (br)Corel; **88** (t)AP/Wide World Photos, (b)Mammoth Cave National Park; **90** (t)Bert Krages/Visuals Unlimited, (b)Jim Sugar Photography/CORBIS; **91** Doug Martin; **92** (t)David Muench/CORBIS, (b)Peter & Ann Bosted/Tom Stack & Assoc.; **93** (l)William A. Blake/CORBIS, (r)Michael Dwyer/Stock Boston; **95** C.C. Lockwood/Earth Scenes; **98–99** Warren Bolster/Getty Images; **100** (l)Norbert Wu/Peter Arnold, Inc., (r)Darryl Torckler/Stone/Getty Images; **102** Cathy Church/Picturesque/PictureQuest; **105** Bob Daemmrich; **106** (t)Darryl Torckler/Stone/Getty Images, (b)Raven/Explorer/Photo Researchers; **110** Jack Fields/Photo Researchers; **111** Tom & Therisa Stack; **112** (top & bottom) Stephen R. Wagner, (cl)Spike Mafford/PhotoDisc, (cr)Douglas Peebles/CORBIS; **113** Arnulf Husmo/Stone/Getty Images; **114** (l)Groenendyk/Photo Researchers, (r)Patrick Ingrand/Stone/Getty Images, (b)Kent Knudson/Stock Boston; **118** (t)Mark E. Gibson/Visuals Unlimited, (b)Timothy Fuller; **119** Timothy Fuller; **120** (bkgd)Chris Lisle/CORBIS, (inset)Sovfoto/Eastfoto/PictureQuest; **121** (l)Carl R. Sams II/Peter Arnold, Inc., (r)Edna Douthat; **126–127** Bill Curtsinger/Getty Images; **128–129** The Floor of the Oceans by Bruce C. Heezen and Marie Tharp, (c) 1977 by Marie Tharp. Reproduced by permission of Marie Tharp; **127** Mark Burnett; **130** Woods Hole Oceanographic Institution; **131** Thomas J. Abercrombie/National Geographic Society; **132** (t)J. & L. Weber/Peter Arnold, Inc., (bl)Arthur Hill/Visuals Unlimited, (br)John Cancalosi/Peter Arnold, Inc.; **133** (t)Instutute of Oceanographic Sciences/NERC/Science Photo Library/Photo Researchers, (b)Biophoto Associates/Photo Researchers; **135** Fred Bavendam/Minden Pictures; **137** Nanct Sefton/Photo Researchers; **138** (t)Manfred Kage/Peter Arnold, Inc., (b)M.I. Walker/Science Source/Photo Researchers; **139** (l)Nick Caloyianis/National Geographic Society, (c)Herb Segars/Animals Animals, (r)Norbert Wu; **140** Fred Bavendam/Peter Arnold, Inc.; **141** (1)Lloyd K. Townsend, (2)Fred Whitehead/Animals Animals, (3)Andrew J. Martinez/Photo Researchers, (4)Andrew J. Martinez/Photo Researchers, (5)Hal Beral/Visuals Unlimited, (6)Michael Abbey/Photo Researchers, (7)Gerald & Buff Corsi/Visuals Unlimited, (8)Anne W. Rosenfeld/Animals Animals, (9)Andrew J. Martinez/Photo Researchers, (10)Zig Leszczynski/Animals Animals, (11)Gregory Ochocki/Photo Researchers, (12)Peter Skinner/Photo Researchers; **142** James H. Robinson/Photo Researchers; **145** (tl)C.C. Lockwood/Earth Scenes, (bl)C.C. Lockwood/DRK Photo, (others)David Young-Wolff/PhotoEdit, Inc.; **146** NASA; **147** David Young-Wolff/PhotoEdit, Inc.; **148** (t)Jim Nilsen/Stone/Getty Images, (b)Fred Bavendam/Minden Pictures; **149** (t)Fred Bavendam/Minden Pictures, (b)Jeff Rotman/Peter Arnold, Inc.; **150** (t)Rick Price/CORBIS, (b)Emory Kristof/National Geographic; **151** Fred Bavendam/Minden Pictures; **156** PhotoDisc; **158** Tom Pantages; **162** Michell D. Bridwell/PhotoEdit, Inc.; **163** (t)Mark Burnett, (b)Dominic Oldershaw; **164** StudiOhio; **165** Timothy Fuller; **166** Aaron Haupt; **168** KS Studios; **169** Matt Meadows; **170** Clyde H. Smith/Peter Arnold, Inc.; **173** Amanita Pictures; **174** Bob Daemmrich; **176** Davis Barber/PhotoEdit, Inc.

PERIODIC TABLE OF THE ELEMENTS

Columns of elements are called groups. Elements in the same group have similar chemical properties.

Element — Hydrogen
Atomic number — 1
Symbol — H
Atomic mass — 1.008
State of matter

Gas
Liquid
Solid
Synthetic

The first three symbols tell you the state of matter of the element at room temperature. The fourth symbol identifies elements that are not present in significant amounts on Earth. Useful amounts are made synthetically.

	1	2	3	4	5	6	7	8	9
1	Hydrogen 1 **H** 1.008								
2	Lithium 3 **Li** 6.941	Beryllium 4 **Be** 9.012							
3	Sodium 11 **Na** 22.990	Magnesium 12 **Mg** 24.305							
4	Potassium 19 **K** 39.098	Calcium 20 **Ca** 40.078	Scandium 21 **Sc** 44.956	Titanium 22 **Ti** 47.867	Vanadium 23 **V** 50.942	Chromium 24 **Cr** 51.996	Manganese 25 **Mn** 54.938	Iron 26 **Fe** 55.845	Cobalt 27 **Co** 58.933
5	Rubidium 37 **Rb** 85.468	Strontium 38 **Sr** 87.62	Yttrium 39 **Y** 88.906	Zirconium 40 **Zr** 91.224	Niobium 41 **Nb** 92.906	Molybdenum 42 **Mo** 95.94	Technetium 43 **Tc** (98)	Ruthenium 44 **Ru** 101.07	Rhodium 45 **Rh** 102.906
6	Cesium 55 **Cs** 132.905	Barium 56 **Ba** 137.327	Lanthanum 57 **La** 138.906	Hafnium 72 **Hf** 178.49	Tantalum 73 **Ta** 180.948	Tungsten 74 **W** 183.84	Rhenium 75 **Re** 186.207	Osmium 76 **Os** 190.23	Iridium 77 **Ir** 192.217
7	Francium 87 **Fr** (223)	Radium 88 **Ra** (226)	Actinium 89 **Ac** (227)	Rutherfordium 104 **Rf** (261)	Dubnium 105 **Db** (262)	Seaborgium 106 **Sg** (266)	Bohrium 107 **Bh** (264)	Hassium 108 **Hs** (277)	Meitnerium 109 **Mt** (268)

The number in parentheses is the mass number of the longest-lived isotope for that element.

Rows of elements are called periods. Atomic number increases across a period.

The arrow shows where these elements would fit into the periodic table. They are moved to the bottom of the table to save space.

Lanthanide series	Cerium 58 **Ce** 140.116	Praseodymium 59 **Pr** 140.908	Neodymium 60 **Nd** 144.24	Promethium 61 **Pm** (145)	Samarium 62 **Sm** 150.36
Actinide series	Thorium 90 **Th** 232.038	Protactinium 91 **Pa** 231.036	Uranium 92 **U** 238.029	Neptunium 93 **Np** (237)	Plutonium 94 **Pu** (244)